# Energizing Green Cities in Southeast Asia

**DIRECTIONS IN DEVELOPMENT**
Environment and Sustainable Development

# Energizing Green Cities in Southeast Asia

## Applying Sustainable Urban Energy and Emissions Planning

Dejan R. Ostojic, Ranjan K. Bose, Holly Krambeck, Jeanette Lim, and Yabei Zhang

**THE WORLD BANK**
Washington, D.C.

# Contents

## Boxes

## Figures

## Maps

## Tables

# Foreword

Cities currently account for about two-thirds of the world's annual energy consumption and about 70 percent of the greenhouse gas (GHG) emissions. In the coming decades, urbanization and income growth in developing countries are expected to push cities' energy consumption and GHG emissions shares even higher, particularly where the majority of people remain underserved by basic infrastructure services and where city authorities are underresourced to shift current trajectories. These challenges are facing many cities and hundreds of millions of people in the East Asia and Pacific (EAP) region, which is experiencing unprecedented rates of urbanization, as the region's urban population grows almost twice as fast as the world's urban population.

This report lays out the challenges and proposes strategies for Sustainable Urban Energy and Emissions Planning (SUEEP) and development. It shows that the above challenges also present a unique opportunity for EAP cities to become the global engines of green growth by choosing energy efficient solutions for their infrastructure needs and avoiding locking in energy-intensive infrastructure that has accompanied economic growth in the past.

The SUEEP studies in the three pilot cities—Cebu City (the Philippines), Surabaya (Indonesia), and Da Nang (Vietnam)—show a clear correlation between the scaling up of energy efficiency in all major infrastructure sectors and economic growth. This relationship is recognized by the municipal governments in the three pilot cities and has been incorporated into their visions of green urban development. Achieving this vision requires institutional reforms and capacity building, including strengthening energy governance at the municipal level. Furthermore, to ensure effective implementation of their green growth plans, municipal governments will have to foster alliances and closely collaborate with a coalition of actors from the national, state, and local levels, and from civil society and the private sector, who share a commitment to advance the green economy.

The SUEEP framework presented here is designed to facilitate such collaboration and the development of capacity-building programs to strengthen energy governance and maximize energy efficiency across municipal sectors, as well as to help define actions and prioritize investments in energy efficient infrastructure. For this purpose, this volume includes as part III the "Sustainable Urban Energy and Emissions Planning Guidebook: A Guide for Cities in East

Asia and Pacific," which provides step-by-step guidance to help a city develop its own energy and emissions plan and link its aspirations to actionable initiatives to improve energy efficiency and reduce emissions.

The World Bank is committed to providing support to EAP cities for sustainable urban energy and emissions planning and for mobilizing financing for priority investments in green infrastructure. We look forward to working hand in hand with cities to facilitate capacity building and public and private investments in programs that help them achieve their green growth objectives and a sustainable future for the generations to come.

John Roome
*Director*
*Sustainable Development*
*East Asia and Pacific Region*
*The World Bank Group*

# Preface

This volume is a product of the Australian Agency for International Development (AusAID)–supported Sustainable Urban Energy and Emissions Planning (SUEEP) program in the East Asia and Pacific (EAP) region. The SUEEP program seeks to help EAP city governments formulate long-term sustainable urban energy and low-carbon development strategies that can be integrated into existing development plans.

The first phase of SUEEP was implemented in three Southeast Asian pilot cities—Cebu City, the Philippines; Surabaya, Indonesia; and Da Nang, Vietnam. Part I of this book synthesizes the information and lessons of the first phase. Part II provides detailed background on the energy and emissions profiles of each of the pilot cities along with recommendations for making the cities "greener." Part III uses the knowledge gained from the pilot cities to formulate guidelines for energy and emissions planning in other developing cities in the EAP region. This book, including the Toolkit featured in part III, can be found at http://www .worldbank.org/eap/energizinggreencities.

Part I stresses the unique opportunity in the EAP region provided by rapid urbanization and growing standards of living—EAP cities can become global engines of green growth by choosing modern, energy-efficient solutions to their infrastructure needs and by refusing to lock in the energy-intensive infrastructure of yesterday. However, mainstreaming energy efficiency on a citywide scale and introducing low-carbon policies require city governments to reform institutions, build capacity, and strengthen energy planning and governance.

Part II provides the details behind the recommendations in part I, as gleaned from the experiences in the three pilot cities. Part II demonstrates that cities can strive for the goal of sustainable urban energy using a wide variety of approaches.

The Guidebook in part III aims to help cities establish and implement a road map for achieving a sustainable energy future. Its comprehensive framework and indicative step-by-step guidance are intended to support cities' efforts to develop their unique energy and emissions plans. The World Bank will be using this Guidebook as a tool to support cities in sustainable urban energy and emissions planning. The Guidebook will be revised after subsequent phases of the SUEEP process are implemented and the lessons learned are reviewed. We look forward to providing support to EAP cities that are keen to embark on the journey toward sustainable development for the benefit of future generations.

This report was produced by the Infrastructure Unit of the Department for Sustainable Development in the EAP region of the World Bank under the guidance of John Roome, Director, and Vijay Jagannathan, Sector Manager.

The preparation of the report was undertaken by a team led by Dejan R. Ostojic, Energy Sector Leader, EAP, and comprising Ranjan K. Bose, Holly Krambeck, Jeanette Lim, and Yabei Zhang.

The team would like to give special recognition to the following World Bank staff members for their help with the peer review process and for providing insightful feedback: Jan Bojo, Feng Liu, Jas Singh, Monali Ranade, and Om Prakesh Agarwal. The team also wishes to acknowledge Luiz T. A. Maurer of the International Finance Corporation as well as David Hawes, Adviser, AusAID, for his contribution as external peer reviewer.

Important comments and suggestions were also received from Dean Cira, Arish Dastur, Franz Gerner, Franz Drees Gross, Ky Hong, Paul Kriss, Nguyet Anh Pham, Dhruva Sahai, Victor Vergara, Xiaodong Wang, Paul Wright, and Yijing Zhong, to whom we would like to express our appreciation.

During the course of the project, the team gained considerable knowledge and benefited greatly from a wide range of consultations in Indonesia, the Philippines, and Vietnam and wishes to thank the participants in these consultations, who included government officials, nongovernmental organization and civil society representatives, and the private sector. We would like to express particular appreciation and gratitude to municipal officials in Cebu City, Da Nang, and Surabaya for their active participation in the pilot studies and numerous invaluable comments and suggestions.

The team acknowledges the valuable contributions of the Energy Sector Management Assistance Program (ESMAP, jointly sponsored by the World Bank and the United Nations Development Programme) and Happold Consulting.

The team would also like to thank Sherrie Brown for editing the document, Laura C. Johnson and Michael Alwan for providing the graphic design work, and Laurent Durix for his contribution on the dissemination process.

The team would like to acknowledge the continued generous support from the government of Australia, which funds the EAP Energy Flagship Report series. This series includes *Winds of Change: East Asia's Sustainable Energy Future* (2010); *One Goal, Two Paths: Achieving Universal Access to Modern Energy in East Asia and the Pacific* (2011); and the present volume, *Energizing Green Cities in Southeast Asia: Applying Sustainable Urban Energy and Emissions Planning*.

# Abbreviations

ADB        Asian Development Bank
ASEAN      Association of Southeast Asian Nations
AusAID     Australian Agency for International Development
BEEC       building energy efficiency code
BERDE      Building for Ecologically Responsive Design Excellence
BRT        bus rapid transit
BTU        British thermal unit
C          centigrade
CDM        Clean Development Mechanism
CDP        Carbon Disclosure Project
CER        Certified Emission Reduction
CFL        compact fluorescent lamp
$CO_2$     carbon dioxide
$CO_2e$    carbon dioxide equivalent
CW         characteristic weight
DAWACo     Da Nang Water Supply Company
DEPW       Department of Engineering and Public Works
DKP        Dinas Kebersihan dan Pertamanan, or Cleansing and Park Department
DOC        Department of Construction
DOE        designated operational entity
EAP        East Asia and Pacific
EB         Executive Board
ESCO       energy services company
ESMAP      Energy Sector Management Assistance Program
EU         European Union
F          Fahrenheit
GDP        gross domestic product
GEF        Global Environment Facility
GHG        greenhouse gas

| | |
|---|---|
| GSO | General Services Office |
| GTZ | German Technical Cooperation |
| HFC | hydrofluorocarbons |
| HPS | high pressure sodium |
| ICLEI | Local Governments for Sustainability |
| IEA | International Energy Agency |
| IMCCC | Inter-Ministerial Committee on Climate Change |
| IPCC | Intergovernmental Panel on Climate Change |
| JICA | Japan International Cooperation Agency |
| KPI | Key Performance Indicator |
| kW | kilowatt |
| kWh | kilowatt-hour |
| kWhe | kilowatt-hour equivalent |
| LED | light-emitting diode |
| LEED | Leadership in Energy and Environmental Design |
| LPG | liquefied petroleum gas |
| M | million |
| MCWD | Metropolitan Cebu Water District |
| MRF | materials recovery facility |
| MRV | Measurement, Reporting, and Verification |
| MW | megawatt |
| MWh | megawatt-hour |
| NCCS | National Climate Change Secretariat |
| NGO | nongovernmental organization |
| OBO | Office of the Building Official |
| OECD | Organisation for Economic Co-operation and Development |
| PFC | perfluorocarbons |
| PhP | Philippine peso |
| PJ | petajoule |
| PLN | Perusahaan Listrik Negara (state-owned electricity company) |
| PMEB | Barcelona Energy Improvement Plan |
| PoA | programs of activity |
| PPP | public-private partnership |
| PRG | partial risk guarantee |
| PS | Project Score |
| PV | photovoltaic |
| REF | reference (scenario) |
| ROI | return on investment |
| Rp | Indonesian rupiah |

SAR        Special Administrative Region
SEAP       Sustainable Energy Action Plan
SED        sustainable energy development (scenario)
SET        Sustainable Energy for Tshwane
SUEEP      Sustainable Urban Energy and Emissions Planning
t          ton
toe        ton of oil equivalent
TRACE      Tool for Rapid Assessment of City Energy
TWPS       Total Weighted Project Score
UNFCCC     United Nations Framework Convention on Climate Change
URENCO     Urban Environment Company
US         United States
USAID      United States Agency for International Development
VECo       Visayan Electric Company
WG         working group

# Overview

*Fast-growing cities in the East Asia and Pacific (EAP) region will define the region's energy future and its greenhouse gas (GHG) footprint. Rapid urbanization and growing standards of living offer a major opportunity to EAP cities to become the global engines of green growth by choosing energy efficient solutions to suit their infrastructure needs and by avoiding locking in energy-intensive infrastructure. The underlying studies in three EAP pilot cities show a clear correlation between investments in energy efficient solutions in all major infrastructure sectors and economic growth—by improving energy and GHG emissions efficiency, cities not only help the global environment, but they also support local economic development through productivity gains, reduced pollution, and more efficient use of resources.*

Mainstreaming energy efficiency on a citywide scale and introducing low-carbon policies requires city governments to reform institutions, build capacity, and strengthen energy planning and governance. The report reveals that a common barrier to implementing cross-sectoral urban energy efficiency and emissions mitigation programs is the absence of institutional mechanisms that support coordinated energy project evaluation, planning, and investment. However, the need for institutional reforms and capacity building is increasingly being recognized across the region, as demonstrated by the governments in the three pilot cities, Cebu City (the Philippines), Da Nang (Vietnam), and Surabaya (Indonesia), which are strongly committed to implementing their visions of green urban development. To ensure effective implementation of their green growth plans, city governments will have to foster alliances and close collaboration with a coalition of actors from the national, state, and local levels, and from civil society and the private sector, which share a commitment to advance the green economy, placing it among the top strategic priorities for the city.

It is within the power of cities to develop policies and establish institutions to support citywide energy efficiency programs and green urban development. With the right institutional framework, cities can develop and implement policies supporting the next generation of urban infrastructure, which must be more efficient, smarter, and socially and environmentally sustainable. Such green infrastructure requires changing the way systems are designed and decisions are made, as well as the application of advanced infrastructure solutions conducive to long-term energy efficient and low-carbon development paths.

The cross-cutting nature of energy efficiency offers a unique platform for the identification and prioritization of green investments in modern infrastructure across all infrastructure subsectors, as demonstrated through the case studies in this report. The sustainable urban energy and emissions planning (SUEEP) process and methodology used in this study and outlined in the SUEEP Guidebook (part III) are designed to help city leadership formulate long-term urban green growth strategies and identify and evaluate investments in energy efficient infrastructure, thereby maximizing return on investment, relative green impact, and contributions to other social and economic development goals. The result is a high-quality pipeline of green investment projects across all key infrastructure subsectors that can be effectively communicated to local stakeholders, private investors, financing institutions, and the international donor community.

Investments in green infrastructure require financing from both public and private sources in a coordinated manner. Investments in the transportation and buildings sectors present some of the largest opportunities for scaling up energy efficiency at the city level. Financing these large investments in fast-growing EAP cities will require partnerships and coordination between public and private investors. SUEEP can help foster such public-private partnerships (PPPs) and mobilize green infrastructure financing by helping to prioritize and coordinate investment projects, as well as through systematic monitoring, reporting, and verification of the impact of the projects on the overall efficiency of energy use and GHG emissions at the city level.

The World Bank Group is committed to providing support to EAP cities in sustainable urban energy and emissions planning, as well as to mobilize financing for priority investments in green infrastructure. The World Bank Group has accumulated global knowledge and experience in supporting institutional development and building capacity for planning and implementing green infrastructure investments in cities around the world, including successful urban development projects in the EAP region. Thus, the Bank is well positioned to assist municipal governments in building institutions, creating policies, developing long-term green growth plans that will attract financial support and investments from both the private sector and the donor community, and linking efficiency and low-carbon programs to international concessional financing and funding, as well as to the private sector investors who will play an important role in achieving green growth objectives.

## Urbanization, Economic Growth, Energy, and Emissions Trends

Cities around the world account for about 70 percent of global gross domestic product (GDP), 67 percent of global energy consumption, and nearly 70 percent of world GHG emissions (IEA 2008). These figures are trending ever upward, and about 80 percent of global growth in urban energy use and 89 percent of growth in GHG emissions are expected to come from developing countries,

**Figure O.1  Urbanization Rate**

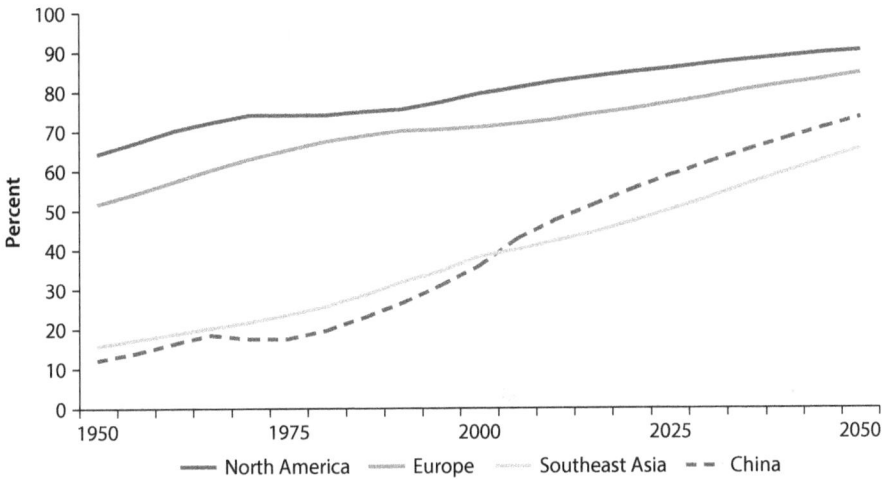

*Source:* UN 2010.

and from EAP cities in particular. As rapid urbanization in EAP countries continues (see figure O.1 for urbanization rates for Southeast Asia and China), national and city authorities will have to make decisions that will fundamentally define how cities will source and use their energy for decades to come.

Urban consumers require more energy as the economy grows and their standards of living rise. Urban growth and increased energy use are strongly linked to the economic growth required for cities to meet their diverse energy needs. In Southeast Asia, which is one of the world's least urbanized regions but whose population is expected to grow 1.75 times faster than the worldwide urban population (Yuen and Kong 2009), the rapid pace of urbanization is posing huge challenges for city governments to meet increasing energy demand in a sustainable manner. Given this, the report focuses on Southeast Asia.

Annual real GDP growth in Indonesia, Malaysia, the Philippines, Singapore, Thailand, and Vietnam (which collectively account for close to 95 percent of GDP and 86 percent of the population for all Association of Southeast Asian Nations or ASEAN countries)[1] is projected to average 6 percent between 2011 and 2015. Income growth in urban areas is also leading to increased demand for new services, particularly those that use electricity. City sprawl, which has increasingly led to the development of areas not easily served by public transportation and has discouraged pedestrians, has resulted in an explosion of personal motor vehicles like cars and motorized two-wheelers. In addition, the lifestyles, and corresponding energy use profiles, of urban middle- and upper-income residents in the developing world increasingly mimic those of their counterparts in the developed world. These factors, coupled with EAP countries' low per capita energy consumption (see figure O.2), ensure continued strong increases in urban energy demand across the region—demand that is expected to double during the next two decades.

**Figure O.2  Changes in Annual Energy Consumption per Capita, 1990 and 2009**

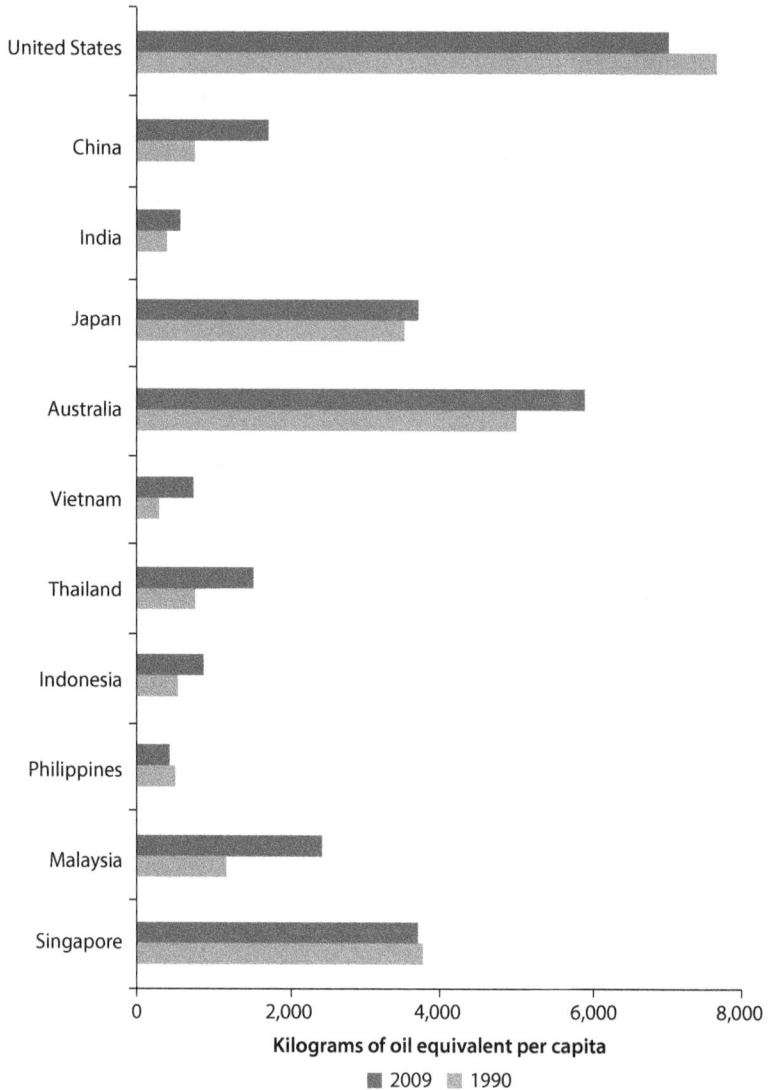

Kilograms of oil equivalent per capita

■ 2009  ▨ 1990

*Source:* IEA 2012.

Urbanization and economic growth in Southeast Asia have not only resulted in a continued increase in energy consumption, but also a profound shift in the energy mix. Although oil and gas will remain major sources of primary energy supply in Southeast Asia (accounting for 35 percent and 16 percent, respectively, in the ASEAN energy mix by 2030), coal is expected to have the fastest annual growth rate, 7.7 percent on average for the next 20 years in the business as usual scenario. This rate of increase would double the share of coal in the ASEAN energy mix from about 15 percent in 2007 to 30 percent by 2030. With a movement toward more carbon-intensive fuels, carbon dioxide ($CO_2$) emissions per unit of energy consumption will increase from 0.49 ton of $CO_2$ equivalent per ton of oil equivalent ($tCO_2$ per toe) in 2007 to 0.63 $tCO_2$ per toe in 2030 in a business as usual scenario. Even in an alternative policy scenario considered by ASEAN countries to mitigate the rise of GHG emissions, the carbon intensity of energy consumption is expected to increase to 0.59 $tCO_2$ per toe by 2030. The main reason for the rapid increase in the share of coal and the carbon intensity of energy consumption is the quickly growing demand for electricity in the expanding urban areas—which are supplied by coal-fired power plants.

The region's governments and city authorities are capable of maintaining economic growth, improving environmental sustainability, and enhancing reliability of energy supply. Because energy efficiency and GHG emissions are influenced directly and permanently by urban form and density, cities' planning and infrastructure investment choices will have a substantial impact on energy and emissions trends (World Bank 2005). Furthermore, large EAP cities are increasingly vocal and influential in formulating national policies that will shape the energy future and the ways in which the cities source and use energy. Finally, cities will be the main arena for economic transformation and mainstreaming of energy efficiency policies and practices, which are the backbone of sustainable energy development in the region—a sustained improvement in energy efficiency under an alternative energy scenario (ACE 2011) can provide a reduction of energy intensity of GDP from 580 toe per million US dollars in 2007 to 408 toe per million US dollars in 2030, compared with 501 toe per million US dollars under the business as usual scenario in 2030.

This alternative energy path requires a paradigm shift to new low-carbon development models and lifestyles. EAP cities need to avoid the carbon-intensive path and pursue sustainable lifestyles for their citizens by promoting novel urbanization models (World Bank 2010) focusing on compact city design, enhanced public transportation, green buildings, clean vehicles, and distributed generation. Smart urban planning—higher density, more spatially compact, and more mixed-use urban design that allows growth near city centers and transit corridors to prevent urban sprawl—can substantially reduce energy demand and $CO_2$ emissions and help cities become greener and more prosperous (World Bank 2009). With support from the Australian Agency for International Development (AusAID), the World Bank Group initiated a regional program—East Asia and Pacific SUEEP Program—to provide support and guidance to city governments in the EAP region to formulate such long-term urban energy strategies within cities' overall development plans.

## Understanding the Cities: Energy Use and GHG Emissions

The first phase of SUEEP was implemented in three Southeast Asian pilot cities—Cebu City, the Philippines; Surabaya, Indonesia; and Da Nang, Vietnam. Pilot city key statistics are summarized in table O.1.

The rapid population increase and rising standards of living in the three pilot cities are causing a considerable increase in city energy consumption. Da Nang is currently experiencing 11.7 percent yearly increases in energy consumption,

**Table O.1  Summary of Structural and Economic Data for the Pilot Cities**

| Parameter | Cebu City | Da Nang | Surabaya |
|---|---|---|---|
| Population (m) | 0.8 | 0.9 | 2.8 |
| City area (km²) | 291 | 1,283 | 327 |
| Population density (per km²) | 2,748 | 711 | 8,458 |
| GDP/cap/year ($) | 5,732 | 1,627 | 8,261 |
| Economic structure (%) | | | |
|    Services | 73 | 56 | 50 |
|    Industry | 19 | 42 | 32 |
|    Agriculture/other | 8 | 2 | 18 |

*Source:* Phase I pilot study.
*Note:* GDP = gross domestic product; km² = square kilometer; m = millions.

which will lead to a doubling of energy demand in six years' time. The increases in Surabaya and Cebu City's yearly energy use are 4.9 percent and 4.3 percent, respectively, still notably high. The main drivers of the strong energy demand are the transportation sector and industry, which account for about 87 percent of energy consumption in Cebu City, about 66 percent in Da Nang, and 68 percent in Surabaya. Consequently, transportation and industrial emissions account for more than 53 percent of emissions for all pilot cities. Transportation alone is responsible for more than 40 percent of emissions in both Cebu City and Da Nang (see chapter 3 for more details).

Understanding the city energy balance and carbon footprint is the first step in formulating a long-term sustainable urban energy development strategy. The SUEEP approach uses three different city-level diagnostic tools to assess the city energy profile. These include (a) the energy balance to analyze energy sources and uses across all sectors and categories of consumers, (b) the GHG emissions inventory to determine the main sources of GHG emissions from energy use, and (c) the Tool for Rapid Assessment of City Energy (TRACE) to evaluate energy efficiency opportunities in city sectors and identify priority areas for further investigation and intervention. An example of a GHG emissions inventory based on fuel source and sector for the pilot cities is shown in figures O.3 and O.4.

Most EAP cities, including the three pilot cities, have relatively low energy consumption and GHG emissions per capita. Figure O.5 illustrates that the current level of energy consumption and associated GHG emissions is relatively modest in the three pilot cities when compared with many cities around the world and also suggests that countries can follow different paths in energy consumption and GHG emissions as GDP per capita increases.

For example, the paths taken by Seoul and Tokyo offer a greener alternative that calls for a significant reduction of energy intensity of economic activities (energy use per unit of GDP) and improvements in energy efficiency. Only Cebu City is currently at the level of energy intensity that makes such a path relatively straightforward, whereas Da Nang and Surabaya have longer paths to reduce the energy intensity of their economies.

The fuel used for electricity generation is a key determinant of the intensity of GHG emissions from energy. In Surabaya, electricity generated from coal-fired

**Figure O.3  GHG Emissions by Fuel Source**

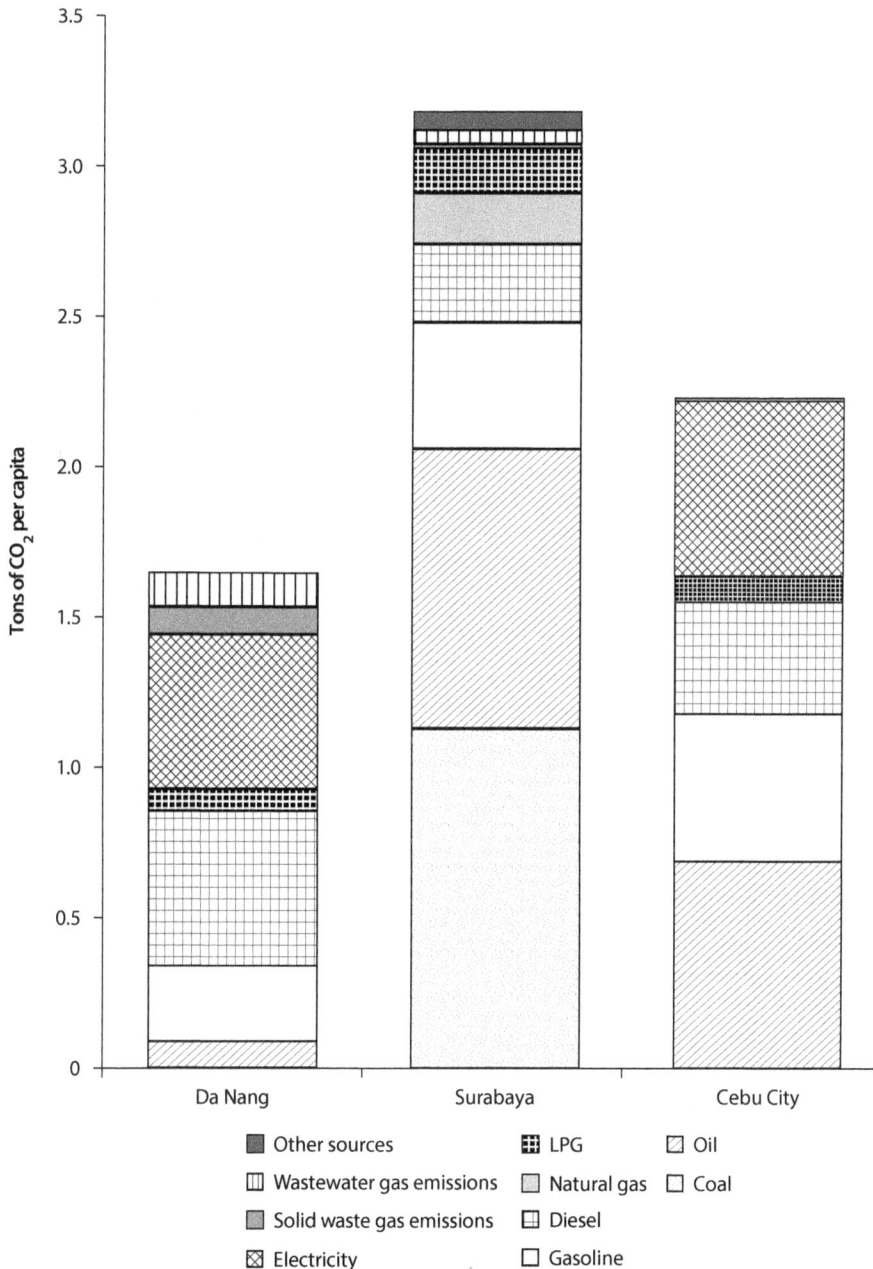

Legend:
- Other sources
- Wastewater gas emissions
- Solid waste gas emissions
- Electricity
- LPG
- Natural gas
- Diesel
- Gasoline
- Oil
- Coal

*Source:* Phase I pilot study.
*Note:* $CO_2$ = carbon dioxide; GHG = greenhouse gas; LPG = liquefied petroleum gas.

**Figure O.4   GHG Emissions by End Use**

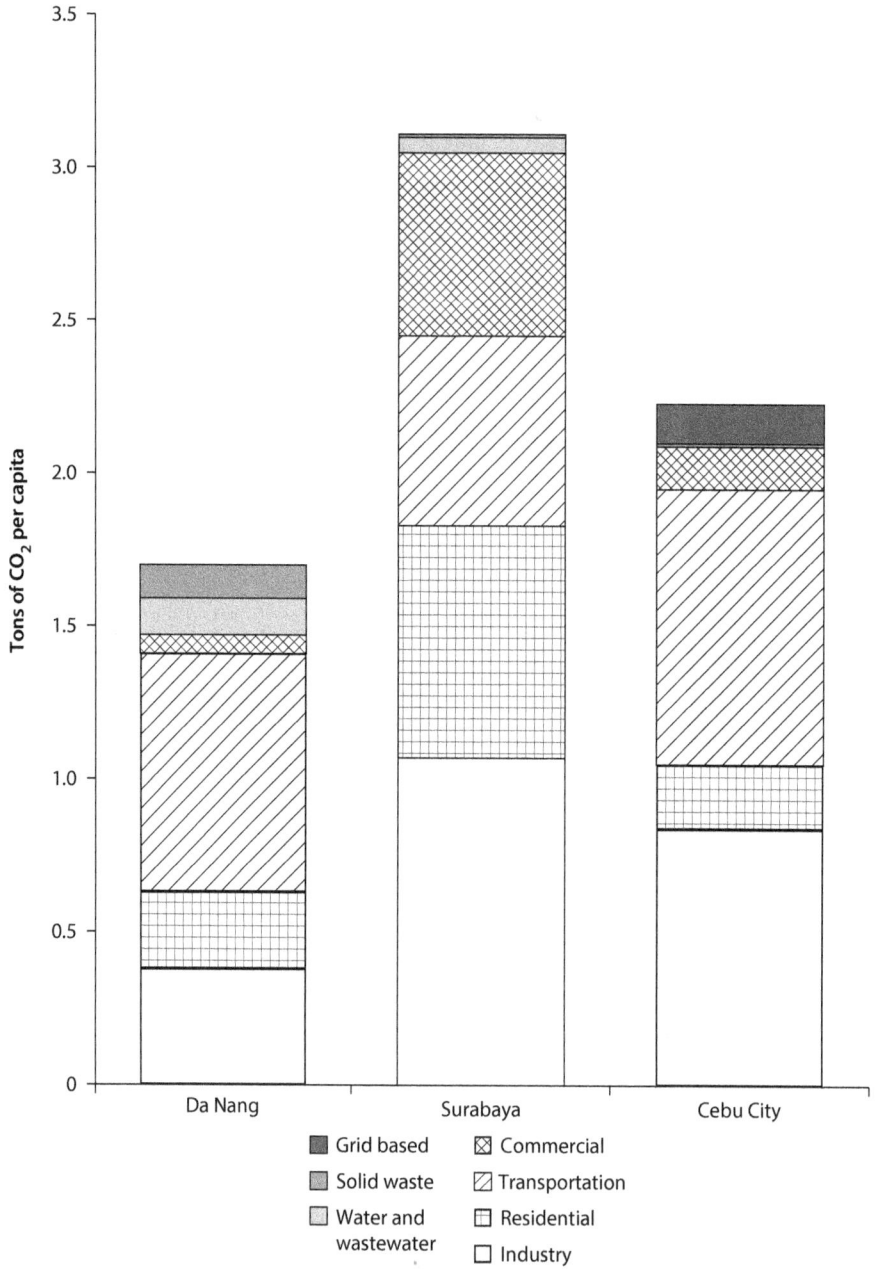

*Source:* Phase I pilot study.
*Note:* Grid based refers to electricity internally generated by the city. This is a separate category for Cebu City because data on end-use sector for electricity generated within the city were not available. $CO_2$ = carbon dioxide; GHG = greenhouse gas.

**Figure O.5  Energy Intensity and GDP per Capita in Select Cities**

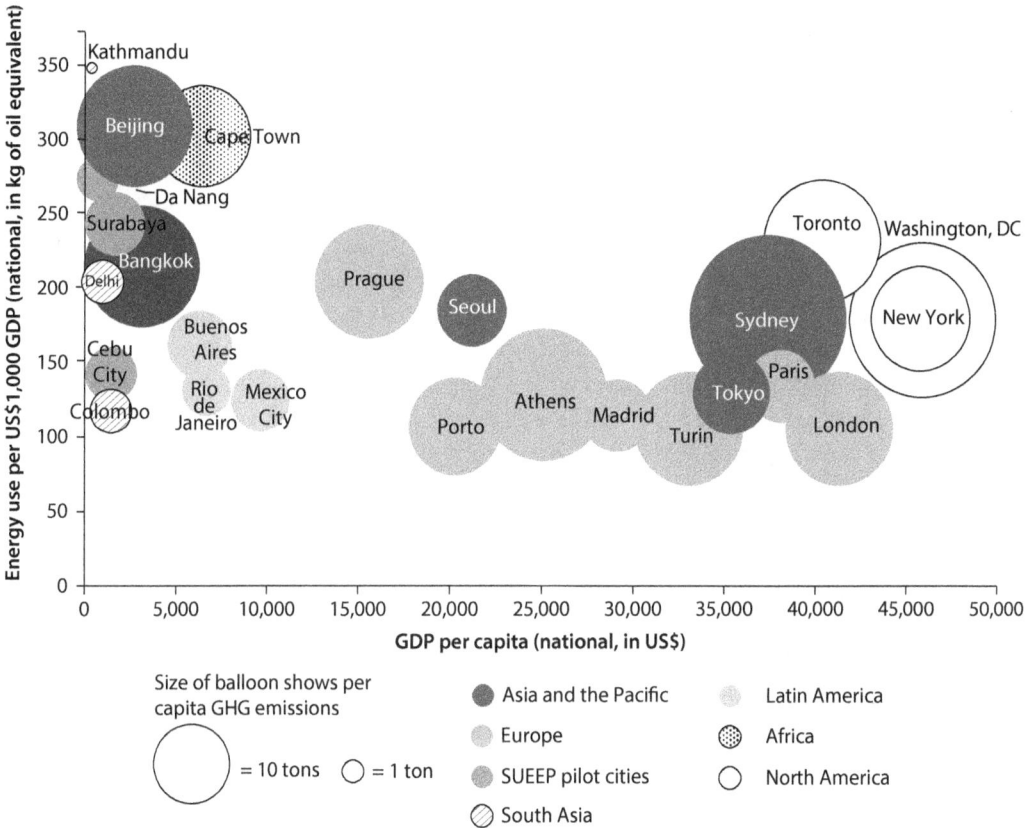

*Sources:* IEA statistics (http://www.iea.org/stats/index.asp); OECD National Accounts data; and World Bank national accounts data.
*Note:* GDP = gross domestic product; GHG = greenhouse gas; kg = kilogram; SUEEP = Sustainable Urban Energy and Emissions Planning.

power plants is responsible for 36 percent of the city's emissions. Cebu City and Da Nang both have significant amounts of renewable electricity generation (49 percent of electricity in Cebu City is generated by geothermal and hydro-power, and 30 percent of electricity in Da Nang comes from hydropower). Thus, GHG emissions in Cebu City and Da Nang are significantly lower (per capita) and are caused mainly by diesel and gasoline fuels used for transportation and local (diesel-based) electricity generation.

All three cities source the vast majority of their electricity from the national grid. The use of distributed renewable energy production (such as solar photovol-taic) is at a very early stage in all three pilot cities. Apart from local power produc-tion, which is particularly prominent in Cebu City (11 percent of the electricity supply), most electricity is supplied by the national power grid. Therefore, the cities have limited influence over the choice of primary fuel and over the effi-ciency of energy conversion processes associated with the production and delivery of electricity. Furthermore, electricity tariffs and pricing policies (such as subsidies), which are key tools for demand-side promotion of energy efficiency, are outside city control. Thus, close collaboration between the city and national authorities is needed for developing an optimal approach to meet fast-growing urban electricity needs in a reliable, efficient, and environmentally sound manner.

Breakdowns of GHG emissions roughly match the three cities' energy use patterns but with some variation. Surabaya's energy balance (figure O.6) illustrates the importance of fuel choice and conversion efficiencies in GHG emissions and the relative importance of energy efficiency in the industrial and buildings sectors. All three pilot cities will face challenges for achieving their economic development aspirations while managing growing energy demand and developing local generation capacity. More consistency in the cities' approaches to energy planning and better coordination across departments would solidify the basis for improving energy efficiency and reducing GHG emissions on a citywide scale. Therefore, EAP cities will benefit from establishing strong sustainable urban energy and emissions planning approaches that take a comprehensive view of energy needs to future proof against unsustainable increases in energy consumption and GHG emissions.

There is a strong need in the pilot cities for further development of energy efficiency governance and capacity. The planning and management of energy is usually a multi-agency function, but none of the pilot cities demonstrated a cohesive approach that encouraged communication among the relevant agencies. In addition, some cities lacked any means of, or concentration on, coordination between national- and city-level initiatives. If conflicts between national and local policies, programs, and initiatives exist, cities are advised to proactively lobby the national government for action in areas that are outside of the city's control or in which the government has promised action, but delivery is absent or ineffective in the city. The ability to influence the national government effectively requires cities to build energy governance into their institutional fabric and use it to lead and guide the planning and implementation of citywide energy efficiency programs.

TRACE, developed by the World Bank Group's Energy Sector Management Assistance Program (ESMAP),[2] offers a quick diagnosis of energy efficiency performance across a city's systems and sectors. It prioritizes sectors and presents a range of potential solutions along with implementation guidance and case studies. TRACE is a software platform for assessing the energy efficiency performance of six city sectors or services: urban passenger transport, city buildings, water and wastewater, public lighting, solid waste, and power and heat. As shown in figure O.7, TRACE consists of three principal components: an energy benchmarking tool that compares key performance indicators among peer cities, a sector prioritization process that identifies sectors with the greatest potential for energy efficiency improvements, and a "playbook" of tried and tested energy efficiency recommendations that helps in the selection of appropriate interventions. The TRACE deployment is a three-month assessment process that includes several weeks of upfront data gathering and benchmarking, sector meetings, and preparation of a final report. Based on TRACE results, city governments can identify early wins in key sectors and start developing citywide energy and emissions strategies.

**Figure O.6  Surabaya Sankey Diagram**

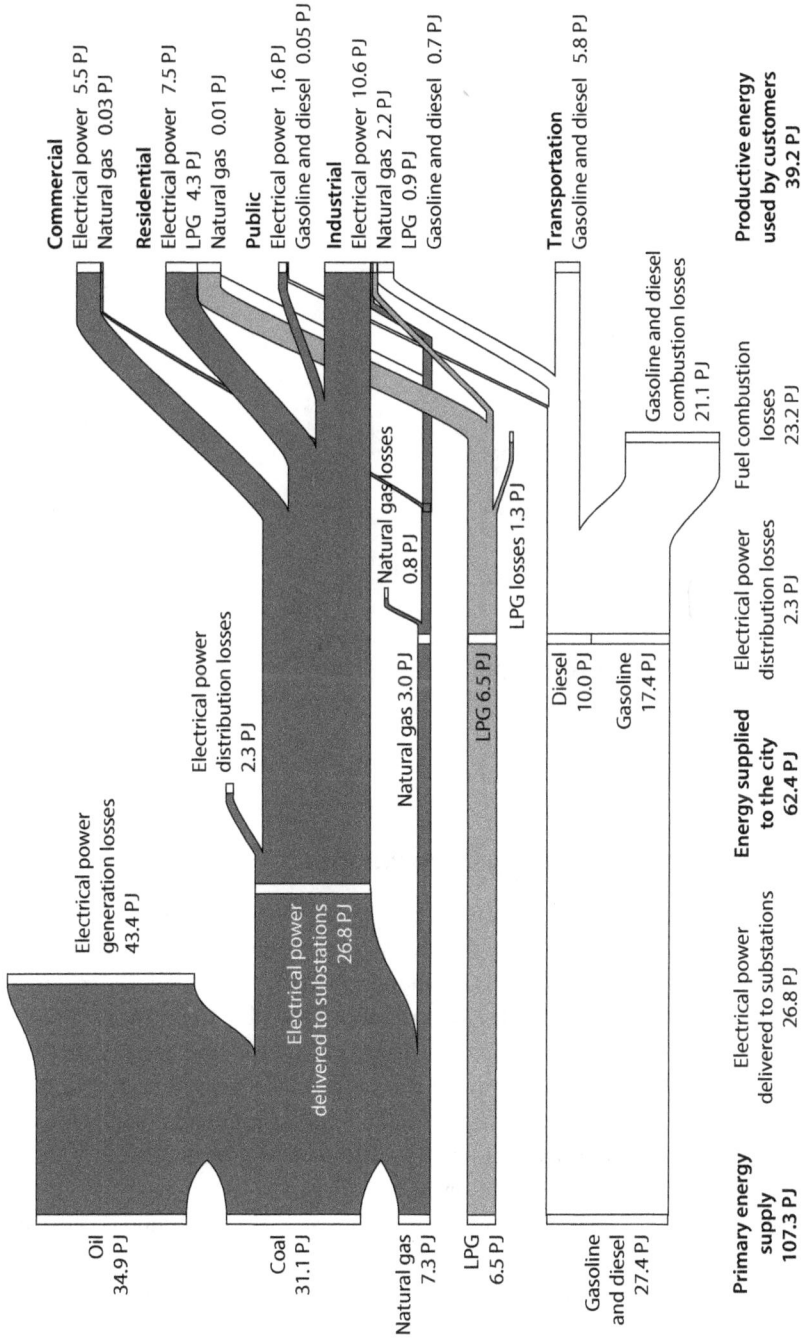

Oil
34.9 PJ

Coal
31.1 PJ

Natural gas
7.3 PJ

LPG
6.5 PJ

Gasoline
and diesel
27.4 PJ

**Primary energy
supply
107.3 PJ**

Electrical power
generation losses
43.4 PJ

Electrical power
delivered to substations
26.8 PJ

**Electrical power
delivered to substations
26.8 PJ**

Electrical power
distribution losses
2.3 PJ

Natural gas 3.0 PJ

LPG 6.5 PJ

Diesel
10.0 PJ

Gasoline
17.4 PJ

**Energy supplied
to the city
62.4 PJ**

Natural gas losses
0.8 PJ

LPG losses 1.3 PJ

Electrical power
distribution losses
2.3 PJ

Gasoline and diesel
combustion losses
21.1 PJ

Fuel combustion
losses
23.2 PJ

**Commercial**
Electrical power   5.5 PJ
Natural gas   0.03 PJ

**Residential**
Electrical power   7.5 PJ
LPG   4.3 PJ
Natural gas   0.01 PJ

**Public**
Electrical power   1.6 PJ
Gasoline and diesel   0.05 PJ

**Industrial**
Electrical power   10.6 PJ
Natural gas   2.2 PJ
LPG   0.9 PJ
Gasoline and diesel   0.7 PJ

**Transportation**
Gasoline and diesel   5.8 PJ

**Productive energy
used by customers
39.2 PJ**

*Source:* Phase I pilot study.
*Note:* LPG = liquefied petroleum gas; PJ = petajoule. "Public" includes the end-use energy of city buildings, street lighting, city vehicles, water, wastewater, and solid waste management.

11

**Figure O.7   Cover Page of TRACE Web-Based Tool**

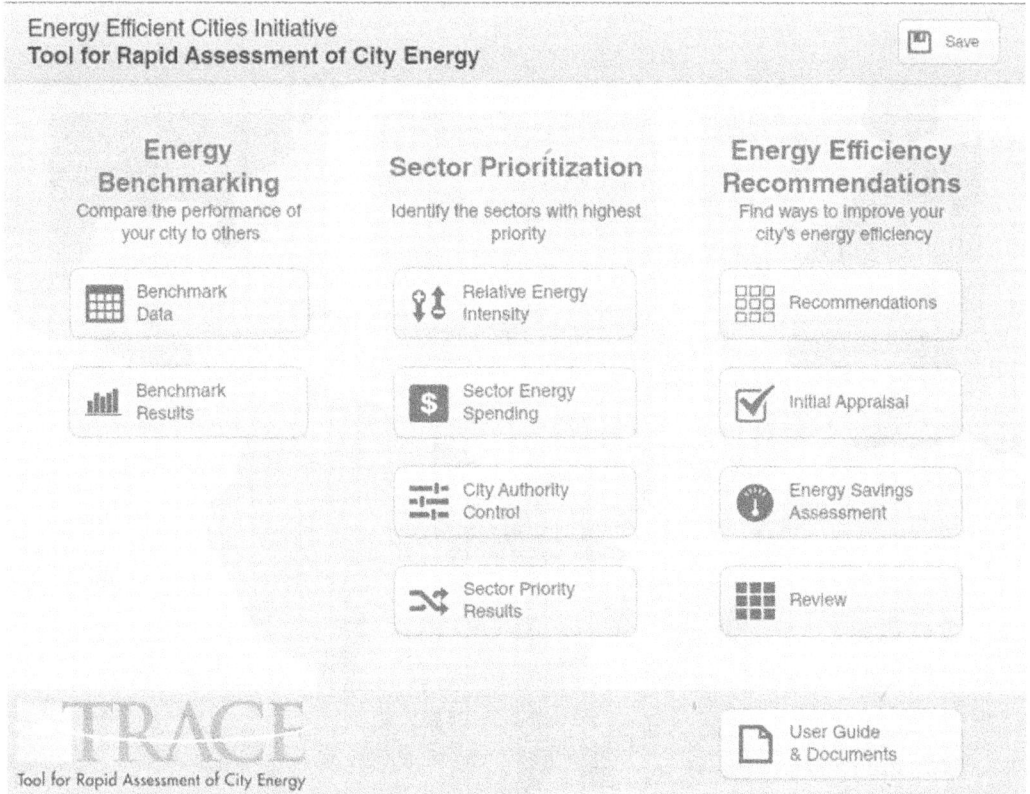

Energy Efficient Cities Initiative
**Tool for Rapid Assessment of City Energy**                                    💾 Save

| **Energy Benchmarking** | **Sector Prioritization** | **Energy Efficiency Recommendations** |
| Compare the performance of your city to others | Identify the sectors with highest priority | Find ways to improve your city's energy efficiency |

Benchmark Data

Benchmark Results

Relative Energy Intensity

Sector Energy Spending

City Authority Control

Sector Priority Results

Recommendations

Initial Appraisal

Energy Savings Assessment

Review

TRACE
Tool for Rapid Assessment of City Energy

User Guide & Documents

*Source:* Tool for Rapid Assessment of City Energy (TRACE).

## Sector Diagnostics: Identifying Opportunities

The transportation and buildings sectors present the largest opportunities for scaling up energy efficiency at the city level. The buildings sector was found to have the most potential for success of energy efficiency measures based on city authorities' high degree of control or influence, and the transportation sector accounted for significant energy consumption and GHG emissions (figures O.8 and O.9). The three pilot cities showed the potential to benefit from integrated transportation planning and the deployment of green building codes. However, public lighting has the most impact on city authorities' budgets; therefore, energy efficiency improvements in this sector are a priority.

A comprehensive approach to integrated transportation planning is needed in all three pilot cities. The transportation sector in all pilot cities is responsible for significant energy consumption and GHG emissions and is thus a key target for action. The transportation sector is the single largest user of energy in Cebu City (51 percent), Da Nang (45 percent), and Surabaya (40 percent) and contributes significantly to GHG emissions in each of the three cities (Cebu City 40 percent, Da Nang 46 percent, and Surabaya 20 percent). Plans to implement public

**Figure O.8 Level of Influence of City Governments in Various Sectors**

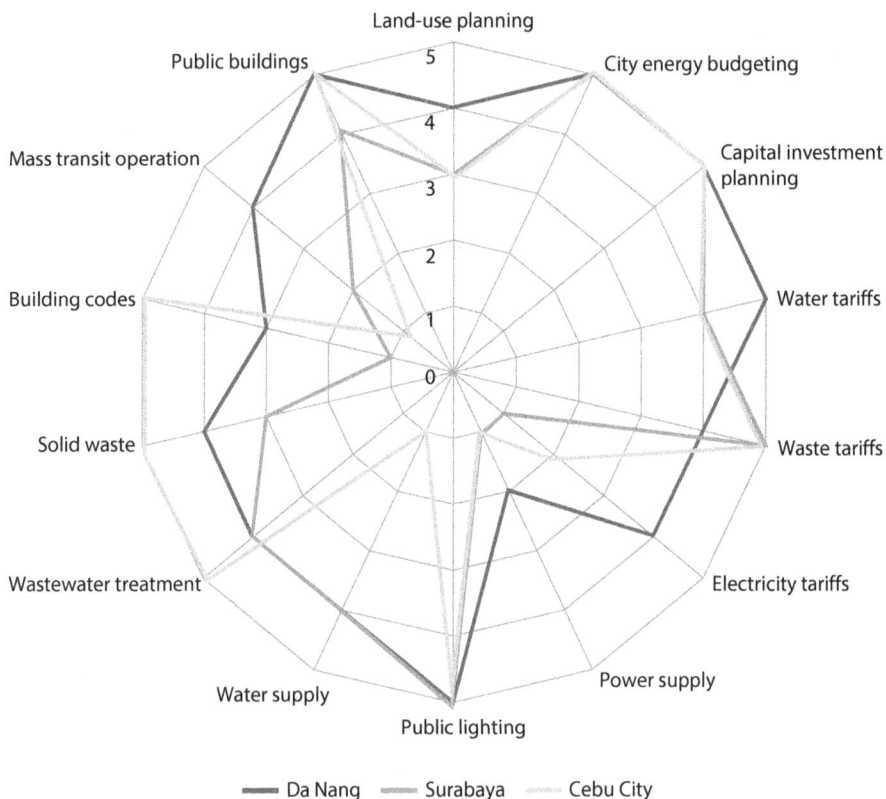

Source: Phase I pilot study.
Note: 0 = No influence; 5 = Maximum influence.

transportation systems in the pilot cities were generally not aligned with wider city-level planning strategies, giving rise to the potential for unforeseen challenges affecting the overall energy performance of the sector. Furthermore, trends in the three cities show a shift from nonmotorized and public transportation to private vehicles. Compounding the problem of growing demand for private transportation, all three cities experience problems with the low fuel efficiency of their current vehicle fleets. Public transportation has the potential to reduce energy consumption and GHG emissions as well as to alleviate growing congestion and pollution problems, but the quality of public transportation in each of the three cities is deteriorating. High-capacity transit systems are absent in all pilot cities.

Building stock is set to double in the region during the next 20 years, and the sector's energy consumption is projected to grow by 30 percent under a business-as-usual scenario (IEA 2010). EAP cities have low building energy intensity (kilowatt-hour per square meter or kWh per m$^2$) because the existing building stock consists predominantly of smaller, low-rise buildings with basic lighting, air conditioning, and appliances. However, new buildings are

**Figure O.9 Primary Energy Consumption by Sector**

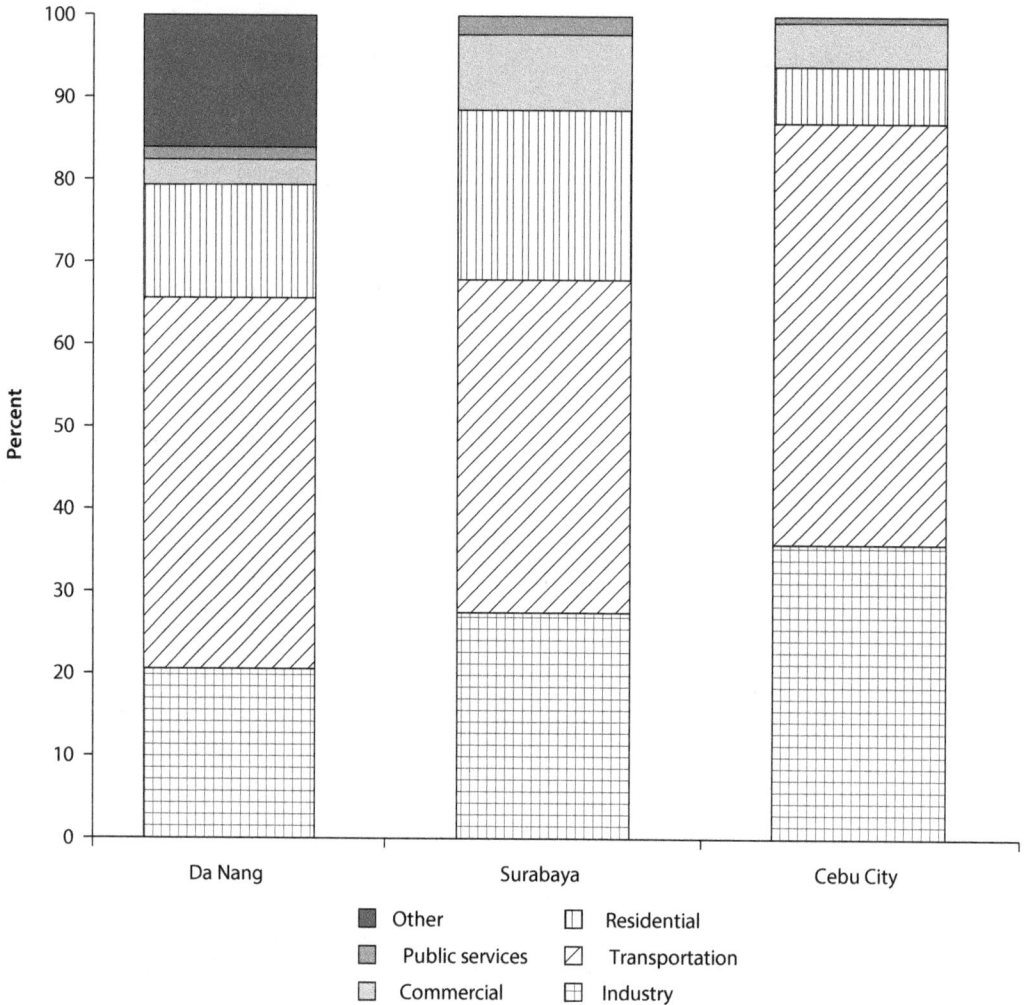

*Source:* Phase I pilot study.

responsible for a rapid increase in energy consumption in the sector because they have larger floor areas, and increased air conditioning, ventilation, lighting, appliances, and computers. Therefore, of all energy-consuming sectors, the buildings sector holds the most promise for cost efficient energy reduction. Voluntary green building codes, which exist in all three pilot cities, reflect this opportunity. However, none of the green building codes have been implemented within city boundaries yet because most developers and financiers still perceive the associated market risks to be too high.

EAP cities are recycling waste through informal methods, but action is needed to capitalize on energy recovery and composting potential. As household income grows within the region, waste generation is expected to increase, underscoring the future need for formalized methods of waste processing. However, landfill gas capture projects appear to be difficult for the pilot cities—proposals for old landfills in two of the three pilot cities were not successfully concluded largely as a result of technical and institutional challenges. Cities need to be better equipped to take advantage of specialized procurement and funding arrangements such as public-private partnership opportunities and carbon finance.

The water and wastewater sectors face numerous challenges, including high leakage rates, lack of city-scale infrastructure, and low demand-side efficiency awareness. EAP cities can take action to improve or develop their centralized infrastructure by prioritizing energy efficient water resources, upgrading pumps, and addressing the high leakage rates that characterize the region. On the demand side, lack of awareness is a major challenge. Some EAP cities are implementing awareness campaigns, but as personal wealth and demand for resources increase, more aggressive demand-side management initiatives will be required.

Managing electricity peak load and increasing the efficiency of electricity supply are major challenges for the three pilot cities, and also provide opportunities for the application of smart grid technologies. Within the power sector, direct city-driven interventions to affect energy efficiency performance are limited to the demand side and to the provision of decentralized renewables, underscoring the need for cities to leverage their influence systematically at the national level to shape this sector. Cities can, however, successfully manage the demand side to reduce the extent of capacity expansion and can level off peak loads, mitigating the risk of rolling blackouts. The recent advances in smart metering and grid-control technologies offer opportunities for cities to work with electricity suppliers to promote and harness demand-side energy efficiency and reduce distribution losses.

Many cities in Southeast Asia, including the three pilot cities, have the capability to increase the use of renewable energy and distributed generation. However, none of the pilot cities has a renewable energy master plan that can guide a scale-up of renewable energy production at the city level. Another way in which cities can improve efficiency of energy conversion and reduce GHG emissions is by increasing the use of natural gas and liquefied petroleum gas, including for distributed generation and other applications where it can cost-effectively replace other fuels and coal-fired grid-based power generation. However, these options are fairly limited at the moment because subsidies (at the national level) favor large power producers.

In the public lighting sector, EAP cities can take advantage of potential energy efficiency gains by replacing existing metal halide bulbs with more efficient technologies such as high pressure sodium bulbs or light-emitting diodes. In addition, optimized operation (on and off times) and maintenance regimes can further reduce energy consumption by this sector. The public lighting sector is characterized by a high level of city government influence and accounts for a considerable proportion of overall city energy expenditure. However, limited awareness of technology benefits and available financing are major barriers. All cities displayed some confusion about the benefits of energy efficient lighting technologies and concern about the capital funds required for replacement programs. Cities can address these issues through pilot studies to demonstrate improved energy performance and financial viability.

## The Role of Institutions

The institutional and policy environment is emerging, but EAP cities still lack cohesive, citywide regulatory frameworks in the energy sector. The degree of regulation and government oversight in the energy sector varies by country, but national and regional governments are always critical for energy sector management and regulation, and the role of city government in setting energy sector policies and regulations is limited. However, this limitation does not prevent cities from planning and deciding what, where, and how urban energy infrastructure should be built. Furthermore, cities can take steps to influence national policies while continuing to influence local behavior through voluntary programs and incentives. Because city governments are intimately involved with every aspect of urban development and management and wield power and influence over urban energy demand, they have the unique ability to tie urban energy supply to demand.

Promoting sustainable urban energy development requires cities to build institutions, strengthen energy governance, and create conducive regulatory frameworks. Despite their significant regulatory powers, most cities in Southeast Asia have not effectively pursued sustainable energy planning and management practices. Several development gaps need to be bridged, including improvement of legal and fiscal frameworks, land-use planning, and development practices; leveraging of emerging technologies and innovations; and improvement of the institutional structures for effective monitoring, reporting, and management of city energy use and emissions. Filling the institutional gaps will help EAP cities to ensure that urban energy supply is reliable, efficient, and affordable, and that energy demand and emissions are efficiently managed. Cities can optimize operating costs, improve air quality, and improve quality of infrastructure services, while concurrently supporting economic development and climate change mitigation objectives. Given energy and GHG emissions' cross-sectoral nature, city governments need to evaluate options holistically. Within the urban planning framework, options must be assessed across sectors as well as across time because sustainability initiatives do not always have immediate impacts or quick roll-out strategies.

During Phase I of SUEEP, an institutional mapping exercise was undertaken in each city to establish the principal agencies and actors involved in the delivery and management of services that affect the city's energy efficiency and GHG emissions profile. The outcome, presented in figure O.8, shows that each pilot city generally has control of investment, budgets, and several local energy efficiency activities, as well as of certain aspects of public lighting, public buildings, and wastewater treatment and water tariffs. The main areas in which the city authorities' influence is limited relate to power supply and electricity tariffs.

Significantly, in most other energy-related areas there is a mix of influence concentrated at the city level, with coordination between local agencies more likely if facilitated by the city government. This situation reinforces the need for city governments to be aware of their local circumstances so as to tailor their plans to their own needs and governance structures. The institutional mapping exercise is also useful for understanding the range of stakeholders that are involved in energy planning, delivery, and management, and serves to communicate to all parties the relevant responsibilities, ensuring that all understand why each stakeholder is involved. Furthermore, the institutional mapping exercise provides a means for understanding the interplay between agencies.

The limited number of policies enacted at the city level stems from the requirement to coordinate and institutionalize energy policies at a national level as part of the sustainable energy planning process. Implementing energy policies at the national level is more efficient and effective and ensures consistency in application throughout the country. It is in the national government's best interests to work with individual cities on national-level policies and plans, because the efforts of city governments will be essential in the effort to achieve national energy and emissions policy goals and targets. National governments should provide clear guidance to cities on the direction they will take on sustainable development to enable cities to plan and, where possible, cooperate on issues for which national and city goals are aligned. In these areas, policies implemented by both the city and national governments can serve to reinforce each other, thus making efforts to develop the city sustainably more effective. The city government's key task would be to take the lead in energy and emissions planning and to implement and advocate for changes that contribute to the advancement of its goals. City governments must recognize the level of influence they have in implementing policies at the city level and be fully responsible for sectors over which they have significant control (for example, street lighting, public buildings, and wastewater treatment). For sectors in which national policies affect the city, the city government should work closely and establish a strong dialogue with the national government. The city can seek support or financing for measures that are compatible with national goals and can ensure that city policies are not negatively affected by national ones. City governments should also work with national agencies and departments to coordinate activities and develop a mutual understanding of responsibilities and expectations.

A comprehensive sustainable urban energy and emissions plan should demonstrate a clear understanding of which of its components can be aided through national programs, and its leaders should engage with national agencies from the outset to determine the level of support that may be forthcoming. In discussing areas of common interest between the city and the national governments, cities can use the opportunity to secure support or financing from the national government to implement national policies. For example, in Da Nang several programs with significant energy efficiency components were implemented at the city level with coordination and cofinancing by a national agency (see chapter 8).

## Governance Mechanisms

SUEEP results in a strategic plan guided by a "vision" and a set of objectives that city authorities seek to achieve. Reliability, efficiency, and affordability of energy supply; reduction of GHG emissions; and the city's adaptation to climate change should be strongly featured as strategic objectives in this process. An integrated strategic planning process enables providers of public services, from mass transportation to wastewater treatment, to contribute and identify opportunities that would lead to greater energy efficiency, cost savings, and reduced GHG emissions, while helping to define investment programs that respond to future demand projections.

To maintain momentum and traction after development of the plan, an institutional governance mechanism is required to formalize and govern its implementation. One of the principal institutional recommendations is for each of the pilot cities to establish a citywide energy and emissions task force to improve coordination and establish an integrated approach to energy planning and management. In all three cities, a committee or like group has already been entrusted with the governance of various aspects of energy and emissions management. Existing committees can be used by extending their mandates through broader terms of reference and enhanced powers (if necessary) to take on the role of the energy and emissions task force.

Because existing urban planning processes in the pilot cities rarely include a comprehensive citywide assessment of and plan for energy needs and systems, the pilot cities could consider proposing the plan for formal review and agreement by an executive authority to embed the energy and emissions plan into a long-term citywide strategy. By doing so, projects would no longer be at the mercy of political cycles, and the responsibility to follow through on the plan must be taken seriously. The pilot cities will have to consider the appropriateness of this route based on their respective governance structures.

Monitoring and reporting systems are crucial for providing credibility to energy and emissions management; therefore, policies need to be implemented carefully to ensure that they do not increase costs unnecessarily and that they address the needs of investors who may, at some point, become involved in financing SUEEP components. The establishment of a city-level platform for monitoring and managing energy and emissions is one of the first practical steps in implementing the vision of sustainable urban energy development and is a prerequisite for mobilizing "green financing," which requires verification and certification of reductions in GHG emissions.

Of the pilot cities, Da Nang had the most readily available energy data from a variety of government agencies; collection of energy data was much less routine in Cebu City and Surabaya. Even when data were available, sharing of data across departments was limited in all three cities. Sharing of information, experiences, and knowledge will be a major opportunity for improvement of energy efficiency for all three cities. For example, analysis of trending information on population, vehicle use, traffic, economic growth, and industry would help all city agencies to make better decisions with regard to energy consumption and energy policy.

Effective monitoring and reporting systems should be structured and formalized in a way that demonstrates a high level of integrity and reliability and should cover the following:

- *Institutional arrangements*—assign roles and responsibilities to individual agencies and personnel.
- *Boundaries*—define what the energy and emissions inventories have included and excluded.
- *Sources of data*—include data from national agencies to city government collection arrangements and so forth. How data are collected and reported (for example, through surveys or meters), and the frequency with which data are collected, should all be defined clearly.
- *Data collation methods*—provide transparency to the way in which raw data are processed. This is particularly important if proxy data or extrapolations have been used or where data are incomplete.
- *Quality assurance processes*—demonstrate that the methods, processes, and sources used have been adequately audited or reviewed to identify gaps, omissions, and potential improvements to the monitoring and reporting process.

Monitoring and reporting systems should strive for continual improvement to make them accurate, reliable, consistent, transparent, and complete.

## Next Steps

The results of the three pilot city studies demonstrated that there are significant opportunities for energy management and GHG mitigation in EAP cities, but challenges to achieving improvements remain. Lack of coordination and planning and deficiencies in technical know-how, funding, and procurement capability all impede progress. Cities can address these challenges through improved energy governance and a robust energy efficiency and GHG planning framework.

Institution building enables cities to deploy effective energy efficiency programs by improving oversight and coordination of various energy-related initiatives across sectors. All three pilot cities had begun creating appropriate governance structures, but it did not appear that they had adopted a comprehensive approach by, for instance, actively engaging a broad range of stakeholders and legislatively formalizing the energy planning bodies. Cities need to ensure that energy governance is adequately structured to allow effective communication, coordination, and action.

The pilot cities also demonstrated a need for a SUEEP process to achieve improved energy management and GHG mitigation. The SUEEP process provides a comprehensive approach to planning to maximize energy efficiency

Energizing Green Cities in Southeast Asia  •  http://dx.doi.org/10.1596/978-0-8213-9837-1

across sectors, with the intent of helping cities to develop their own initiatives using different mechanisms. The SUEEP process also defines governance systems for implementation and for monitoring and reporting, which are important outcomes because they improve energy governance in the city and create a common platform for collaboration between the city and donors, civil society, and the private sector. The SUEEP process also provides a framework enabling city governments to prepare a pipeline of investments in energy efficient infrastructure (from mass transportation to wastewater treatment) as well as to mobilize "green financing" support. Despite the benefits of such a holistic approach, none of the pilot cities had established an integrated approach to energy planning or GHG mitigation, and the study revealed significant conflicts between various city plans, for example, transportation planning and land-use planning.

Although a comprehensive approach to planning, like the SUEEP process, is ideal, cities have different levels of capacity, resources, and priorities. Given this, city governments could engage with energy planning at three different levels. At the first level, the pilot cities could undertake a high-level, rapid assessment of its energy efficiency measures (for example, TRACE). The second level entails deeper sectoral engagements in selected areas (for example, PPPs and sector-wide interventions). A city that is fully committed to comprehensive planning could approach it from the third level—implementation of the full SUEEP guidelines.

City governments are in a unique position to achieve SUEEP goals through their own activities. Although their institutional arrangements and capacity are still works in progress, city governments perform several roles, including policy making, regulation and enforcement, and leading and facilitating across a broad range of stakeholders. A city government is the only body with this unique blend of roles and relationships, reinforced by the credibility and tenure to facilitate citywide, comprehensive, and strategic energy and emissions planning. City governments should also engage in strong dialogue with national governments, and where appropriate, align SUEEP with national and regional programs. This would attract broader support for SUEEP, which could be channeled to locally led initiatives and specific projects driven by the priorities and unique set of circumstances in each city.

SUEEP is a continuous process that evolves through institutional reform, capacity building, and implementation of priority investments. To develop a successful SUEEP process, cities must establish sustained city government commitment, create a baseline based on energy and emissions diagnostics, articulate a vision and goals, prioritize projects, develop an implementation plan, and regularly monitor and report on implementation progress. SUEEP has many benefits, including the following:

- Identification of the principal energy and emissions issues facing the city
- Establishment of what can be achieved by the city government, as well as with the aid of other agencies, and how it can be achieved
- Integration of energy and emissions issues into wider city planning processes
- Coordination across sectors and agencies and the establishment of shared goals

- Establishment of governance, monitoring, and reporting processes that are essential to future management of the issues and that are prerequisites for third-party involvement in project and carbon finance

Building on the three-city pilot work completed in Phase I, the project team developed an SUEEP Guidebook and Toolkit (see http://www.worldbank.org/eap/energizinggreencities) that cities can use to facilitate the development of their institutional capacity-building programs and sustainable energy and emissions plans. The Guidebook (part III of this volume) and Toolkit are designed to give cities a starting point to begin planning for a more energy efficient development path, guiding leadership through each key step of the process, including crafting a vision statement, forming a task team, communicating with stakeholders, measuring urban energy consumption and emissions, setting green targets, preparing a sustainable urban energy and emissions plan, implementing and financing the plan, and ongoing monitoring and reporting (see figure O.10).

The SUEEP Guidebook will be tested by the pilot cities and refined to support regionwide replication. In addition to assisting pilot cities to develop their own SUEEP processes and create and start implementing their own sustainable urban energy and emissions plans, the next phase of the program will be the

**Figure O.10  SUEEP Process**

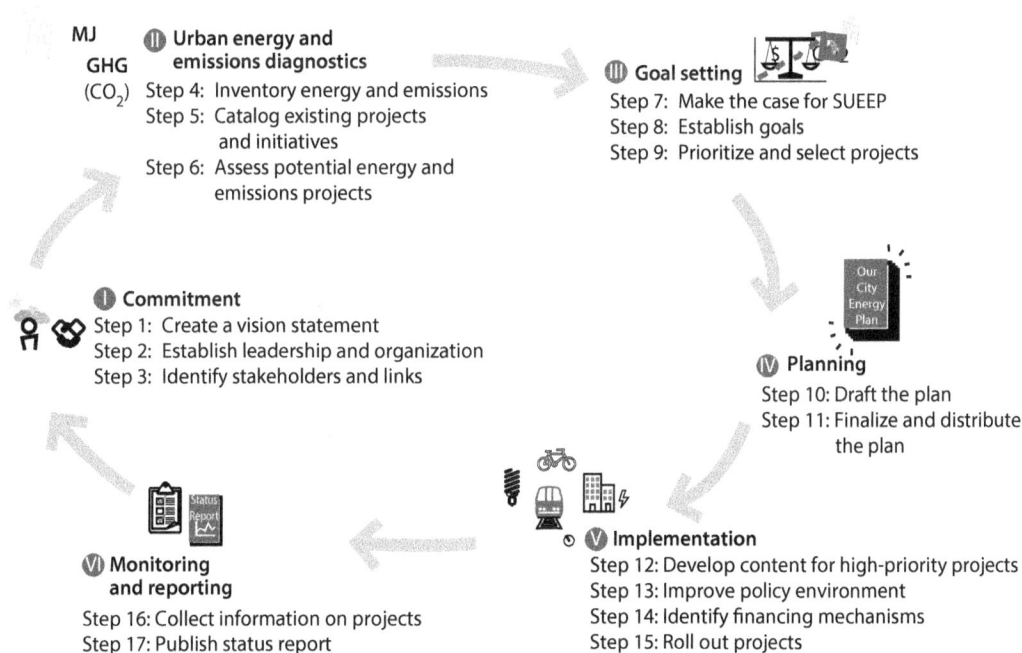

MJ
GHG
($CO_2$)

**II Urban energy and emissions diagnostics**
Step 4: Inventory energy and emissions
Step 5: Catalog existing projects and initiatives
Step 6: Assess potential energy and emissions projects

**III Goal setting**
Step 7: Make the case for SUEEP
Step 8: Establish goals
Step 9: Prioritize and select projects

**I Commitment**
Step 1: Create a vision statement
Step 2: Establish leadership and organization
Step 3: Identify stakeholders and links

**IV Planning**
Step 10: Draft the plan
Step 11: Finalize and distribute the plan

**VI Monitoring and reporting**
Step 16: Collect information on projects
Step 17: Publish status report

**V Implementation**
Step 12: Develop content for high-priority projects
Step 13: Improve policy environment
Step 14: Identify financing mechanisms
Step 15: Roll out projects

*Source:* Phase I Pilot Study.
*Note:* $CO_2$ = carbon dioxide; GHG = greenhouse gas; MJ = megajoules; SUEEP = Sustainable Urban Energy and Emissions Planning.

creation of a web-based platform that cities can use to measure and report energy consumption and GHG emissions, which is essential for the creation of carbon assets (such as GHG emissions reduction credits) and mobilization of carbon financing support.

Looking ahead, the attainment of long-term, sustainable urban energy and emissions development is not a goal easily defined or achieved. But the SUEEP process and its partners bring EAP cities one step closer through three critical contributions briefly described next.

### Institutional Development and Capacity Building

The SUEEP process introduces a number of key foundation-building activities required to support long-term urban green growth strategies. The SUEEP guidelines bring clarity and international best practices to the institutional reform, policy development, and stakeholder outreach processes necessary to achieve targets. The SUEEP process also includes accounting tools that cities can use to quantify their energy consumption and GHG emissions for use in target-setting, as well as for ongoing monitoring and reporting of results and implementation progress.

### Creation of a High-Quality Pipeline of Green Investments

Policy and institutions alone will not create green growth outcomes—investments in energy efficiency improvements and GHG mitigation activities will also play an important role. Through the SUEEP process, city leadership can evaluate investments comprehensively, based not only on their fiscal return, but also on their relative green impact and contribution to other social and economic development goals. The result is a well-defined pipeline of green investment projects that can be communicated not to just local stakeholders and financing institutions, but also to the international donor community and potential partners, including private investors.

### Mobilization of Financing

The international donor community's interest in supporting sustainable infrastructure for green growth in rapidly developing EAP cities is substantial. However, there have been many challenges: defining green city goals; identifying those activities that would optimally support green growth goals; ensuring local governments have the capacity and institutional structures needed to support both construction and maintenance of green investments; and identifying means to measure success. The SUEEP process attenuates these challenges by (a) building an institutional and policy foundation for supporting green investments; (b) setting up a quantitative system of indicators for identifying green growth targets and monitoring and reporting progress over time; and (c) creating a long-term green growth plan and a well-defined, thoroughly evaluated pipeline of bankable investments that can be easily communicated to potential investors and financiers.

## Notes

1. The Association of Southeast Asian Nations, or ASEAN, consists of 10 member countries: Brunei Darussalam, Cambodia, Indonesia, the Lao People's Democratic Republic, Malaysia, the Republic of the Union of Myanmar, the Philippines, Singapore, Thailand, and Vietnam (OECD 2010).
2. For further details please visit http://www.esmap.org/esmap/node/235.

## References

ACE (ASEAN Center for Energy). 2011. "The 3rd ASEAN Energy Outlook." http://www .energycommunity.org/documents/ThirdASEANEnergyOutlook.pdf.

IEA (International Energy Agency). 2008. "Energy Use in Cities." In *World Energy Outlook 2008*. Paris: Organisation for Economic Co-operation and Development/IEA.

———. 2010. *World Energy Outlook*. Paris: IEA.

———. 2012. World Energy Balances, IEA World Energy Statistics and Balances (database). http://www.iea.org/stats/.

OECD (Organisation for Economic Co-operation and Development). 2010. *Southeast Asian Economic Outlook 2010*. Paris: OECD.

Tool for Rapid Assessment of City Energy (TRACE). Energy Sector Management Assistance Program (ESMAP), World Bank, Washington, DC. http://esmap.org/TRACE.

UN (United Nations). 2010. "World Urbanization Prospects: The 2009 Revision: Highlights." Population Division, Department of Economic and Social Affairs, United Nations, New York.

World Bank. 2005. *The Dynamics of Global Urban Expansion*. Washington, DC: Transport and Urban Development Department, World Bank.

———. 2009. *World Development Report 2010: Development and Climate Change*. Washington, DC: World Bank.

———. 2010. *Winds of Change: East Asia's Sustainable Energy Future*. Washington, DC: World Bank.

Yuen, Belinda, and Leon Kong. 2009. "Climate Change and Urban Planning in Southeast Asia." *Cities and Climate Change* 2 (3).

# PART I

# Urban Energy Use and Greenhouse Gas Emissions in East Asia and Pacific

# CHAPTER 1

# Introduction and Background

## Urbanization, Economic Growth, and Impact on Energy and Greenhouse Gas Emissions

*The movement of people from rural areas to cities is a prominent feature of the development process and a defining feature of the twenty-first century. The trend is likely to continue. The United Nations projects that approximately 60 percent of the world's population will live in cities by 2030 as a consequence of the rapid urbanization in developing countries (see figure 1.1).*

Cities, which account for about 70 percent of global gross domestic product (GDP), are engines of economic growth.[1] Their rapid expansion is shaping energy demand and greenhouse gas (GHG) emissions. According to the International Energy Agency (IEA), cities accounted for about 67 percent of global energy requirements and 71 percent of world GHG emissions in 2006 (IEA 2008). These figures are forecasted to grow—by 2030, cities' share of energy demand and associated GHG emissions are projected to increase to about 73 and 76 percent, respectively (UN 2010). IEA further estimates that approximately 80 percent of the growth in energy use and 89 percent of the growth in GHG emissions by cities is expected to occur in developing countries, where the vast majority of people remain underserved by basic infrastructure services and where city authorities are underresourced to shift current trajectories.

Urban growth and increased energy use are fundamentally linked to the economic growth that cities need to develop sustainably. Global urban energy use patterns show an increasing trend in energy intensity, measured by energy used to produce one unit of GDP. According to IEA, global city energy use will grow by 1.9 percent per year, from approximately 7,900 million tons of oil equivalent (Mtoe) in 2006 to 12,374 Mtoe in 2030, compared with an expected overall global growth rate of 1.6 percent in energy use in the same period. Because of high levels of energy intensity in urban areas, the pattern of energy use in cities will increasingly shape global energy use.

The share of natural gas (82 percent) and electricity (76 percent) consumed in cities in 2006 was disproportionate to the average consumption of all fuels (see figure 1.2), and much higher than the share of the world's population

**Figure 1.1  Urbanization, 1950–2050**

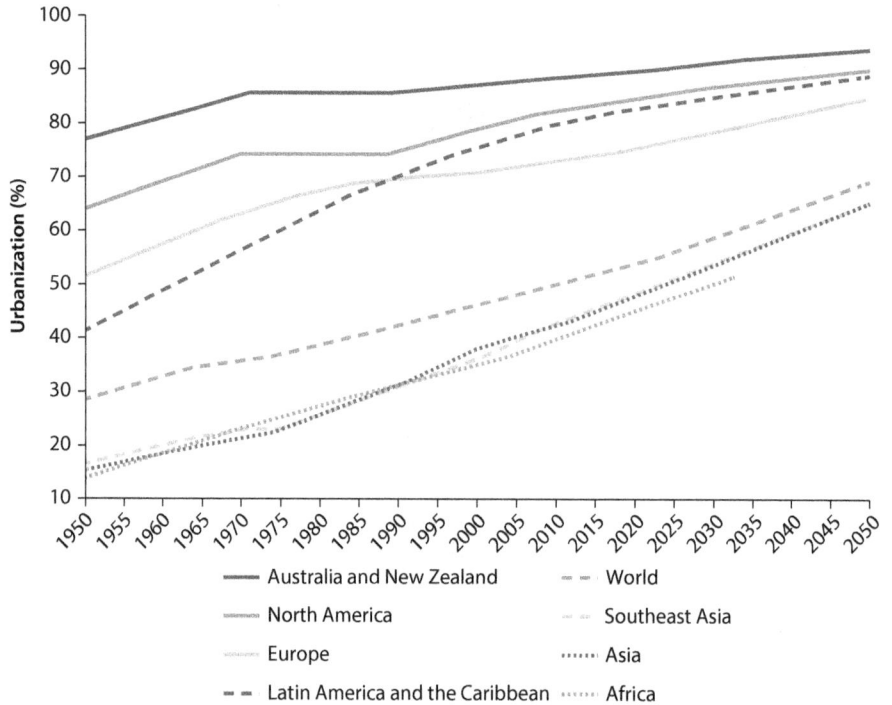

*Source:* UN 2010.

living in cities. This imbalance results from the more extensive infrastructure in cities for energy distribution and higher appliance ownership rates in developing-country cities relative to rural areas. Coal consumption in cities also accounts for 76 percent of the global total, mainly from coal-fired power generation. Figure 1.2 also illustrates that the share of oil consumed in cities (63 percent) is smaller than the average of all fuels, resulting from the higher penetration of electricity for heating and cooking and wider use of motor vehicles for mobility of goods and people in urban areas. The share of biomass and waste consumption, at 24 percent, is much lower in cities than in rural areas. Use of renewables in cities is higher (72 percent), mainly for power generation in Organisation for Economic Co-operation and Development countries.

Given the close links between rapid urbanization, economic growth, and energy and GHG emissions growth, the way cities are planned, financed, and managed will be a determinant of sustainability outcomes and of the lives and livelihoods of city residents today and into the future, thereby placing cities at the forefront of the climate change and sustainability agenda. Therefore, policy makers have a vital role in directing the sustainable development of their cities through the identification of the main drivers of energy consumption and GHG emissions—and the implementation of policies and measures that aim to minimize the growth of these factors while enabling economic growth.

**Figure 1.2  World Energy Demand in Cities, 2006**

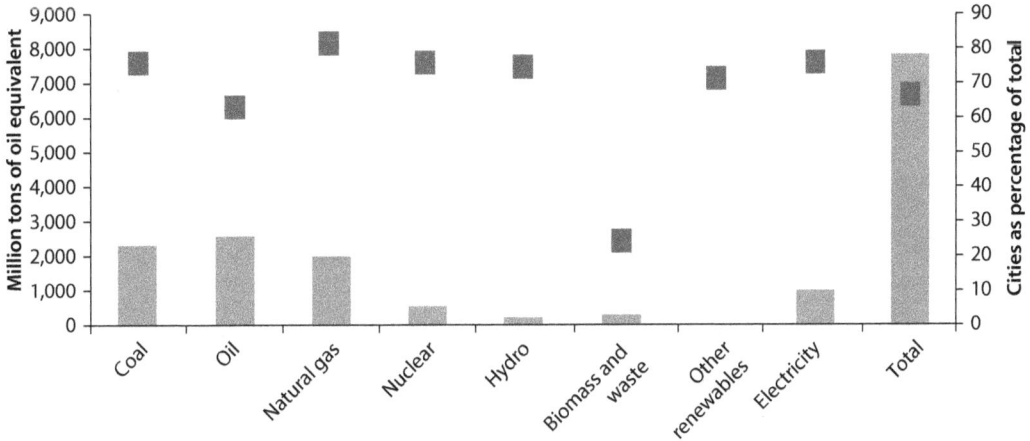

*Source:* IEA 2008.
*Note:* Blue (left scale) is the worldwide energy consumed by fuel type. Orange (right scale) is the proportion of that fuel that is consumed by cities.

The rest of the chapter focuses on Southeast Asia. It is one of the world's least urbanized regions, but its urban population is growing rapidly—the rate of population growth is 1.75 times faster than worldwide urban population growth (Yuen and Kong 2009). This rapid urbanization rate means that Southeast Asia faces huge challenges in meeting increasing energy demand in a sustainable manner.

### Southeast Asia: Trends in Economic Growth and Urbanization

According to the *Southeast Asian Economic Outlook 2010*, economic indicators point toward steady economic growth in the region based on strong growth of exports, domestic consumption, and private investment, supported by improved business sentiment (OECD 2010). The average economic growth rate of the six major countries in Southeast Asia (Indonesia, Malaysia, the Philippines, Singapore, Thailand, and Vietnam) reached 7.3 percent in 2010, compared with 1.3 percent in 2009. These six countries constituted 95 percent of total GDP and 86 percent of total population in Association of Southeast Asian Nations (ASEAN) countries in 2009.

Real annual GDP growth in the six Southeast Asian countries is projected to average 6 percent during 2011–15 (see figure 1.3). The region's steady growth will be led by above-average growth rates in Indonesia and Vietnam, supported by economic growth that is likely to rely on domestic consumption and investment.

Economic growth in Southeast Asia is inducing the migration of people seeking work, thereby resulting in an increase in urban population. Between 1950 and 2010, the urban population in the region (currently about 241 million people) grew from 15.5 percent to 41.4 percent and is expected to increase to 50 percent by 2025.

**Figure 1.3  Real GDP Growth Rates in Southeast Asia**

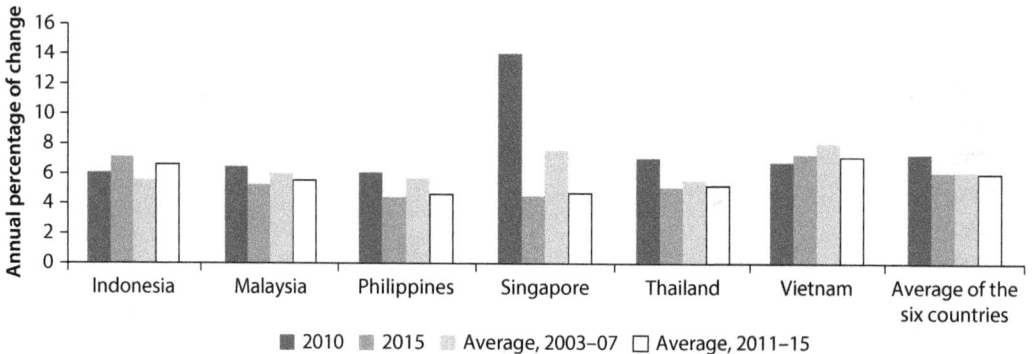

Source: OECD 2010.
Note: GDP = gross domestic product.

The effects of urbanization include a significant increase in an educated urban middle class that enjoys greater disposable incomes and uses its newfound wealth to purchase goods other than basic necessities. Stability of income and employment and increasing levels of home ownership are also characteristics of increased urbanization. Most important, consumerism in the East Asia and Pacific (EAP) region has become a significant force in the urban economy.

### Effects of Urbanization on Energy Demand and Energy Type

With urbanization and economic growth, energy consumers are expected to shift from carbon dioxide ($CO_2$)-neutral energy sources (such as biomass and waste) to $CO_2$-intensive energy sources (fossil fuels), leading to increasing GHG emissions in cities (IEA 2008). This trend is apparent in Southeast Asia. Urbanization and economic growth in Southeast Asia have not only resulted in a continued rise in the proportion of global energy consumed by cities, but have also caused profound impacts on the type of energy used. Oil and gas dominate the total primary energy supply mix of the five major economies of Southeast Asia (Indonesia, Malaysia, the Philippines, Thailand, and Vietnam), with coal accounting for a smaller portion of the energy supply mix (see figure 1.4).

Although coal accounts for a smaller portion of the energy supply mix than do oil and gas in these five economies, dependence on coal has substantially increased during the past three decades. Coal-exporting countries such as Indonesia and Vietnam have coal reserves and have significant incentives to continue increasing the share of coal in their energy mix. For coal importers— Malaysia, the Philippines, and Thailand—dependence on imports to meet coal demand does not raise energy security concerns because of the relatively low share of coal in their energy mixes and the less volatile nature of coal prices (compared with oil) in the international market. For these reasons, the countries in Southeast Asia are likely to increase their use of coal significantly unless the environmental costs are internalized sufficiently to reverse this trend.

Southeast Asia's dependence on oil as an energy source has declined during the past three decades, although the extent of this decline has varied appreciably

**Figure 1.4  Oil and Gas Dominate Southeast Asia's Primary Energy Supply Mix, 2009**

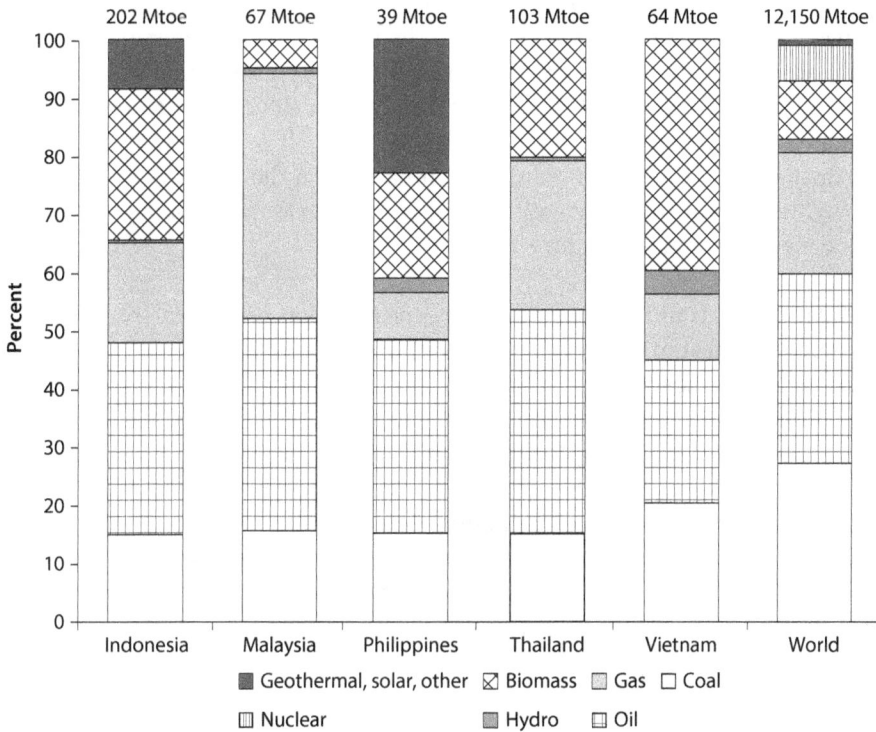

*Source:* IEA 2010.
*Note:* Mtoe = million tons of oil equivalent.

among the countries. Vietnam increased its share of oil in total primary energy consumption from 1980 to 2006, but Indonesia, Malaysia, and Thailand reduced their shares of oil to 40–50 percent from 80–90 percent in the same period.

The five countries have significantly increased their dependence on natural gas as an energy source over the past three decades. Fuel switching from coal to gas has been particularly notable in electricity generation. In the five countries, coal's share of power generation is only 21 percent, much smaller than the 57 percent share of gas, indicating that there is still considerable room for the expansion of gas use in Southeast Asia in response to energy security and environmental concerns.

Oil and gas will remain major sources of primary energy supply in Southeast Asia (accounting for 35 and 16 percent, respectively, in the ASEAN energy mix by 2030), but coal is expected to have the fastest annual growth rate, at 7.7 percent in the next 20 years in the business-as-usual scenario. This will double the share of coal in the ASEAN energy mix from about 15 percent in 2007 to 30 percent by 2030. With a movement toward more carbon-intensive fuels, $CO_2$ emissions per unit of energy consumption will increase from 0.49 ton of $CO_2$ equivalent per ton of oil equivalent ($tCO_2e$ per toe) in 2007 to 0.63 $tCO_2e$ per toe by 2030 in a business-as-usual scenario. Even in an alternative policy scenario that is being considered by ASEAN countries to mitigate the rise of GHG emissions, the carbon intensity of energy consumption is expected to increase to 0.59 $tCO_2e$ per toe by 2030. The main reason for the rapid increase in the share of coal and the carbon intensity of energy consumption is the rapidly growing demand for electricity originating largely in the expanding urban areas that are supplied by coal-fired power plants.

An analysis of the energy consumption patterns in the five major economies of Southeast Asia over two different periods—1990 and 2009—brings the following observations to light:

- Total energy consumption and energy consumption per capita are still small fractions of that of developed countries (see figure 1.5).
- Vietnam has substantially reduced its energy intensity although there is much room for improvement because of its higher energy intensity levels in comparison with the EAP region (see figure 1.6).

### Implications of Urbanization for Energy Use

Energy is intrinsic to urban settlements, embedded in the built environment; directly used to power socioeconomic activity, transport, and communications; and enabling the provision of city government services. Urbanization facilitates the transition from traditional energy sources (oil and coal) to alternative sources

**Figure 1.5  Changes in Annual Energy Consumption and Energy Consumption per Capita**

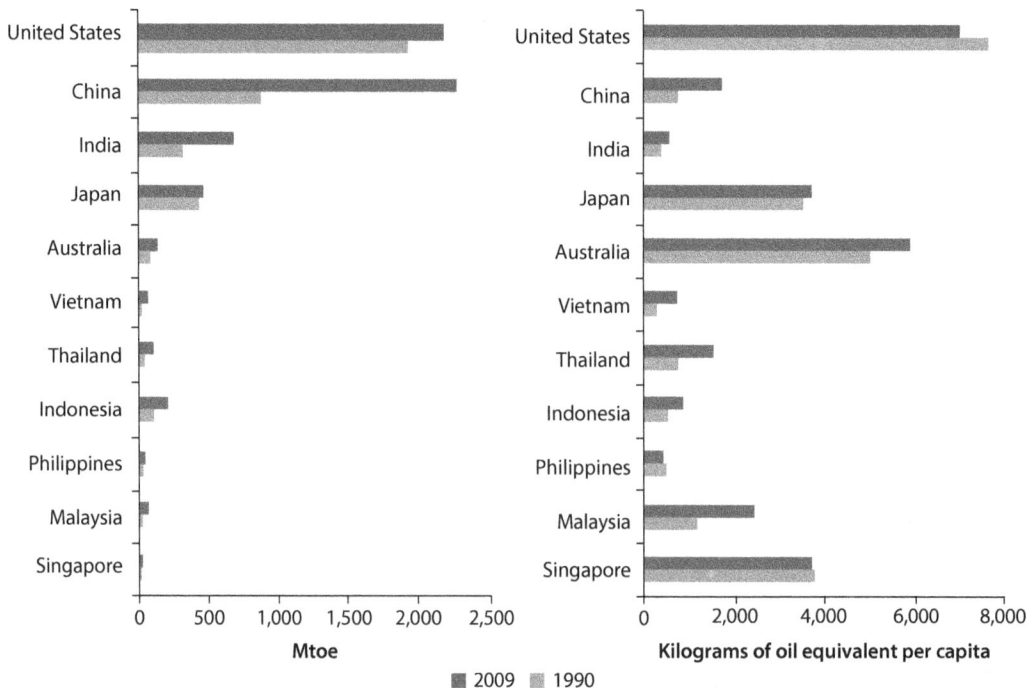

Source: IEA 2012.
Note: Mtoe = million tons of oil equivalent.

**Figure 1.6 Changes in Annual Energy Intensity**

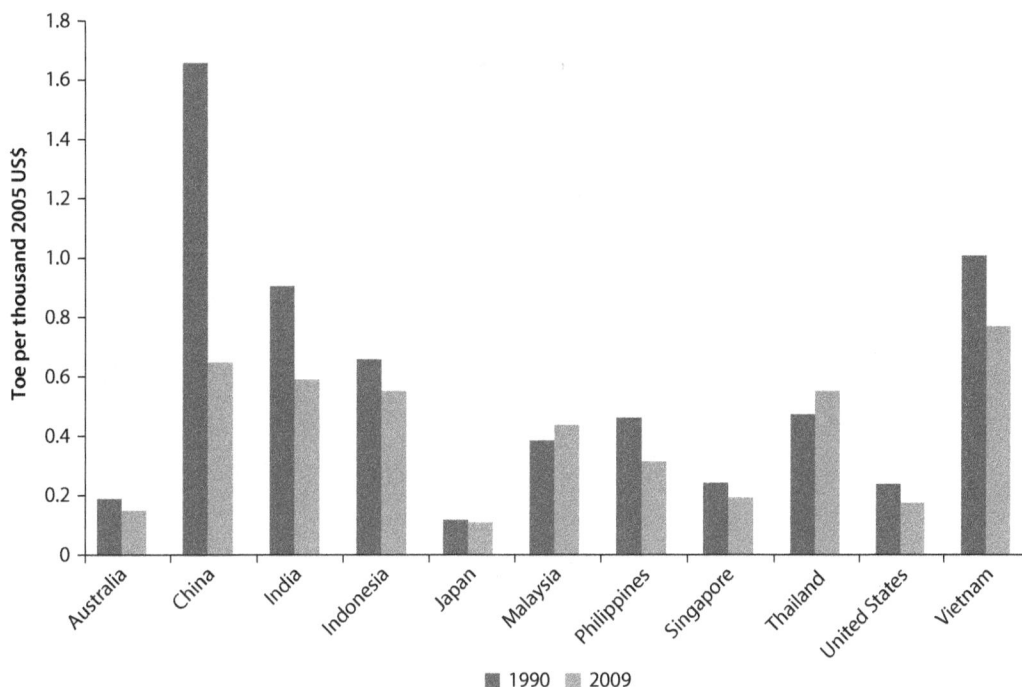

*Source:* World Bank 2010a.
*Note:* Toe = ton of oil equivalent.

(gas and electricity). The urban poor rely on biofuels to meet a major portion of their energy needs, resulting in pressure on such resources.

The growth in income in urban areas is leading to greater demand for new services, particularly services that use electricity. City sprawl, which has increasingly led to the development of areas that are not easily served by public transportation and which discourages pedestrians, has resulted in an explosion of personal motor vehicles—cars and motorized two-wheelers. In addition, the lifestyles and corresponding energy use profiles of urban middle- and upper-income residents in the developing world progressively mimic those of their counterparts in the developed world.

Despite these trends, energy usage in Southeast Asian cities is still generally low because of the relative poverty of the majority of the population. Because energy use efficiency is influenced directly and permanently by urban form and density, cities' investment choices with regard to urban infrastructure (transportation, water, energy, and so forth) and capital will have a major impact on both energy demand and associated GHG emissions.

### Implications of Urbanization for GHG Emissions

Southeast Asia contributed 5,187 million tons of $CO_2$ equivalent (MtCO$_2$e) (12 percent) to global GHG emissions in 2000, of which 59 percent was derived from Indonesia, 6 percent from Thailand, 4 percent from the Philippines,

**Figure 1.7  GHG Emissions in Southeast Asia, 1990–2000**

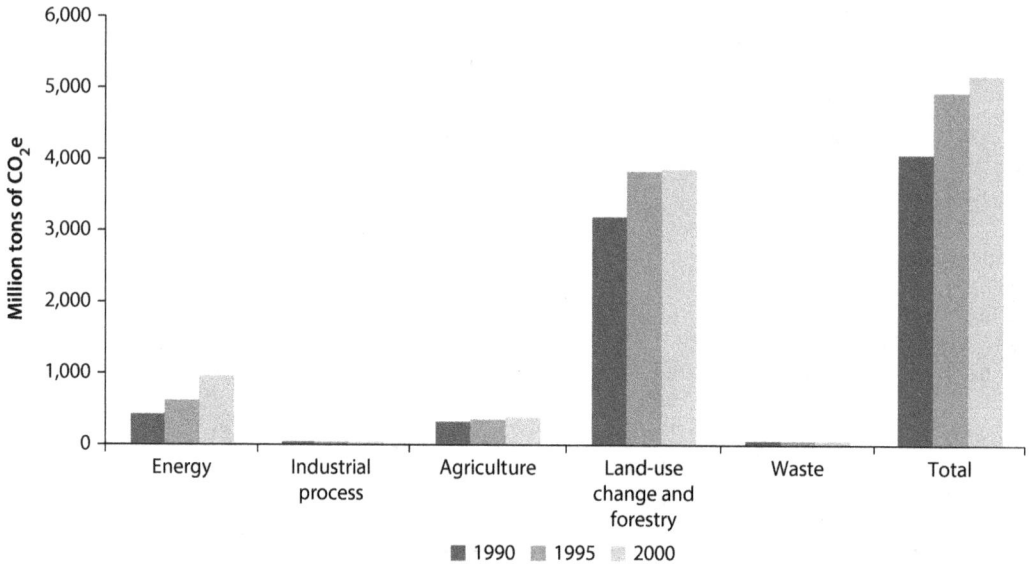

*Source:* ADB 2009.
*Note:* $CO_2$e = carbon dioxide equivalent; GHG = greenhouse gas.

2 percent from Vietnam, and 1 percent from Singapore. The energy sector was the second-largest contributor of GHG emissions in the region[2] (see figure 1.7).

Per capita GHG emissions in Southeast Asia are higher than the global average, but still low compared with developed countries. But $CO_2$ emissions are expected to double by 2030, which means that more action should be taken to enable Southeast Asian cities to manage energy use effectively to minimize the impact of $CO_2$ emissions.

Southeast Asia has 173,251 kilometers (km) of coastline and high concentrations of population and economic activity in coastal areas. Climate change will expose them to increases in sea level, river flooding, and volatile weather conditions. The poor will be most affected because they live in the most vulnerable locations (low-lying flood-prone areas, marshlands, steep slopes) and lack the resources to protect themselves.

### Increasing Concerns about Urban Environmental Sustainability

The rapidly growing cities in Southeast Asia are encountering mounting issues with environmental sustainability. They face a range of urban environmental challenges, from rapidly expanding but poorly planned cities to polluted air and water; inadequate water supply, sanitation, and energy; waste management; deficient drains; and flooding. Some of the growing concerns related to environmental sustainability include the following:

*Inadequate sewerage systems.* Piped sewer systems reach only 15 percent of the urban population of Manila (the Philippines). Domestic waste of some

192,000 tons yearly enters drains and groundwater and receives only minor treatment in unmaintained septic tanks. In Indonesia, the rate of sewerage and sanitation coverage is very low. Domestic sewerage, industrial effluents, agricultural runoff, and solid waste are polluting surface water and groundwater.

*Polluted rivers.* Rivers in Vietnam's major cities are severely polluted by untreated industrial wastewater. Lakes, streams, and canals serve as sinks for domestic sewerage and municipal and industrial wastes.

*Poor air quality.* Combustion of fossil fuels for power and transportation is causing significant deterioration in local air quality in many of the region's largest cities. The United Nations Environment Programme has ranked Jakarta, Indonesia, and Bangkok, Thailand, among the world's most polluted megacities. Poor outdoor air quality in many Southeast Asian cities has been linked to severe health impacts. Extreme levels of urban air pollution result in about 800,000 deaths per year in the world, out of which 65 percent occur in the Asia Pacific region.

*Inadequate solid waste sector.* Local governments face huge problems disposing of waste in urban areas. Common disposal methods for solid waste in Southeast Asia, such as open dumping and landfills, are becoming limited because urban local governments encounter greater and greater difficulties in finding suitable sites within city boundaries. Incineration (which has its own negative environmental impacts) is costly in the region because much of the waste consists of organic matter with a high moisture content, requiring high temperatures.

Cities can turn these environmental challenges into opportunities by promoting sustainable energy solutions across urban sectors. Good environmental stewardship in energy planning and management is essential to mitigate local, regional, and global environmental impacts that affect cities' long-term well-being. Action is required on three interconnected fronts: increasing eco-efficiency and greening of urban development; increasing inclusiveness and equity; and increasing resilience to climate change and other shocks and crises. These actions are cross-sectoral and would require fundamental changes to the current development paradigm.

### Enhancing the Institutional and Policy Environment

In general, national and regional legislators and governments are responsible for enacting energy sector policies and regulations. The degree of regulation and government oversight in the energy sector varies by country. In many large economies, the energy sector is governed by numerous policies and regulations and is influenced by a mix of government institutions arising from concerns about energy security, market competition, social and environmental issues, and other considerations. Fuel pricing is also often subject to government intervention through taxes and subsidies. Energy sector policies and regulations used to be supplycentric, but this orientation has changed substantially since the first oil crisis in 1973. Many countries now implement regulations and standards requiring minimum energy performance in energy-consuming equipment, appliances, and building components. Governments also initiate special policies and programs to create incentives for the adoption of renewable energy and energy-efficient appliances and equipment.

The multitiered and multifaceted nature of energy sector management and regulation leads to complicated institutional interactions. Institutional and regulatory settings for urban energy planning and management in the Southeast Asian region are currently not conducive to addressing challenges and capitalizing on opportunities.

National and regional governments play critical roles in energy sector management and regulation. National and regional energy policies, legislation, and regulations influence the transparency, consistency, and predictability of modern energy supply systems in cities and address common social and environmental issues. For example, national and regional governments can establish general provisions that give cities incentives to adopt sustainable energy practices. These provisions could include renewable energy feed-in tariffs that mandate electricity utilities to purchase wind- or solar-generated electricity at set prices and energy performance standards that set minimum energy efficiency levels for new appliances and new buildings. Conversely, national and regional regulations may hinder achievement of the sustainable energy measures of cities. For example, in most countries, the prevailing regulations on electric utilities discourage demand-side management and installation of distributed generation facilities, including renewable technologies.

The ability of the city government to set broad energy sector policies and regulations is limited. However, this should not prevent cities from planning and deciding what, where, and how urban energy infrastructure should be built. Cities can take steps to influence national policies, while seeking to influence local behavior through voluntary programs and incentives. Because city governments are intimately involved with every aspect of urban development and management and wield power and influence over urban energy demand, they have the unique ability to tie urban energy supply and demand together. Cities are among the most effective actors in pursuing sustainable energy solutions. Despite this, most cities in Southeast Asia have not vigorously pursued sustainable energy planning and management practices.

Several development gaps need to be bridged, including inadequate legal and fiscal frameworks; lack of financing for sustainable development solutions; inadequate land-use planning and urban land development; insufficient design approaches and methodologies, technologies, and innovations; and lack of institutional structures and capacity building for effective monitoring, reporting, and management of city energy use. Solutions to bridge these gaps are discussed below.

*Strengthening Legal and Fiscal Frameworks*
Regulatory and fiscal measures can both be implemented to internalize environmental and social costs. Legislative measures could be introduced to change market incentives and signals and bring about changes in consumers' and producers' behavior. Examples include the requirement that public vehicles in Delhi (India) use compressed natural gas, the requirement that individual automobiles in downtown Jakarta follow passenger occupancy limits, and the ban on plastic bags in Dhaka (Bangladesh).

Fiscal measures need to be introduced to reflect the true cost of providing natural resources, particularly water and energy. Costs can be more accurately reflected through measures such as progressive pricing, which actively subsidizes the poor and penalizes overuse and wastage. For example, a reasonable level of water necessary to meet basic needs could be provided for free or at a subsidized level, if needed, for social developmental reasons, and more intensive use or wastage could be priced at a progressive rate. Similar policies could be enacted in the provision of electricity. In the same vein, congestion pricing for automobiles entering downtown areas is a fiscal measure to internalize environmental costs. Such policies have been introduced in many Southeast Asian cities but need to be strengthened and more effectively enforced.

Penalizing the overuse and wastage of natural resources, particularly water and energy, would encourage businesses and households to conserve resources and adopt efficiency measures. Actively making urban infrastructure, particularly water, transportation, and buildings, more energy efficient would also reduce the carbon footprint of the city. Governments would have to take a long-term view on issues and implement policies that may be costly in the initial phases or may not result in short-term benefits—but have sustained benefits over the long term.

*Financing Sustainable Urban Development Solutions*
How urban development is financed is crucial to the sustainability of cities. Key challenges include reforming existing city government finance systems to make them more effective, accessing new external sources of finance, and building stronger links between the formal urban development finance system and financing systems for the urban poor.

Although functions and executive authority in many countries in Southeast Asia have been devolved to a local level, fiscal authority often remains at higher levels of government, and local governments are not able to spend sizable funds on urban development without authorization from higher levels. Even in countries in which local governments may have such authority, many cities lack the capacity to make full use of their theoretical powers. Reforming the city government finance sector and building local government capacity are therefore important development strategies.

Making cities sustainable will require the infusion of funds to finance investment in urban infrastructure and services such as mass transit systems, water and sanitation systems, and the like. Although internalizing environmental costs (for example, through the implementation of a carbon tax) will increase funds available for investment in urban infrastructure, additional funds will be required, at least in the short term. New sources of finance need to be tapped to provide funding for sustainable urban development.

### Improving Land-Use Planning and Urban Development

Cities in Southeast Asia have dense cores with extended suburban areas that tend to grow along transportation corridors. New urban developments are often unplanned and haphazard—with closely intermingled industrial, residential, commercial, and agricultural land uses—and lack adequate infrastructure and services. Moreover, because development occurs along transportation corridors, large tracts of land farther away from the transportation corridor are not developed, resulting in "ribbon" or "strip" development that is environmentally unsustainable and resource intensive. Urban development often extends across several provincial and jurisdictional boundaries, posing new economic, social, and environmental challenges that require a rethinking of urban planning, management, and governance approaches and institutions. Strengthening and extending planning laws to cover suburban areas is crucial for sustainable urban development. Examples include local laws and regulations in China, Vietnam, and Pakistan that create local intergovernmental agencies that cover both urban and suburban areas, empowering them to strengthen urban planning laws and regulations in these areas.

A major area for policy reform is urban land use. The commoditization of urban land must be balanced by the recognition that land is a public and environmental good. Planning and land-use restrictions need to be imposed to increase the social and environmental functions of land, such as the provision of housing to the urban poor and the development of urban green spaces, parks, and mass transit corridors.

### Promoting Innovation and Technologies

Cities are major consumers of energy and emitters of GHGs, but they also have the potential to create solutions to urban problems because they are centers of knowledge and of technological and process innovation. The EAP region is bursting with innovative technologies, approaches, and practices that could lead the way toward an inclusive and sustainable future.

Recent advances in smart metering and grid-control technologies offer opportunities for cities to work with electricity suppliers to promote and harness demand-side energy efficiency and reduce distribution losses. The challenge is to systematically identify, document, analyze, adapt, and scale up this technology along with other innovative solutions. To do so requires the creation of a fiscal, regulatory, and institutional environment that allows individuals, businesses, communities, and civil society organizations, and even government agencies, to find innovative solutions. It would also require research and training institutes to identify and analyze the reasons for the success of certain technologies and to assist governments in scaling up such technologies and innovations.

Introduction of new technologies, production processes, and innovative practices often means that those with access to information, knowledge, and capital benefit whereas those who do not have access are left behind. New technologies often mean higher prices—given the need to factor in costs associated with development—and the poor often lose out. Therefore, policies that minimize the impact on the poor would have to be developed and implemented. For example, the cost of grid-control technologies could be recovered from consumers other than those in nonurban slums, whose usage of electricity is at the most basic level.

### Strengthening Institutional Structures and Building Capacity

To improve the resilience of Southeast Asian cities to climate change impacts, institutional strengthening and capacity building are required. Knowledge about "green growth" and climate change needs to be embedded in institutions and imparted to the people who staff them so they can make effective decisions, allocate resources efficiently to reduce escalating energy costs, and manage risk and implementation of programs. To enable this, local and global data would have to be made available, and expertise in energy sector analysis, international policy best practices, and carbon finance flows will have to be developed.

Sustainable urban energy planning and practices are elusive without realistic metrics to quantify performance (using indicators) and to measure progress (using benchmarks). Indicators and benchmarks do not merely reveal gaps; they also inspire actions to better manage services without reducing affordability or compromising the environment. Developing metrics is a crucial but difficult task because every city is unique in its energy use and levels of service. Hence, indicators need to be selected carefully to avoid distortions. A small set of key energy indicators will allow the most meaningful cross-city comparisons. An energy accounting framework should be established, supported by a structured data collection template and a baseline against which progress will be measured. This framework will help city authorities effectively carry out monitoring, reporting, and management of energy uses and carbon emissions to ensure the city remains on a low-carbon growth path.

The issue of energy use cuts across many agencies, but fostering interagency collaboration is a big challenge, particularly if the costs and benefits are distributed unevenly. The functional areas within agencies—represented by technical staff members, environmental officers, budget teams, procurement personnel, and so forth—also bring unique biases, expertise, incentives, and constraints to efforts to improve energy efficiency. These constraints are especially apparent in cities in which infrastructure provision and services lag behind demographic growth and geographic expansion. These challenges are compounded by the cross-cutting nature of energy services, which often leads to decisions about their efficient use falling through the cracks. At the local level, some of these issues may be addressed through policies and programs, but strong mayoral leadership is often needed to insist that the parties work together.

### Promoting Sustainable Urban Energy Solutions

Sustainable urban energy—the provision of energy that meets the needs of the present without compromising the ability of future generations to meet their own needs, within a sound urban planning framework—is critical to sustainable urban development. By ensuring that urban energy supply is secure, reliable, and affordable and that energy demand is efficiently managed, cities can optimize operating costs, improve air quality, and improve quality of infrastructure services, while concurrently supporting economic development and climate change mitigation objectives. These objectives require consideration of issues beyond those in the energy sector, to sectors that use energy to produce urban services (notably the transportation and water supply and sanitation sectors). Lifetime cost savings and local and global externalities are key considerations for city governments in prioritizing sustainable energy measures.

Given energy's cross-sectoral nature, city governments need to evaluate options comprehensively. Within the urban planning framework, options have to be assessed across sectors as well as across time, because sustainability initiatives cannot always be rolled out quickly and do not always have immediate impacts. Effective urban land-use planning and land development provide unique opportunities to adopt "systems approaches" to addressing urban energy issues by integrating physical, socioeconomic, and environmental planning and identifying synergies and cobenefits. Land-use planning is also a sound platform for mobilizing additional and concessional financial resources because many financial institutions are looking for opportunities to invest in "green growth" despite uncertainties about the post-Kyoto[3] price of carbon.

Making cities more energy efficient, increasing accessibility to renewable energy supplies, and adopting smart grids all hedge against the risks of higher energy costs if a global agreement were to be reached to reduce GHG emissions drastically. But EAP cities should not try to address all sustainable energy options simultaneously. Pursuing actions on sustainable energy, however cost effective they may be, requires public and private investment, efforts on the part of city governments and citizens, and the strong support of regional and national governments. Cities should tailor their efforts to available resources and pursue initial steps toward sustainable energy practices that generate significant and immediate local benefits.

Adopting sustainable urban energy planning in developing Southeast Asia is a challenge. In recent years, this region has been deluged with warnings, advisories, and information about climate change and energy efficiency—often at the city level. Although city government officials would like to take action, the World Bank Group has found that the officials often do not know where to start. The city officials typically need technical assistance to help them take the first few steps toward achieving cross-sector efficiencies, including evaluating baseline energy consumption and supply options, identifying priority sectors for promoting sustainability, and achieving measurable improvements. These activities are the building blocks for the development of comprehensive, phased urban energy policies and investment strategies.

### Program Objective

With support from the Australian Agency for International Development (AusAID), the World Bank Group initiated a regional program—EAP Sustainable Urban Energy and Emissions Planning (SUEEP)—in January 2011 to provide support and guidance to city governments in the EAP region to formulate long-term urban energy strategies, keeping in mind the cities' overall development plans.

The establishment of SUEEP was a response to the report *Winds of Change: East Asia's Sustainable Energy Future* (World Bank 2010b), which was a joint publication of the World Bank Group and AusAID. This report concluded that, despite the challenges faced in mitigating the impact of climate change, it is possible for East Asia to develop a sustainable energy pathway, maintain economic growth, mitigate climate change, and improve energy security, provided that immediate action is taken to transform the energy sectors in cities toward much higher energy efficiency levels and the widespread use of low-carbon technologies. For this clean energy revolution to take place, major institutional and domestic policy reforms will be required.

### Phased Approach

SUEEP is structured in three phases; this report summarizes the outcomes of Phase I: Three City Pilots, whereby the current status of urban energy use and GHG emissions in three pilot cities in East Asia is established and a range of policy and technical measures are identified to enable the city governments to formulate long-term sustainable urban energy development strategies, which will lead to programmatic investment plans and policies (see part II of this volume).

Phase II uses the lessons learned and outputs from the three pilot city programs to develop a Guidebook and SUEEP Toolkit (see part III of this volume) designed specifically for the EAP region, enabling city authorities to access the tools and resources developed by the program. The Guidebook includes a set of guidelines that cities can use to develop their own SUEEP processes, as well as identifies areas for which cities need support in implementing these strategies.

Phase III of SUEEP is planned for 2013 and will comprise development of detailed SUEEP plans and web-based tools that can be used to help cities monitor and manage energy use and GHG emissions.

### SUEEP Phase I Approach

Phase I involved working closely with three pilot cities in three different countries in the EAP region—Surabaya in Indonesia, Da Nang in Vietnam, and Cebu City in the Philippines, to

- Assess their current energy usage, energy expenditure, and GHG emissions;
- Review existing and proposed urban plans and initiatives that affect energy usage;
- Map the institutional structures relevant to policy and decision making for sustainable urban energy planning;
- Provide information about international best practices that can be adopted by city governments; and
- Provide a framework to allow the identification of priority sectors for energy efficiency improvements, as well as to develop a sustainable energy strategy.

### *Methodological Summary*

To achieve the objectives efficiently and consistently across the three pilot cities, a methodology was established that capitalized on the work carried out by the World Bank Group's Energy Sector Management and Assistance Program (ESMAP) in 2009–10. ESMAP developed a tool (Tool for Rapid Assessment of City Energy, or TRACE) that was used during a three-month period to quickly assess energy used by a city government, benchmark energy use against peer cities, establish priority areas, and develop interventions for improved energy efficiency across a wide range of municipal services.

Alongside the application of TRACE, an energy balance assessment was undertaken in each city to obtain a reasonable understanding of the types of fuel used, the sectors that consumed energy, and the way energy was consumed within each sector. Notably, the breakdown is limited to the sector level and all data and calculations are contained within a single, practical worksheet.

In addition, a GHG inventory was compiled using the Urban Greenhouse Gas Emissions Inventory Data Collection Tool, developed by the World Bank Group in 2011. The tool covers emissions from transport, solid waste management, water and wastewater treatment, and stationary combustion (power generation, building energy usage, industry, and so on).[4] The tool aims to provide a simple, user-friendly, and easily replicable approach, and emphasizes providing descriptions of data sources, years, implied boundaries, and quality of the data.

This approach was adopted to enable a swift, high-level sector analysis for prioritizing projects with significant energy savings, GHG mitigation, or low-carbon energy generation potential. A sector-based approach was chosen because it allows energy use to be mapped to institutional structures, highlighting areas with both strong governmental influence and verified significant and beneficial opportunities. Further details on methodology are provided in appendix A.

### Report Structure

This report is structured to provide the reader with a clear understanding of Phase I of the SUEEP program, of the process undertaken in the pilot cities, and of the themes emerging from the pilot that are of interest to other cities across the EAP region. Following this background and approach chapter, the report is set out as follows:

- Chapter 2: Understanding the Cities: Energy Use and Greenhouse Gas Emissions, provides a brief overview of each of the pilot cities, followed by a description of the energy and GHG emissions profile of each city.
- Chapter 3: Sector Diagnostics: Identifying Opportunities, provides a review of transportation, solid waste, water and wastewater, power, public lighting, and city government buildings sectors; identifies key trends, barriers, and opportunities for energy efficiency and renewable energy in areas under city government control; and discusses how and why respective areas may be prioritized for action by city governments.
- Chapter 4: Governance, discusses the principal governance issues associated with energy planning and management at a city level.
- Chapter 5: Sustainable Urban Energy and Emissions Planning: The Way Forward, sets out how, given the significant body of evidence presented, city governments can establish a road map for a sustainable urban energy path in the future.

Part II provides more details on the studies undertaken in the pilot cities and the recommended courses of action for enhancing energy efficiency in the various sectors.

- Chapter 6: Cebu City, the Philippines
- Chapter 7: Surabaya, Indonesia
- Chapter 8: Da Nang, Vietnam

Part III provides step-by-step guidance to help municipal governments in the EAP region develop their own energy and emissions plans and link their aspirations to actionable initiatives to improve energy efficiency and reduce GHG emissions.

- Chapter 9: Introduction to the Guidebook, reviews why urban energy and emissions planning is important and outlines the structure of the Guidebook.
- Chapter 10: Stage I: Commitment, lays the groundwork for successful energy and emissions management by establishing high-level political buy-in to propel the rest of the process. The chapter discusses the steps for establishing political and stakeholder commitment to energy and emissions planning.
- Chapter 11: Stage II: Urban Energy and Emissions Diagnostics, explains the way in which a city collects the basic data needed to understand its energy and emissions baseline and create an energy balance and GHG emissions inventory, which will help a city identify major trends and opportunities to achieve its sustainability goals.
- Chapter 12: Stage III: Goal Setting, demonstrates how to combine the city's overarching priorities with the findings of Stage II to develop energy and

emissions goals relevant to the city. Establishing a convincing story about the importance of energy and emissions planning and how the city will benefit are crucial.

- Chapter 13: Stage IV: Planning, shows how to bring together all the knowledge and thinking developed thus far into a documented plan that clearly expresses the city's strategic focus on energy, the initiatives that will help the city achieve its goals, and how progress will be monitored.
- Chapter 14: Stage V: Implementation, reviews the steps for developing content for projects, ensuring policies are conducive to successful implementation, and accessing financing.
- Chapter 15: Stage VI: Monitoring and Reporting, takes stock of SUEEP progress and identifies components of the plan that need readjusting, thus providing crucial inputs into each successive SUEEP process.
- Appendix A describes the approach and methodology taken in Phase I of the program.

## Notes

1. "Cities" refers to all urban areas including towns; the terms "city," "urban," and "municipal" are used interchangeably in this document.

2. The largest contributor of greenhouse gas (GHG) emissions in Southeast Asia is land-use change and forestry, but the largest rate of increase in GHG emissions is occurring in the energy sector.

3. The Kyoto Protocol, which was signed in 1997, is an international agreement linked to the United Nations Framework Convention on Climate Change. The major feature of the Kyoto Protocol is the binding targets set for 37 industrial countries and the European Community for reducing GHG emissions equivalent to an average of 5 percent against 1990 levels over the five-year period 2008–12. The "post-Kyoto" period refers to the time after 2012 for which no international agreements have been made addressing the reduction of GHG emissions, which leaves the price of carbon uncertain.

4. Urban forestry considerations will be included in the tool in the future.

## References

ADB (Asian Development Bank). 2009. *The Economics of Climate Change in Southeast Asia: A Regional Review*. Manila: ADB.

IEA (International Energy Agency). 2008. "Energy Use in Cities." In *World Energy Outlook 2008*, 179–93. Paris: Organisation for Economic Co-operation and Development/ International Energy Agency.

———. 2010. *Energy Balances of Non-OECD Countries*. Paris: IEA.

———. 2012. World Energy Balances, IEA World Energy Statistics and Balances (database). http://www.iea.org/stats/index.asp.

OECD (Organisation for Economic Co-operation and Development). 2010. *Southeast Asian Economic Outlook 2010*. Paris: OECD.

UN (United Nations). 2010. "World Urbanization Prospects: The 2009 Revision: Highlights." United Nations, Department of Economic and Social Affairs, Population Division, New York.

World Bank. 2010a. "A City-Wide Approach to Carbon Finance." Carbon Partnership Facility Innovation Series, World Bank, Washington, DC.

———. 2010b. *Winds of Change: East Asia's Sustainable Energy Future*. Washington, DC: World Bank.

Yuen, Belinda, and Leon Kong. 2009. "Climate Change and Urban Planning in Southeast Asia." *Cities and Climate Change* 2 (3).

# CHAPTER 2

# Understanding the Cities: Energy Use and Greenhouse Gas Emissions

## City Overview

The pilot cities were selected for the Sustainable Urban Energy and Emissions Planning (SUEEP) Program because they were representative of "second" (not capital) cities and typify cities across the East Asia and Pacific (EAP) region. Like many other cities, these three cities are experiencing the following trends:

- Increased urban migration is fueling population growth.
- Economies are growing, and living standards are improving.
- The services sector, especially tourism, is a growing component of the economy, although the city is still heavily reliant on manufacturing industries.

Each of these factors alone could reasonably result in an increase in urban energy consumption and greenhouse gas (GHG) emissions. Together, the impact is considerable.

### Cebu City

Cebu City is the capital of Cebu Island and the center of Metro Cebu, which includes the cities of Lapu-Lapu, Mandaue, and Talisay. Cebu City has a population of approximately 800,000, and more than 2.4 million people live in the wider Metro Cebu area. Located on the coast of Cebu Island, flanked by the Cebu Straight, and surrounded inland by mountainous terrain, the city has limited opportunities for physical expansion, maintaining its relatively compact nature. Although the city has a land area of 291 square kilometers ($km^2$), the population density for the city is 2,748 inhabitants per $km^2$ because of the concentration of the population in a dense urban area of 23 $km^2$.

Cebu Island occupies a strategic position in the Philippines, being centrally located and easily accessible by air and water. Cebu City is an important port city and is well connected to the EAP region and to the Philippines capital, Manila. The city's strategic location and seaport support local trade and services, which together employ the majority of Cebu's labor force. Some 73 percent of the population of Cebu City work in trade or related services, such as banking, real estate, insurance, or community services. Another 19 percent are employed in industry, and 8 percent in agriculture and related services. The services sector is growing and is expected to continue to be the biggest contributor to gross domestic product (GDP) in Cebu City. The majority of establishments in the city are considered to be micro or small enterprises with an average capitalization of 1.5 million Philippine pesos or less. More than two-thirds of these business establishments are situated in the central part of the city, and collectively account for more than three-quarters of the city's economy. Cebu City is the second-fastest growing city economy in the Philippines after Manila, and growth rates are anticipated to continue rising for the foreseeable future.

### Da Nang

Da Nang is the largest urban center in central Vietnam, with an area of 1,283 km$^2$ and a population of more than 900,000. Population density is relatively low in comparison with the other pilot cities, at just over 700 inhabitants per km$^2$. Located at the mouth of the Han River and facing the South China Sea, Da Nang is the fourth-largest seaport in Vietnam and forms an important gateway to the Central Highlands of Vietnam as well as to Cambodia, the Lao People's Democratic Republic, the Republic of the Union of Myanmar, and Thailand.

Da Nang's economy has historically been dominated by the industrial and construction sectors. However, in 2006 services became the largest economic sector in the city as measured by gross output. Although economic growth has been concentrated in the road, rail, and seaport hubs, as well as other service-oriented industries, the tourism sector's contribution to GDP is expected to grow. As the city's economy moves from primary and secondary sectors toward tertiary and services sectors, changes in its energy demand can be expected to occur.

Population growth in the early twenty-first century was relatively slow (1.7 percent annually between 2000 and 2007), but Da Nang now appears to be poised for a significant increase in population, primarily because of migration from neighboring rural areas. In 2007, Da Nang's official population was approximately 807,000, with projections that it will increase at a 3.7 percent annual rate through 2015, and at a rate of 4.9 percent per year between 2015 and 2020, resulting in a 70 percent increase in population by 2020. Counting unregistered migrants, the population may reach nearly 1.65 million people by 2020. Da Nang's economy is also growing rapidly; between 2000 and 2007, Da Nang's gross regional domestic product grew at 12.3 percent annually and in 2009 totaled US$1.48 billion. Household income has also risen rapidly, increasing from 187,000 Vietnamese dong (D) per month in 1996 to 853,000 D/month in 2006. Despite the steep increase, household income levels are still less than the national average for the same period, due to the preponderance of low-value-added production in the city and "brain drain" to other parts of the country.

**Table 2.1  Summary of Structural and Economic Data for the Pilot Cities**

| Parameter | Cebu City | Da Nang | Surabaya |
|---|---|---|---|
| Population (m) | 0.8 | 0.9 | 2.8 |
| City area (km²) | 291 | 1,283 | 327 |
| Population density (per km²) | 2,748 | 711 | 8,458 |
| GDP/cap/year ($) | 5,732 | 1,627 | 8,261 |
| *Economic structure (%)* | | | |
|   Services | 73 | 56 | 50 |
|   Industry | 19 | 42 | 32 |
|   Agriculture/other | 8 | 2 | 18 |

*Source:* Phase I pilot study.
*Note:* m = meter; km² = square kilometer.

### Surabaya

Surabaya is a large city with a population of about 2.8 million, which is more than three times the populations of Cebu City or Da Nang. Much of the city's center is densely populated—population density can be as high as 8,000 persons per km² in some areas. The city occupies coastal terrain with a land area of 327 km², and is located in the Brantas River Delta at the mouth of the Mas River, an area that confronts a high flooding hazard. Surabaya is the capital of the Indonesian province of East Java.

Surabaya is a key node in air, water, and land transportation. The city is served by Juanda International Airport and Perak Port, one of Asia's largest and busiest seaports. Hence, Surabaya serves as an important gateway to the East Java region for both passengers and goods. Surabaya has an annual population growth rate of 0.65 percent. However, GDP grew 6.3 percent in 2008, compared with a national GDP growth rate of 6.1 percent. Surabaya's main industries are the trade, hotel, and restaurant sectors (which together account for 36 percent of GDP) and the manufacturing sector (which accounts for 32 percent of GDP). Other significant contributors to Surabaya's GDP are transportation and communications, construction, financial services, and the services sector.

See table 2.1 for a summary of the pilot cities' characteristics and map 2.1 for their relative locations.

## City Energy Profile

To understand the cities' energy profiles, an energy balance assessment was undertaken in each city to obtain an overview of energy demand and supply. In addition, a number of key performance indicators were collected and benchmarked against peer cities using the Tool for Rapid Assessment of City Energy (TRACE; see chapters 1 and 3 for a further description).

### Energy Demand

Electricity consumption per capita for the three cities is low in comparison with other Asian cities in the TRACE benchmarking tool (see figure 2.1), but energy

**Map 2.1  Locations of the Pilot Cities**

*Source:* World Bank.

intensity, or energy consumption per unit of GDP, is high (see figure 2.2). Given this fact, these cities will have to become more energy efficient to dampen sharp increases in energy demand as economic growth occurs.

Not only is electricity demand growing year on year in the three pilot cities (see figures 2.3a–2.3c), but it is outstripping population growth, a trend playing out in other EAP cities as well. Da Nang's annual growth in electricity demand is significantly higher than that in Cebu City and in Surabaya, as is its population growth, which will result in a doubling of energy demand in six years if demand growth continues unabated. These growth rates are likely to be replicated across much of the EAP region and will present a challenge to local infrastructure, utility services providers, and city governments to supply sufficient energy to satisfy growing demand. The EAP region should therefore use this as an opportunity to integrate low-carbon generation technologies into the supply mix through long-term SUEEP.

### Energy Demand by Sector
Energy use in different sectors in each city was analyzed, and clearly demonstrated that the two major energy-using sectors are transportation and industry (see figure 2.4).

**Figure 2.1 Electricity Consumption per Capita Benchmarked against Other Cities Using TRACE, 2010**

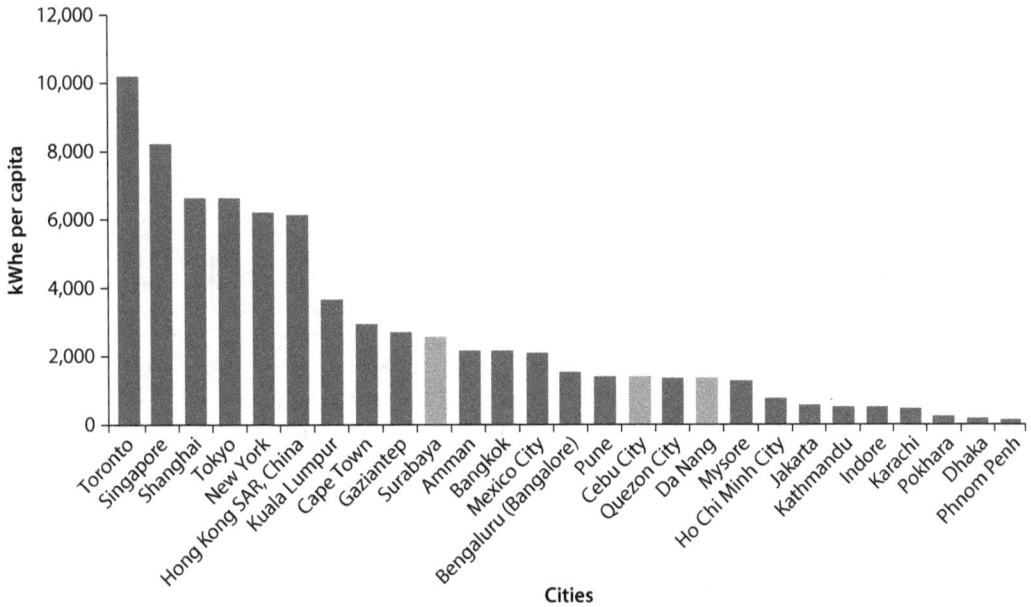

*Source:* Phase I pilot study.
*Note:* kWhe = kilowatt-hour equivalent; TRACE = Tool for Rapid Assessment of City Energy.

**Figure 2.2 Primary Energy Consumption per Unit of GDP Benchmarked against Other Cities Using TRACE, 2010**

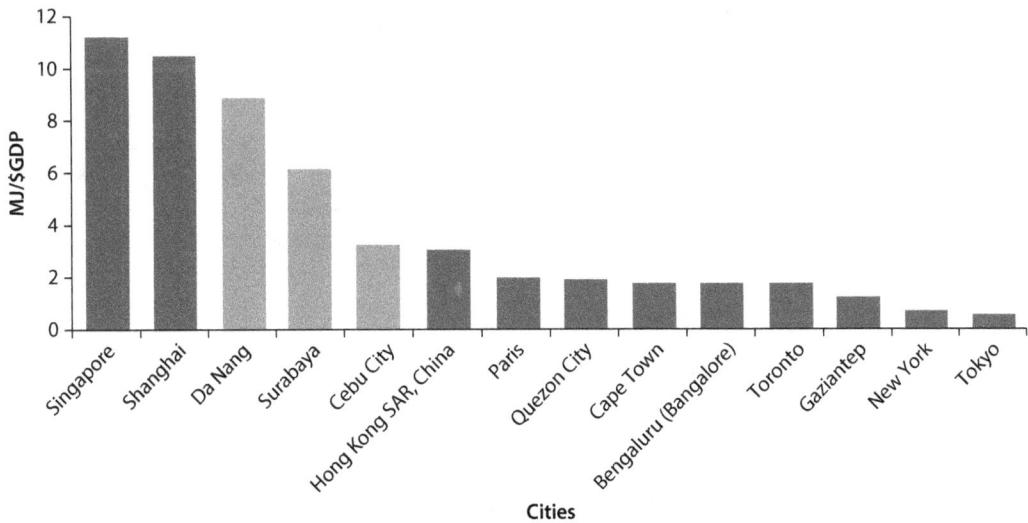

*Source:* Phase I pilot study.
*Note:* GDP = gross domestic product; MJ = megajoules.

**Figure 2.3a  Electricity Consumption in Cebu City**

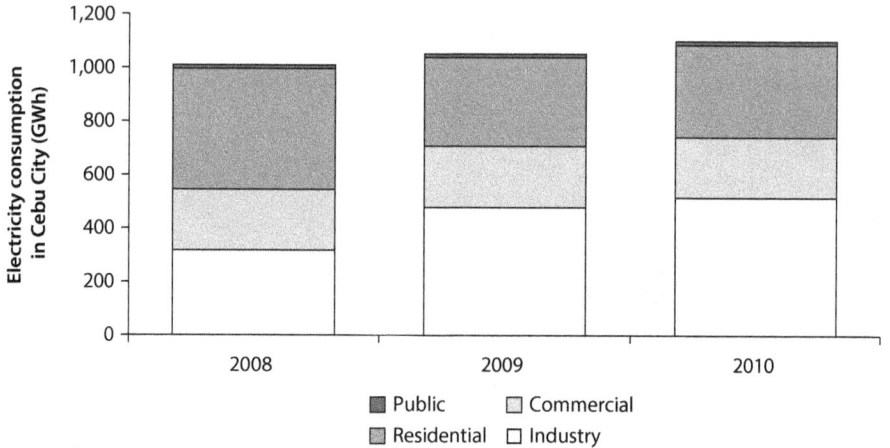

Source: Phase I pilot study.
Note: GWh = gigawatt-hour.

**Figure 2.3b  Electricity Consumption in East Java**

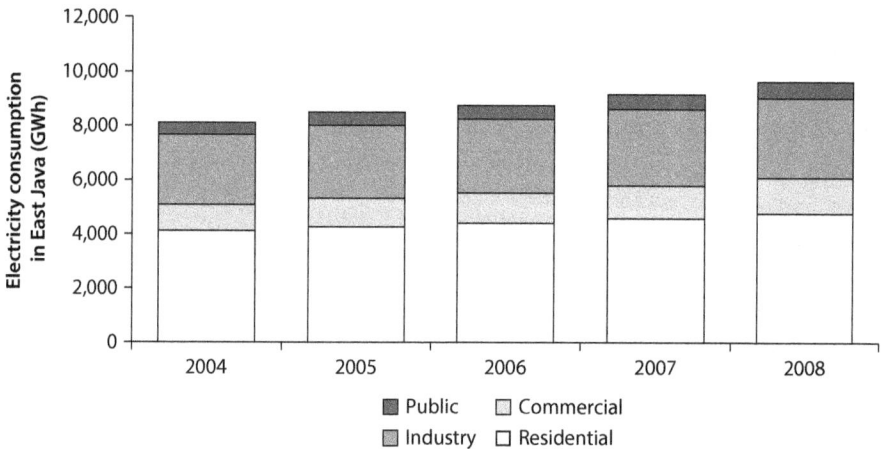

Source: JICA 2009.
Note: Data for Surabaya were not available. Data for East Java are representative and were therefore used as a proxy for Surabaya. GWh = gigawatt-hour.

Figure 2.4 excludes the use of solid fuel for residential cooking. As in many other studies, the energy balance data did not include these solid fuels because of the difficulties with data collection. However, cities should continue to monitor this sector given that every second household in the region still relies on solid fuel for cooking, and uses a traditional stove, which has negative impacts on health, gender equality, poverty reduction, and the local and global environment. This phenomenon affects both rural and urban areas. The use of transition and traditional fuels continues to be high in urban areas that are characterized by

**Figure 2.3c  Electricity Consumption in Da Nang**

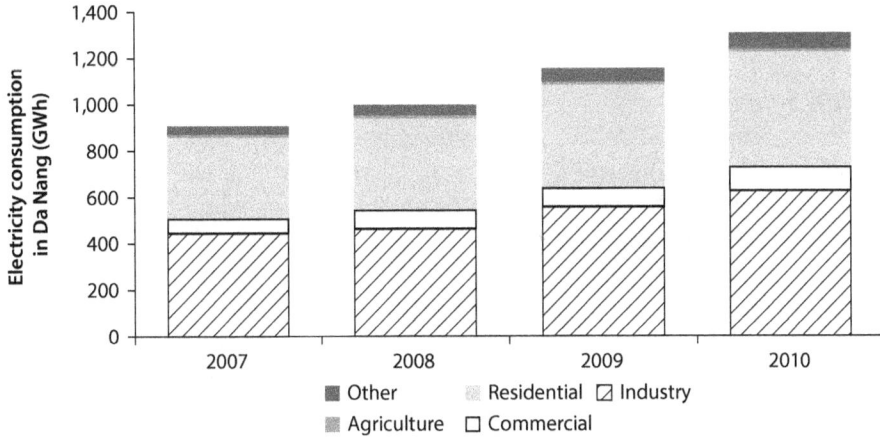

*Source:* Phase I pilot study.
*Note:* GWh = gigawatt-hour.

**Figure 2.4  Primary Energy Consumption by Sector**

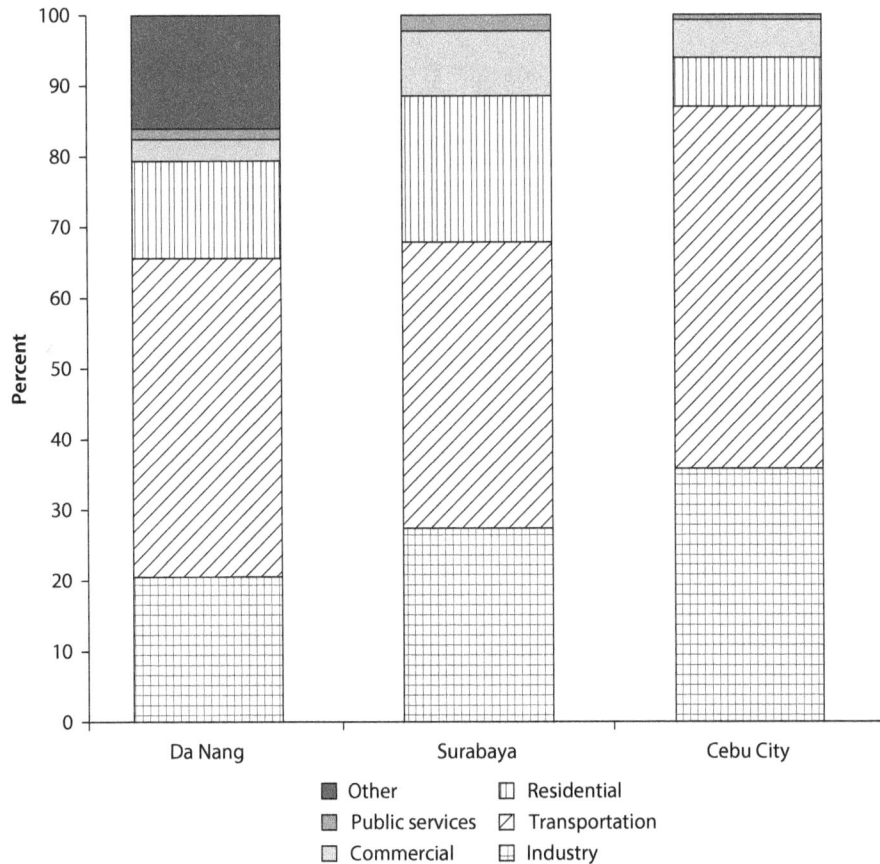

*Source:* Phase I pilot study.

low incomes. In light of this, cities need to actively pursue policies that can achieve universal access to clean cooking solutions to all residents to save energy and improve quality of life.

## Energy Supply

An energy supply profile can provide insights into a city's reliability of energy supply, GHG emissions levels, and economic growth, which will be of value for strategic planning. The outcome of the analysis of energy supply across the pilot cities is broadly discussed below.

### Energy Conversion Is Inherently Inefficient

The generation of electrical energy requires the conversion of one energy source to another, a process that inherently results in losses, although the magnitude of the losses depends on the type of energy or fuel used and the technology deployed. Figure 2.5 presents the typical efficiencies associated with a range of generation technologies, from highly efficient, large hydroelectric plants to still-emerging solar technologies.

Many EAP countries rely heavily on coal and oil for electricity generation. Although newer oil- and coal-fired steam turbine power plants operate at

**Figure 2.5  Efficiencies in Electricity Generation: Principal Technologies**

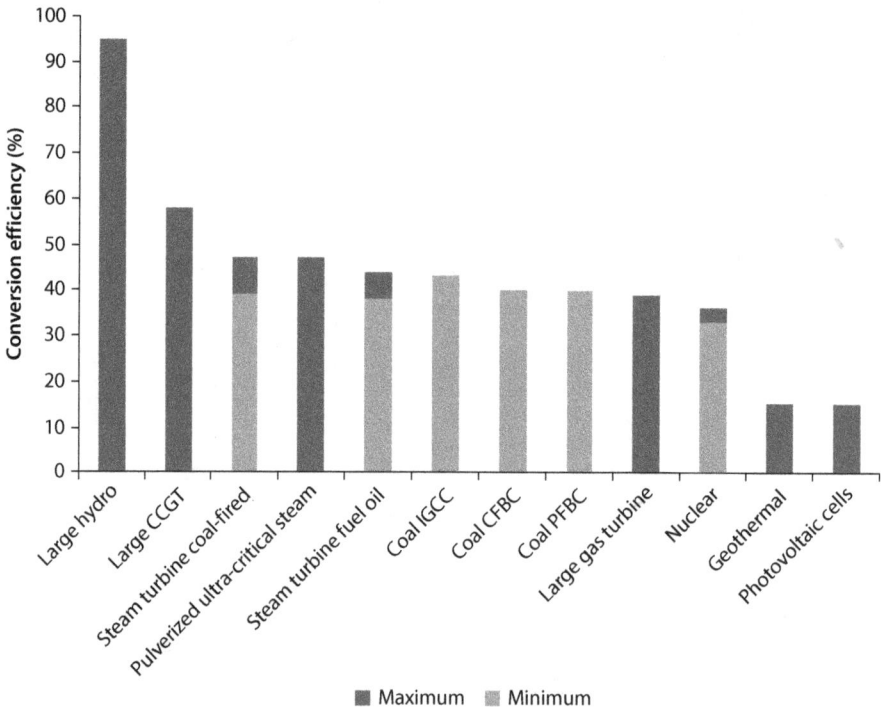

*Source:* Eurelectric 2003.
*Note:* CCGT = combined cycle gas turbine; CFBC = circulating fluidized bed combustion; IGCC = integrated gasification combined cycle; PFBC = preserved fluidized bed combustion.

efficiencies between 39 and 47 percent, older power plants are significantly less efficient. Sankey diagrams, such as the example for Surabaya in figure 2.6, illustrate the efficiencies inherent in energy systems.[1] Surabaya exhibits losses exceeding 60 percent for primary fuel energy and about 77 percent for primary fuel energy for combustion engines (mainly for transportation) that occur in the conversion process. In Cebu City and Da Nang, similar trends are observed. The electrical generation conversion losses are reasonably typical of the generating plants in use (although efficiencies are generally better with modern technologies). In the transportation sector, conversion losses are directly related to engine and transportation infrastructure efficiencies and are thus more dependent on individual city composition and context.

### Power Generation Sources Play an Important Role in a City's GHG Emissions

Cities must address both the choice of generating technology (if the opportunity exists), as well as the manner in which electrical energy is used when considering long-term issues of reliability of energy supply, climate change mitigation, and energy efficiency. The less efficient the generation technology used, the more important it is to address electricity end use—small efficiency gains at the point of use will have an even greater effect on primary energy inputs for electricity generation. This focus is also important because of the long-term and relatively static nature of generation infrastructure, which provides limited opportunities to alter generating efficiencies after a generation plant has been commissioned.

Electricity generation in the pilot cities is fueled by a unique mix of energy sources governed by resource availability in each country. Surabaya's electricity is generated mainly from coal and oil, whereas Da Nang uses a significant proportion of hydropower and gas, and about 49 percent of Cebu City's electricity is generated from hydro and geothermal sources. These electrical generation differences alter the cities' electricity carbon emissions factors, which in turn alter their emissions profiles. The relative carbon intensity of grid-supplied electricity in the three pilot cities is presented in figure 2.7.

Currently, none of the pilot cities generates a significant proportion of its electricity requirements within the city boundaries. They depend heavily on the national or regional electricity grids. The reliance on external sources of electricity has the potential to expose these cities to power shortages that could have major impacts because of the concentration of economic activity in urban areas.

In light of the above, improving energy efficiency and increasing decentralized energy supplies should both be key strategies for enhancing cities' reliability of energy supply. In particular, cities should prioritize the use of local low-carbon energy sources in their energy planning if such opportunities exist within the confines of the city or metropolitan area. City governments have the capacity to promote decentralized low-carbon power generation and fuel switching to lower-carbon, lower-pollution alternatives and should work with utilities to define long-term low-carbon strategies that stimulate local energy production. This effort can be achieved in concert with national and regional plans

**Figure 2.6 Surabaya Sankey Diagram**

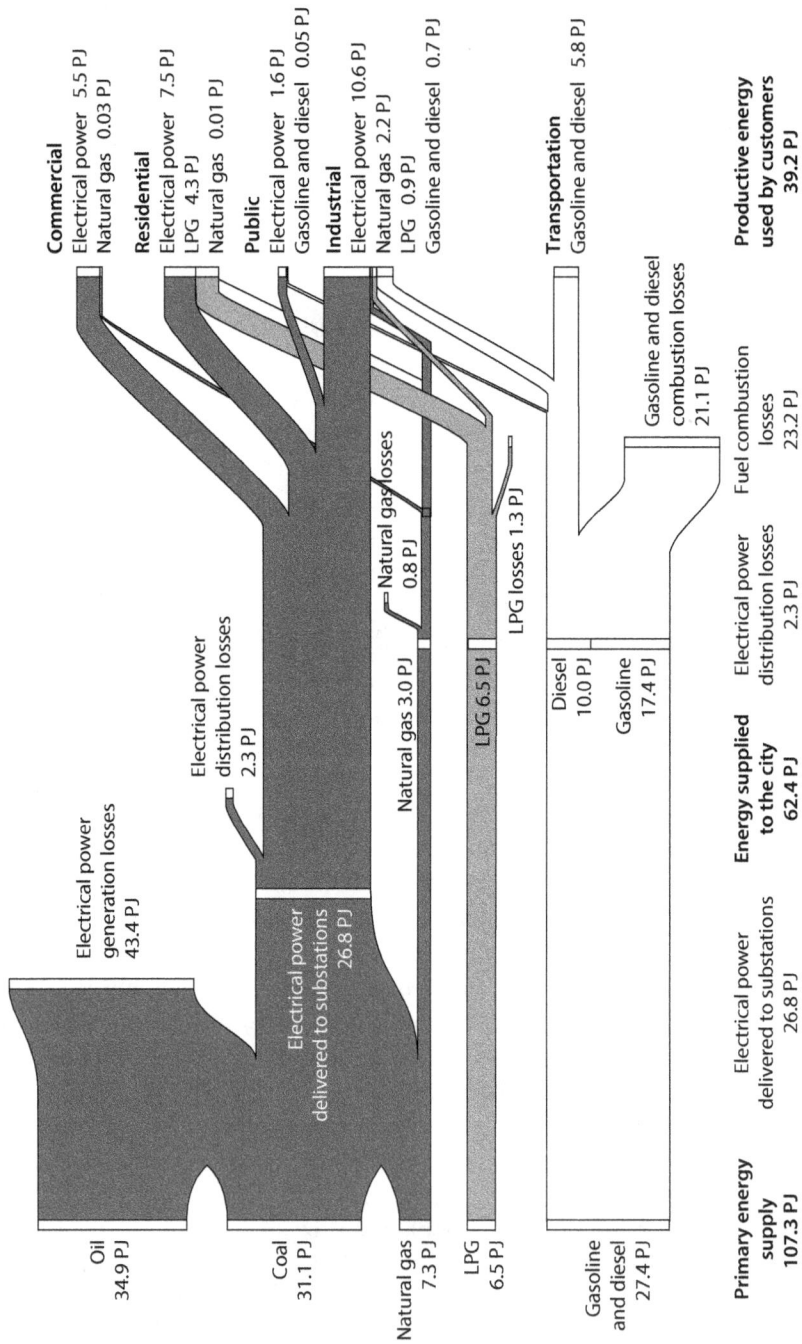

Oil
34.9 PJ

Coal
31.1 PJ

Natural gas
7.3 PJ

LPG
6.5 PJ

Gasoline
and diesel
27.4 PJ

**Primary energy
supply
107.3 PJ**

Electrical power
generation losses
43.4 PJ

Electrical power
delivered to substations
26.8 PJ

**Electrical power
delivered to substations
26.8 PJ**

Electrical power
distribution losses
2.3 PJ

Natural gas 3.0 PJ

LPG 6.5 PJ

Diesel
10.0 PJ

Gasoline
17.4 PJ

**Energy supplied
to the city
62.4 PJ**

Natural gas losses
0.8 PJ

LPG losses 1.3 PJ

**Commercial**
Electrical power   5.5 PJ
Natural gas   0.03 PJ

**Residential**
Electrical power   7.5 PJ
LPG   4.3 PJ
Natural gas   0.01 PJ

**Public**
Electrical power   1.6 PJ
Gasoline and diesel   0.05 PJ

**Industrial**
Electrical power   10.6 PJ
Natural gas   2.2 PJ
LPG   0.9 PJ
Gasoline and diesel   0.7 PJ

**Transportation**
Gasoline and diesel   5.8 PJ

Gasoline and diesel
combustion losses
21.1 PJ

Fuel combustion
losses
23.2 PJ

Electrical power
distribution losses
2.3 PJ

**Productive energy
used by customers
39.2 PJ**

*Source:* Phase I pilot study.
*Note:* LPG = liquefied petroleum gas; PJ = petajoule. "Public" includes the end-use energy of city buildings, street lighting, city vehicles, water, wastewater, and solid waste management.

**Figure 2.7  Carbon Intensity of the Electrical Grid**

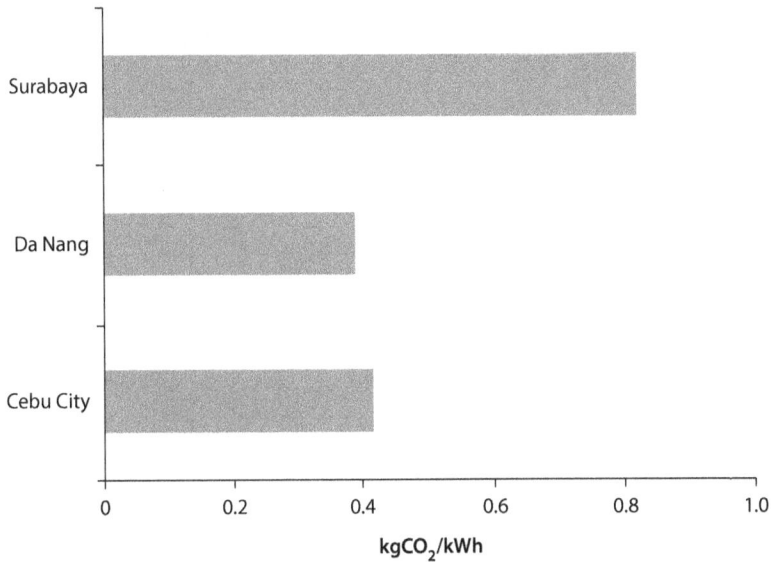

Source: Phase I pilot study.
Note: kgCO$_2$/kWh = kilograms of carbon dioxide per kilowatt-hour.

to decarbonize the grid. The points to emphasize, therefore, are that city governments should engage in strong dialogue with the national government in power generation planning and should seize opportunities at the city level that help stimulate investment and jobs while strengthening resilience to energy shortages and lowering the carbon intensity of grid-supplied power.

### Sustainable Solutions in the Transportation Sector

The current trend in transportation, whereby inefficient public transportation is increasingly replaced by two-wheeled motorized vehicles and then four-wheeled vehicles as household incomes rise, is associated with a range of environmental and economic challenges. These challenges include increased fossil fuel use, higher carbon emissions, local air pollution (and health impacts for inner-city inhabitants), and traffic congestion. This trend, if unabated, will lead to greater expenditures on imported energy, health care, and typically, transportation infra-structure to reduce congestion. By acknowledging the likelihood that the trend will continue unless addressed, city governments open up opportunities to take an alternative path.

It is apparent in the pilot cities that a substantial amount of transportation energy is lost in fuel combustion (in Surabaya, Cebu City, and Da Nang, the figures are 77, 75, and 75 percent, respectively). A range of opportunities exists to increase efficiencies at a city level. An integrated program of regulatory, fiscal, technological, and infrastructure-based reforms and investments can improve energy efficiency. City governments have some level of control over citywide transportation planning and management, and these powers can be used in coordination with railroad or national agencies to bring about greater efficiency in transportation energy use.

There are three types of activities cities can undertake to reduce energy consumption and mitigate GHG emissions from the transportation sector:

- Reduce the carbon intensity of fuels and energy sources (to reduce tons of $CO_2$ per unit of fuel combusted).
- Improve the fuel efficiency of vehicles through vehicle technology improvements and retrofits; improvements in vehicle operations efficiency; and improvements in driving conditions, such as road surface quality, congestion, and the like (to reduce units of fuel combusted per kilometer traveled).
- Reduce systemwide vehicle-kilometers traveled, through a modal shift to high-occupancy transit and nonmotorized transportation (to reduce vehicle-kilometers).

Nearly every activity that can be undertaken by a city requires one or some combination of these three approaches. For example, replacement of a traditional diesel bus system with a network of dedicated bus lanes and compressed natural gas (CNG) hybrid buses would reflect a combination of all three of these activities, in that such a program would (1) affect the carbon content of fuel (replacement of diesel by CNG), (2) improve the fleet fuel efficiency (through new technology, as well as improved drive cycle through use of a dedicated lane), and (3) reduce the activity level per passenger transported (through better transportation management).

As with activities in other sectors to improve energy efficiency, these measures are evaluated with respect to their broader economic, social, and fiscal impacts on the city.

The following are specific examples of cities that have undertaken activities to reduce emissions and improve energy efficiency in urban transportation.

### Egypt Vehicle Scrapping and Recycling Program of Activities

*City:* Cairo, the Arab Republic of Egypt

*Activity category:* (2) Improve fuel efficiency

*Average annual $CO_2$ reductions* (estimated by the Clean Development Mechanism): 250,000 tons

*Description:* The average age of a taxi in Cairo is 32 years. Before the program, more than 48,000 taxis in Cairo were more than 20 years old. Through a scrapping and recycling program managed by the Ministry of Finance, taxi owners were given packages of incentives toward the purchase of new, more fuel-efficient vehicles (in most cases, about 12–15 percent more efficient) in exchange for voluntarily submitting their old vehicles for recycling. To date, more than 35,000 vehicles have been exchanged under the program.

## Transmilenio Bus Rapid Transit (BRT) System

*City:* Bogota, Colombia

*Activity categories:* (1) Reduce carbon intensity of fuel, (2) Improve fuel efficiency, (3) Reduce vehicle activity

*Average annual $CO_2$ reductions* (estimated by the Clean Development Mechanism): 73,000 tons

*Description:* The Transmilenio BRT replaced an aging, inefficient network of microbuses with modern articulated diesel and CNG buses operating on dedicated busways, which feature priority traffic signaling, quick-boarding stations, and frequent headways. The system reduces the carbon intensity of fuel used in the transportation system by carrying more passenger-kilometers on CNG buses; improves fuel efficiency of the public transportation network through use of more efficient technology to move a given number of passengers and by improving the drive cycle through use of dedicated lanes and signal priority; and reduces overall vehicle activity by inducing a modal shift from private passenger modes, such as motorcycles, to buses.

## New Delhi Regenerative Braking Project

*City:* New Delhi, India

*Activity categories:* (1) Reduce carbon intensity of fuel, (2) Improve fuel efficiency

*Average annual $CO_2$ reductions* (estimated by the Clean Development Mechanism): 41,000 tons

*Description:* Through the implementation of electrified rail and locomotives with regenerative braking capabilities, energy typically lost when braking is saved and resupplied to other vehicles. The project reduced the carbon intensity of the system through the switch to electric vehicles and improved fuel efficiency by about 30 percent by reducing losses associated with stopping and starting.

Although cities have mechanisms on hand to enhance the energy efficiency of their transportation systems, the success of their policies will also depend somewhat on national policies. For example, national subsidies for fuel will hinder the promotion of more efficient modes of transportation, as illustrated by the limited take-up of liquefied petroleum gas (LPG) and CNG in Indonesian cities despite the government's promotion of these fuel types in the transportation sector.

## City Emissions Profile

GHG inventories can add a useful layer of analysis to that gained from the city energy balance. The GHG inventory demonstrates the relative impact of different fuels on overall emissions, as well as the contribution of non-energy-related sectors such as agricultural production. The GHG inventories carried out for the pilot cities proved informative, highlighting trends relevant to the wider region.

Perhaps more important, cities that maintain robust systems for monitoring GHG emissions earn access to additional sources of funding and financing for energy efficiency and emissions mitigation programs that may otherwise not be available.

When energy-related GHG emissions are analyzed according to fuel source, the differences in electrical generation in the three pilot cities become apparent. For Surabaya, use of electricity generated mainly from coal is responsible for 36 percent of the city's emissions. For Cebu City and Da Nang, which both use significant renewable electrical generation, diesel and gasoline used for transportation and local electricity generation contribute heavily to the emissions profile. This difference is due to the impact of the carbon factors of grid electricity demonstrated in figure 2.7. An inventory of GHG emissions based on fuel source and end-use sector for the pilot cities is shown in figures 2.8a and 2.8b.

Transportation and industrial emissions account for more than 53 percent of emissions for all pilot cities—sectors that are only partially controlled by the city governments. Transportation alone is responsible for more than 40 percent of emissions in both Cebu City and Da Nang. However, this does not necessarily mean that the transportation sector in Surabaya is any less carbon intensive than in Cebu City and Da Nang.

Urban GHG emissions analysis incorporates processes such as waste management and wastewater treatment. These sources should not be overlooked because they may contribute a fair portion of a city's total emissions. GHG emissions associated with solid waste management in Da Nang account for about 7 percent of the city's emissions.

### *Growing Economic Prosperity and Carbon Emissions*

A large body of evidence supports the assertion that more highly developed and richer nations have greater carbon intensities than less-developed and poorer nations. However, empirical evidence also suggests that more developed nations use energy more efficiently than do developing nations, as illustrated in figure 2.9.

**Figure 2.8a   GHG Emissions by Fuel Source**

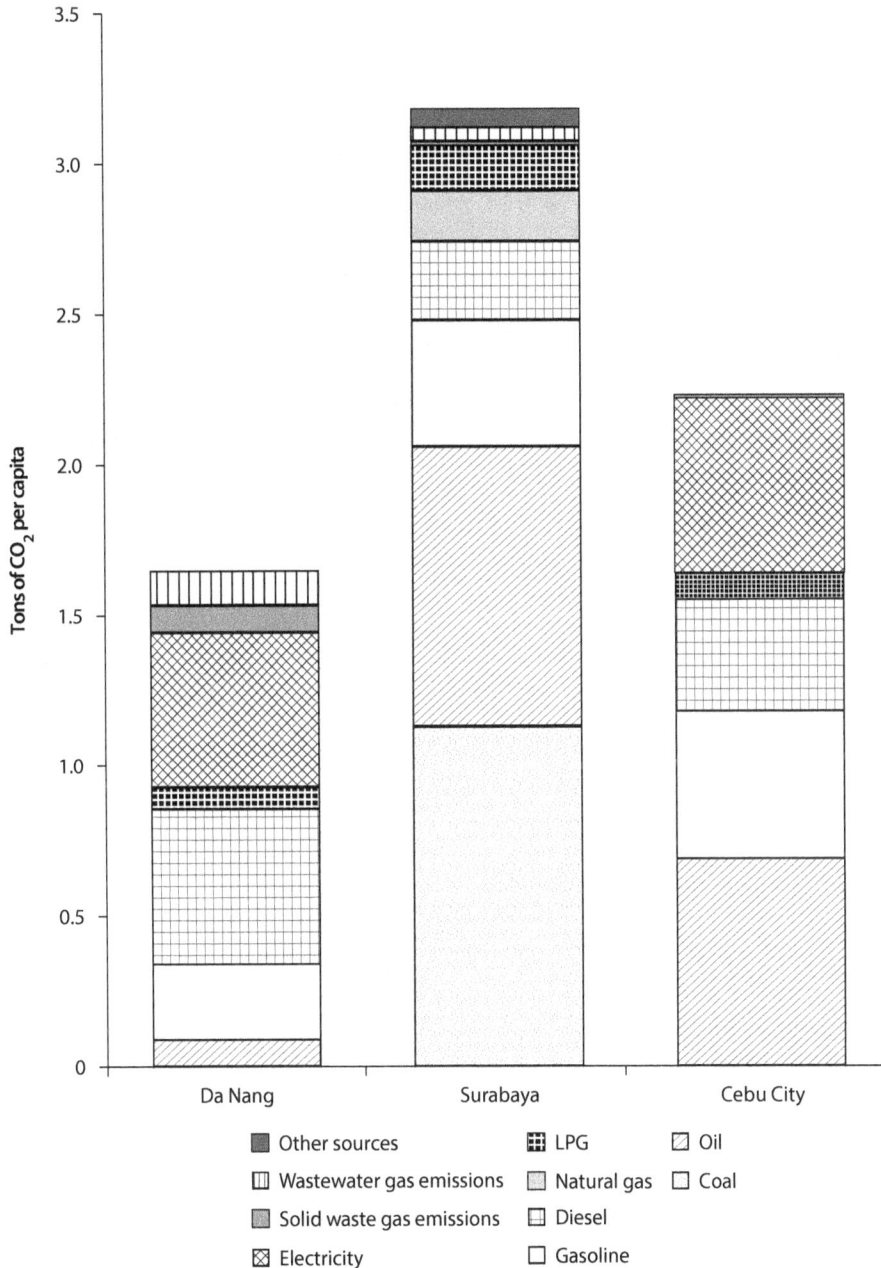

Source: Phase I pilot study.
Note: $CO_2$ = carbon dioxide; GHG = greenhouse gas; LPG = liquefied petroleum gas.

**Figure 2.8b  GHG Emissions by End Use**

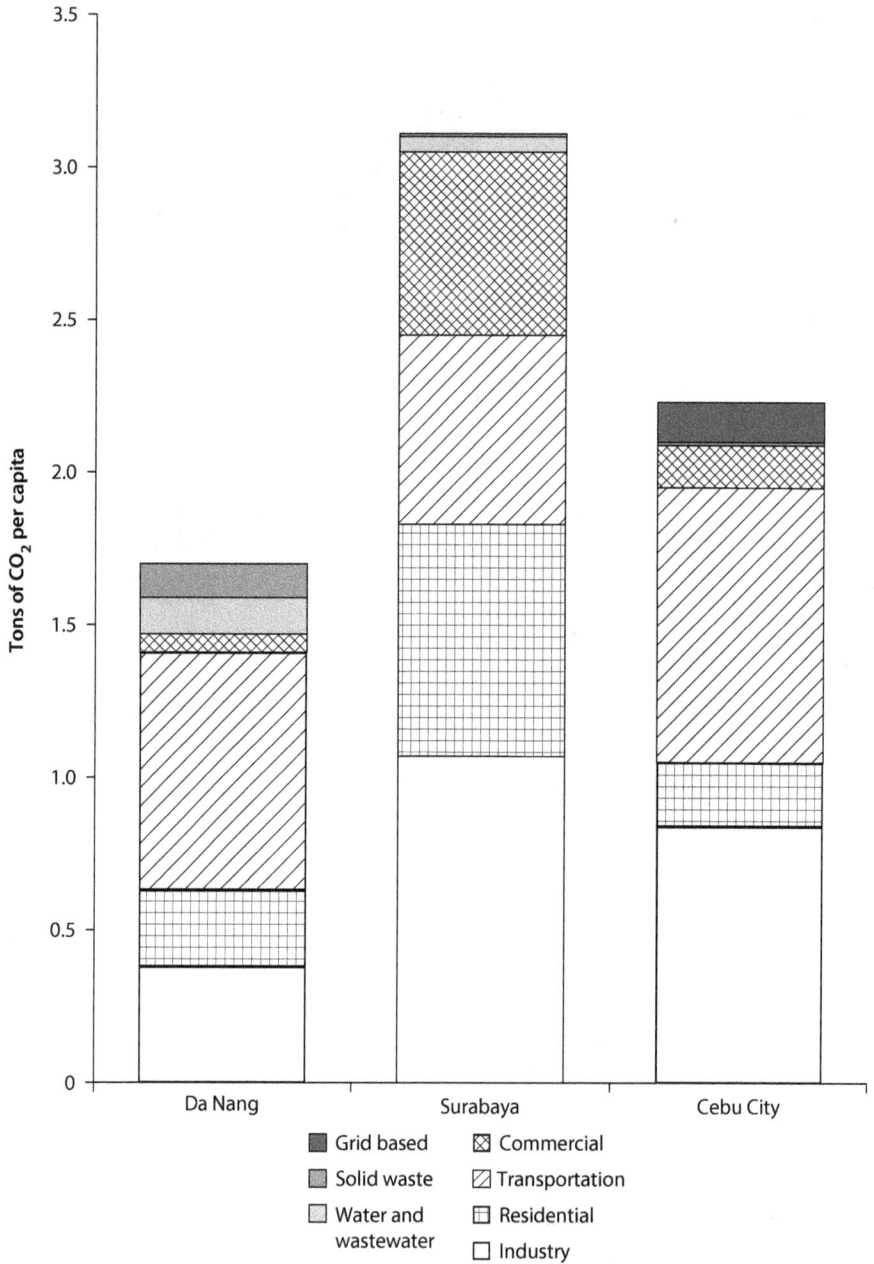

Source: Phase I pilot study.
Note: Grid based refers to electricity internally generated by the city. This is a separate category for Cebu City because data on end-use sector for electricity generated within the city were not available. $CO_2$ = carbon dioxide.

**Figure 2.9  Energy Efficiency and Economic Development, 2008**

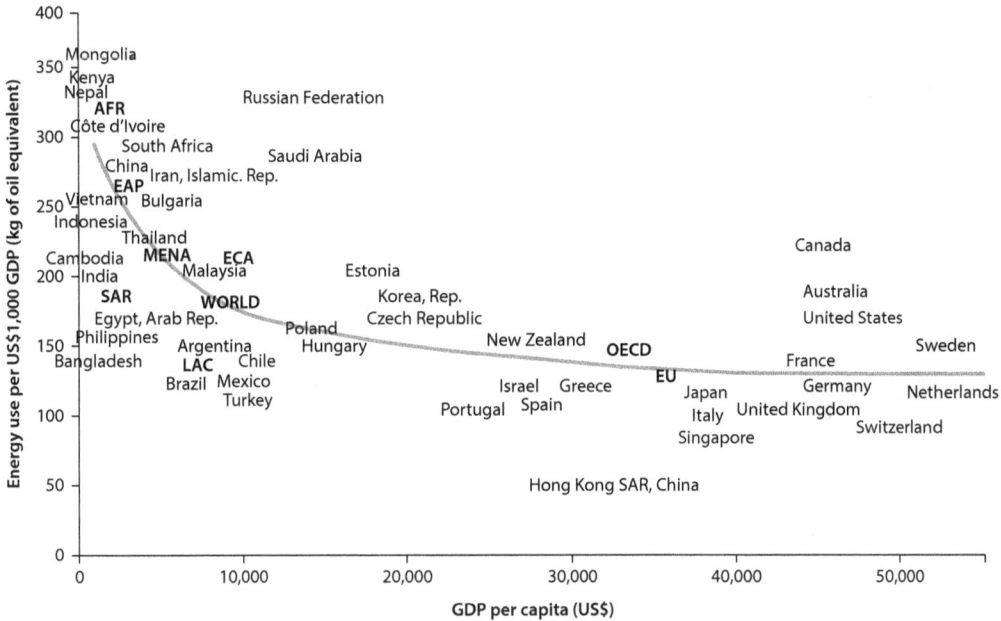

*Sources:* IEA statistics (http://www.iea.org/stats/index.asp); OECD National Accounts data; and World Bank national accounts data.
*Note:* AFR = Africa; EAP = East Asia and Pacific; ECA = Eastern Europe and Central Asia; EU = European Union; GDP = gross domestic product; kg = kilograms; LAC = Latin America and the Carribbean; MENA = Middle East and North Africa; OECD = Organisation for Economic Co-operation and Development; SAR = South Asia Region.

Figure 2.9 demonstrates that as per capita GDP increases, energy use per unit of GDP decreases, plateauing at about 125 kilograms of oil equivalent per US$1,000 of GDP. Hence, there is a positive correlation between economic development (approximated by increases in GDP) and the reduction of energy intensity, that is, energy efficiency improvement.

The key agents of economic development are widely accepted to be investment, employment, and productivity, all of which have a positive correlation with the improvement of energy efficiency. The empirical evidence in figure 2.9 illustrates the positive correlation in EAP cities, which are at the early or middle part of the curve linking economic growth to the energy efficiency improvement rate.

A key message for EAP cities is that energy efficiency improvements lead to

- Increases in investments,
- Jobs creation, and
- Productivity improvements,

all of which, in turn, generate economic growth.

Thus, the case for investment in energy efficiency is clearly demonstrated. However, with respect to carbon intensity, a different story emerges. Figure 2.10 illustrates, using the same indicators as figure 2.9, the relative per capita

**Figure 2.10  Energy Intensity and GDP per Capita in Select Cities**

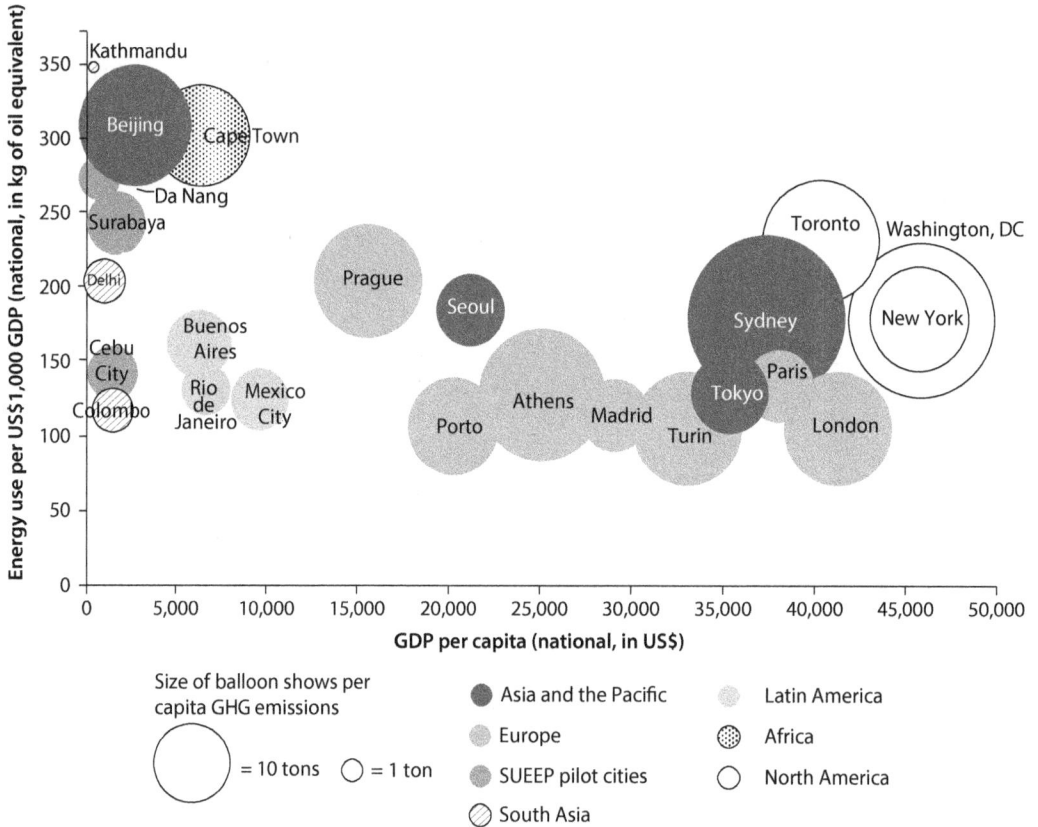

*Sources:* IEA statistics (http://www.iea.org/stats/index.asp); OECD National Accounts data; and World Bank national accounts data.
*Note:* GDP = gross domestic product; GHG = greenhouse gas; kg = kilograms; SUEEP = Sustainable Urban Energy and Emissions Planning.

GHG emissions in each city. The magnitude of per capita emissions is denoted by the size of the balloon (see key in figure).

The figure suggests that although cities may be able to reduce energy intensity, a corresponding reduction in carbon emissions may not occur. Thus, investments in energy efficiency alone are not sufficient to keep cities on a sustainable energy path; for the most developed cities, GHG emissions per capita may plateau or even start increasing with GDP growth unless more comprehensive measures are implemented. Such measures or plans must take into consideration other causes of energy demand and emissions, for instance, land use, public transportation, distributed power generation and its mix, and so forth, which are part of a comprehensive SUEEP process. Hence, cities may quickly run out of "low hanging" energy efficiency opportunities and get stuck in the inherently highly energy-intensive infrastructure and economic models if they fail to address sustainability issues comprehensively.

## Note

1. The Sankey diagrams provide an overview of the primary fuel used upstream to supply energy within the city boundaries. The diagrams also illustrate the magnitude of final energy put to productive use across the city sectors following losses due to conversion inefficiencies. Therefore, the distribution of the total energy supplied to the individual sectors before conversion losses (labeled Energy Supplied to the City) is not apparent from the Sankey diagrams; this distribution of energy supply and consumption on a sector basis is detailed instead in figures 6.3, 7.3, and 8.3.

## References

Eurelectric. 2003. *Efficiency in Electricity Generation*. Brussels, Belgium: Eurelectric; Essen, Germany: VGB PowerTech.

JICA (Japan International Cooperation Agency). 2009. "The JICA Study on Formulation of Spatial Planning for Gerbangkertosusila Zone." Interim report, JICA, Tokyo.

# CHAPTER 3

# Sector Diagnostics: Identifying Opportunities

City government energy expenditure across the three pilot cities accounted for 3.0–4.5 percent of annual city budgets and is expected to increase as cities grow and expand. The increase in city government energy expenditure across the pilot cities was attributed to a combination of increased energy use and increased energy unit costs. Although energy expenditure as a proportion of city budgets in the three cities appears to be small, potential exists for even that level of expenditure to be reduced.

Minimizing energy expenditures in the three cities is crucial, given the expectation of rising energy demand and the need for city governments to demonstrate good practice to wider stakeholders. The importance of the second point should not be underestimated. Many of the levers a city government can use to reduce citywide energy use (such as energy efficient building codes) will affect a wide range of stakeholders. It is therefore imperative that the city government demonstrate good housekeeping in its own affairs and, potentially, use lessons learned to influence the activities of external organizations through mentoring or guidance.

Public lighting and city government vehicles accounted for a significant proportion of the energy budget in the three pilot cities. Wastewater treatment, waste management, and city government buildings also accounted for significant energy spending. This chapter describes the analysis undertaken across the sectors and identifies opportunities for energy efficiency improvements. Proposed projects were developed based on several factors: (a) level of potential for energy efficiency improvement, (b) scope of control or influence of the city government, (c) resource constraints, and (d) compatibility of the projects with other development goals.

## Public Lighting

As cities grow and citizens expect quality of life improvements, city authorities face increasing pressures to enhance the coverage and quality of public lighting.

**Table 3.1  Public Lighting Performance for Three Sustainable Urban Energy and Emissions Planning (SUEEP) Pilot Cities**

| Public lighting | Cebu City | Da Nang | Surabaya |
|---|---|---|---|
| Energy expenditure (%) | 37 | — | 34 |
| Electricity expenditure (%) | 67 | 24 | — |
| Coverage (% of city roads lit) | 95 | 90 | 79 |
| Energy consumption/pole (kWhe/pole) | 656 | 416 | 400 |
| Bulb technologies (%) | — | 55 sodium 18 LED 13 HPS 12 CFL 2 metal halide | 95 HPS |

*Source:* Phase I pilot study.
*Note:* — = not available; CFL = compact fluorescent lamp; HPS = high pressure sodium; kWhe/pole = kilowatt-hour equivalent per pole; LED = light-emitting diode.

Any program to enhance public lighting must consider a range of factors such as cost, public safety, and the aesthetics of public space.

In the pilot cities, public lighting constitutes a considerable proportion of overall city government energy expenditure. In Cebu City, it accounts for about two-thirds of electricity expenditure (see table 3.1). Because public lighting is an activity with significant city government influence in each of the three pilots, awareness of the potential resource and financial gains to be achieved with energy efficiency measures in the sector is high.

### Focus on Bulb Technology

Energy consumption in public lighting systems is controlled by a relatively small range of factors, including the placement of lighting poles, pole height, angle of lighting coverage, and the efficiency of the light source. The installation of new bulbs is a costly procedure, so actions seeking short-term return on investments usually focus on optimization using appropriate light bulb technology. This approach can often reduce energy consumption significantly, reduce maintenance requirements, and increase the life span of light bulbs, the combination of which reduces the associated costs of public lighting for city authorities.

Traditional lighting technologies such as metal halide bulbs are inefficient compared with contemporary technologies. Replacement with more efficient high pressure sodium (HPS) or light-emitting diode (LED) bulbs offers potential in all pilot cities. The energy performance and extended life spans of these bulbs and their ability to withstand fluctuating power conditions common in the region contribute to their usefulness.

### Improving Awareness and Support

Although audit and replacement programs have recently been implemented in each of the three cities, significant potential remains for improvement in the selection and application of bulb technology. A common barrier to selection of appropriate bulbs and fixtures is persistent confusion about the comparative performance of the technologies on the market. Despite awareness of the availability of energy efficient technologies, such as LEDs, these bulbs are rarely deployed by the three cities. An effective means of overcoming such barriers is the implementation of pilot studies to build awareness among the involved stakeholders. Full-scale demonstration projects are usually required to validate manufacturers' promised performance and to secure necessary support from decision makers for public lighting retrofit programs. Once support is received, public lighting procurement guidelines (currently lacking in the cities) would bolster continued energy efficient operation of the public lighting sector.

Lack of upfront investment funds is often cited as a second major barrier to implementation of energy efficient public lighting. However, existing cost-effective financing models can overcome such obstacles. For example, many cities in developing economies in the region have reported success with implementation of energy performance contracting for public lighting retrofits. Energy performance contracting is an innovative financing mechanism, normally offered by energy services companies, that uses cost savings from reduced energy consumption to repay the cost of installing energy conservation measures. These financing mechanisms allow cities to achieve energy savings without up-front capital expenses. Although such a mechanism has yet to be attempted in the three cities, it is worth pursuing actively. Future developments will likely promote financially attractive combinations of photovoltaic power sources with energy efficient technologies such as LEDs for public lighting, as is currently being investigated in Surabaya for both street lights and traffic signals.

### Implementing Strategic Timing or Dimming Programs

Energy consumption in the public lighting sector can be further reduced by strategic timing or dimming programs tailored to the needs in specific areas, while maintaining appropriate levels of lighting for public safety. In Da Nang, the public lighting sector uses twin timing regimes that have yielded significant reductions in the sector's electricity consumption. Under this scheme, every third light in 40 of Da Nang's main streets is switched off from 6:30 pm to 11 pm, after which only every third light is kept on from 11 pm to 5 am. The second regime maintains full use of all lights in the remaining streets through the night.

## Transportation

The transportation sector in all pilot cities is responsible for significant energy consumption and greenhouse gas (GHG) emissions and therefore should be carefully studied. Fuel used for transportation is the single largest energy-consuming sector in Cebu City (51 percent), Da Nang (45 percent), and Surabaya (40 percent), and contributes notably to GHG emissions in each of the three cities (40 percent in Cebu City, 46 percent in Da Nang, and 20 percent in Surabaya). Across the East Asia and Pacific (EAP) region, transportation is both the largest as well as the fastest growing source of GHG emissions.

Energizing Green Cities in Southeast Asia • http://dx.doi.org/10.1596/978-0-8213-9837-1

**Figure 3.1  Total Transportation Energy per Capita**

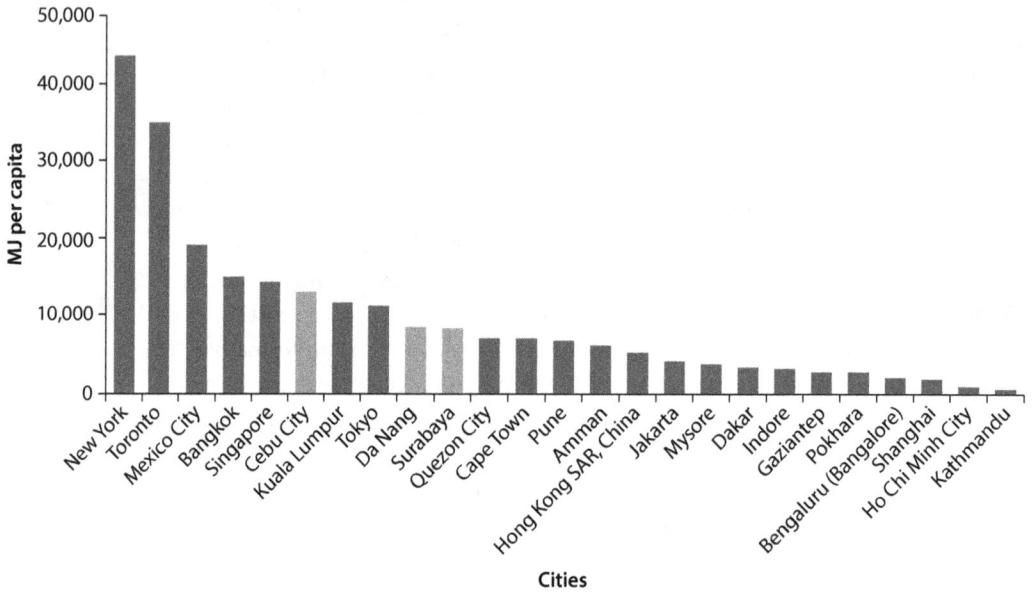

Source: Phase I pilot study.
Note: MJ = megajoules.

Although all three cities currently experience low energy intensity (megajoules per capita, or MJ per capita) in the transportation sector (see figure 3.1), the sector's energy consumption is rapidly increasing as a consequence of trends toward two-wheeled and four-wheeled motorized private transportation. This trend is representative of the sector across the region. City authorities, who are faced with the considerable challenge of influencing the type, length, and frequencies of trips that their citizens take, urgently need to intervene. A comprehensive approach is essential to address travel behavior patterns, modify urban development patterns, and take action to improve the energy efficiency of individual vehicles by increasing the distance traveled per unit of fuel and encouraging adoption of fuels that result in fewer GHG emissions.

### Addressing Motorization

Population expansion and economic growth are rapidly driving demand for enhanced personal mobility. Motorized two-wheeled vehicles are the predominant choice today, offering cheap and readily accessible personal mobility. But four-wheeled modes of transportation are in demand as well. Da Nang, where car ownership and monthly income are highly correlated ($r = 0.846$), and where a progressive shift from public transportation to two-wheeled vehicles to four-wheeled vehicles is occurring, illustrates this trend. In addition, low elasticity in vehicle prices in the three cities means that the market offers few substitutes for private vehicles for personal mobility. Thus, vehicle taxation alone will have only a limited direct impact on the increase in vehicles. Other regulatory and planning interventions will be necessary to decrease fuel consumption. None of the three cities currently places restrictions on the number of new vehicle registrations. Given these factors, the number of vehicles in these cities is set to increase, similar to the trend throughout Asia, where vehicle fleets are estimated to double every five to seven years.

Growing vehicle ownership increases traffic congestion, which is already apparent in some of the pilot cities during peak hours. Increased congestion increases local air pollution, GHG emissions, noise pollution, and personal fuel expenses, and also contributes to reduced health, deteriorating road safety, decreased quality of life, and loss of economic productivity for the city.

None of the three cities has yet implemented mechanisms for internalizing the externalities of personal vehicle transportation. Users of private vehicles currently do not pay the costs of congestion or pollution, and these costs are borne in equal proportion by public transportation passengers. Nonmotorized travel is becoming more difficult and decreasingly socially attractive. Walking in the three cities involves weaving across uneven pavement amid a range of obstructions, including vendor stalls and parked cars. Despite these trends, Surabaya is putting measures in place to encourage the reemergence of pedestrian- and bicycle-friendly areas, increasingly focusing on improving accessibility and services for pedestrians and nonmotorized vehicles, including smaller pedestrianized streets and car-free days in two locations in the city.

The shift toward motorcycles and cars shapes energy performance in the transportation sector. The fuel efficiency of two-wheeled vehicles (0.01–0.02 kilometer per liter, or km per liter) is better than that of four-wheeled vehicles (0.1–0.2 km per liter). High-capacity buses are even more fuel efficient per passenger-km than two-wheeled vehicles when ridership rates are high. A shift toward clean and efficient public transportation therefore results in greater fuel efficiency. In Cebu City, public transportation ridership is high (85 percent of all trips) but the mode is inefficient (see next section). In all pilot cities, the overall fuel economy of vehicle fleets is worsened by growing numbers of idling engines.

### Importance of Improving Fleet Performance

The distribution of vehicle types and performance characteristics produces the difference in energy intensity illustrated by figure 3.1, which shows that Cebu City underperforms in comparison with Da Nang and Surabaya, despite its higher public transportation ridership. The disparity arises from Cebu City's highly inefficient and aging fleet of jeepneys (which have a capacity of 8–12 people per vehicle) with much poorer fuel economy than Surabaya's and Da Nang's predominantly two-wheeled modes of transportation.

Inefficient and aging fleets result in higher emissions levels. In response, the three cities have adopted European emissions standards (ranging from Euro II to Euro IV[1]). Emissions testing for older vehicles is carried out, but enforcement is fragmented and ad hoc. For example, although vehicles in Cebu City are required to be tested when they are first registered, there is no formal repeat testing during the vehicle's life cycle, and they are typically tested only after being apprehended when an officer observes smoke belching from the tailpipe. The situation is exacerbated by a lack of awareness by vehicle owners themselves of the value of regular vehicle maintenance.

Energy use in the transportation sector is currently dominated by petroleum products and the rate of consumption is growing quickly. Some vehicles, such as the angkots (minibuses) in Surabaya, run on highly polluting fuels like kerosene. Currently available vehicle engine and fuel technologies could reduce emissions substantially, and all three cities have introduced liquefied petroleum gas (LPG) infrastructure for their taxi fleets.

### Strengthening Public Transportation

The three cities use different modes of public transportation. Cebu City's public transportation mainly comprises jeepneys, which operate on an ad hoc schedule. In Surabaya, 5,000 registered angkots run city-government-determined routes to regional bus lines served by three major bus stations and one regional train station. Da Nang has regular passenger bus service on six routes traversing the city. Despite the different modes, common threads among the three cities are the deteriorating quality of public transportation services and a lack of clear knowledge on the role of the public versus the private transportation sector and on the (dis)incentives for use of public versus private transportation.

The majority of public transportation trips are handled by the informal sector and operate inefficiently. In light of this, city authorities should initiate a modal shift that promotes lower fuel consumption per passenger-km traveled. Because buses have significantly better energy performance and emissions per passenger-km compared with private transportation, and because it would substantially reduce congestion, city governments should consider putting in place high-capacity public transportation networks. The three pilot cities, all of which have conducted bus rapid transit studies, are moving in the right direction. In Surabaya, plans for a mass transportation system comprising tramway and monorail systems to serve heavy traffic corridors into the city center are being developed, and studies on integrating the rest of its public transportation network with the mass transit system have been initiated.

Although steps are being taken in the three cities to promote more efficient public transportation systems, city governments will continue to face multiple challenges. A key hurdle will be the ability to measure system performance, given the informal nature of the sector. The pilot cities have good vehicle registration data, but detailed data on modal split, passenger- and vehicle-km traveled, origin-destination, and environmental performance are largely nonexistent or, if they do exist, are of variable quality. Another challenge is attributable to the employment of a large number of citizens in the informal public transportation sector. City governments will have to include this group of stakeholders in the planning of future public transportation projects to ensure the efficient and effective operation of the formal sector.

**Table 3.2  Transportation Sector Performance for Three SUEEP Pilot Cities**

|  | Cebu City | Da Nang | Surabaya |
|---|---|---|---|
| Total transportation energy use (MJ/capita) | 13,394 | 8,668 | 8,481 |
| Private transportation energy use (MJ/capita) | 3.70 | 1.20 | 1.47 |
| Public transportation energy use (MJ/passenger-km) | 0.75 | 0.22 | 0.85 |
| Nonmotorized transportation mode split (%) | 5 | 34 | 40 |
| Public transportation mode split (%) | 86 | 1 | 5 |
| Contribution to citywide GHG emissions (%) | 40 | 46 | 20 |

*Source:* Phase I pilot study.
*Note:* GHG = greenhouse gas; km = kilometer; MJ = megajoules; SUEEP = Sustainable Urban Energy and Emissions Planning.

### Planning Integrated Development

Similar to other Asian cities, urbanization at the edges of Cebu City, Da Nang, and Surabaya is high and increasing, exerting pressures on road space, as do the higher density areas and city centers. Accelerating urban sprawl exacerbates motorization trends, so pilot cities need to promote mixed-use urban development that reduces distances from origins to destinations to reduce motorized travel.

Table 3.2 shows that the share of nonmotorized transportation is the determining factor for total transportation energy use. In Cebu City, total transportation energy use (MJ per capita) is still substantially higher than in Da Nang and Surabaya even though it has a much higher public transportation share. This occurs because Cebu City's nonmotorized share is much lower.

Although vehicle ownership will rise with increased incomes, demand for motorized transportation can be managed by integrating land-use planning and transportation planning. A comprehensive approach will ensure that plans for mass public transportation, land use, parking policies, registration pricing, street signaling, pavement upgrades, and other measures affecting transportation systems are integrated and collectively provide an effective strategy for delivering the targeted reductions in private trip lengths and frequencies. Engaging in a strong dialogue with national governments on transportation policies will also ensure that the city's plans are compatible with, or at least not in conflict with, national policies. Minimal efforts have been made so far in the three cities to effectively integrate transportation policy with land-use planning. Much scope remains for developing density, zoning, and development approaches to support pedestrian-friendly, transit-oriented transportation and urban development.

### City Government Buildings

Energy consumption in the buildings sector has grown along with the rapid urbanization that has increased volumes of building stock and with the behavioral changes that accompany economic growth. If prevailing practices and trends continue, global final energy consumption in buildings will grow by 30 percent between 2007 and 2030, with the increase expected to be greatest in fast-growing EAP cities, including Cebu City, Da Nang, and Surabaya. This sector has

**Figure 3.2  City Government Buildings Electricity Consumption**

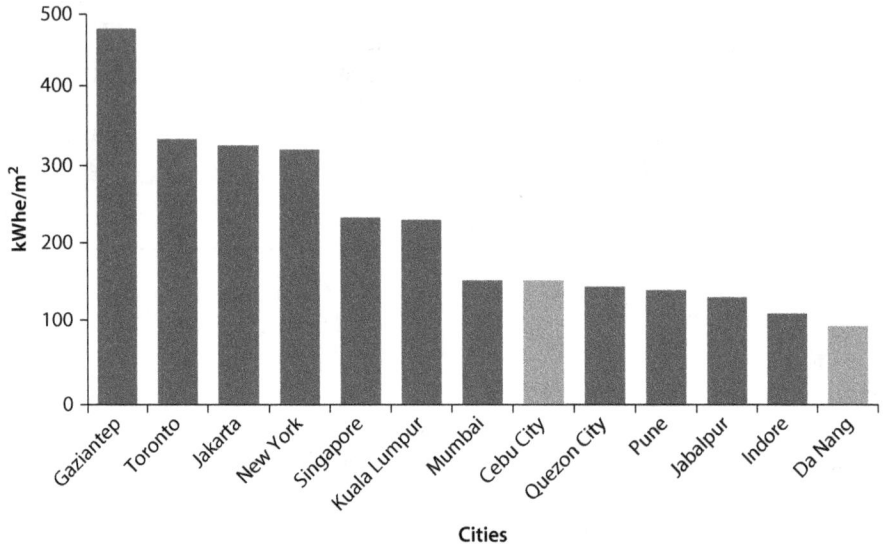

*Source:* Phase I pilot study.
*Note:* Surabaya data not available. kWhe/m² = kilowatt-hour equivalent per square meter.

been identified by the Intergovernmental Panel on Climate Change as having the greatest potential for cost-effective reduction of GHG emissions by 2030 (IPCC 2007).

Because the urban building stock in developing countries is expected to more than double by 2030, there is enormous potential to improve energy performance in the sector by focusing on new construction—preventing the lock-in effect in new buildings—as compared with retrofitting existing buildings to improve energy efficiency. This is relevant for the pilot cities, given the potential for newly constructed buildings to consume higher levels of energy. City buildings in Cebu City and Da Nang currently exhibit low energy intensity (see figure 3.2) because the existing building stock comprises predominantly smaller, low-rise buildings minimally equipped with lighting, air conditioning, and appliances. However, this situation will change with the trend for new buildings to have larger floor space and increased cooling, ventilation, lighting, and plug load intensities. Thus, authorities in the three cities have ranked the buildings sector as having the highest potential for energy efficiency action following the public lighting sector.

Given that electricity is the primary energy source in buildings, two areas can be targeted to reduce energy use and emissions: (a) reducing electrical loads and (b) reducing energy loads through building design and codes.

### Reducing Electrical Loads
Energy use in buildings is relatively low because a large proportion of hallways in city government buildings, schools, and hospitals use natural ventilation and

daylight. Minimal energy is used for air conditioning and for providing hot water to supplement the black water tanks on the roofs in the pilot cities. Newer commercial buildings such as retail malls and office buildings are much larger and require central air conditioning, ventilation, and much more lighting.

As incomes grow, the size of homes and the use of electrical appliances, such as air conditioning, computers, and washing machines, also increase. In Da Nang, air conditioner ownership in homes is correlated with higher monthly incomes ($r = 0.837$). This trend is likely to continue as incomes keep rising in the three cities.

Electrical load reduction can be addressed with targeted policies and programs. In the pilot cities, national-level energy efficiency standards for major appliances and equipment are broadly recognized as necessary interventions. A range of tactical and educational energy efficiency initiatives were also undertaken by the city authorities, including restricted air conditioning operation times, dimming light switches, and compact fluorescent lamp replacement programs.

### Reducing Energy Loads through Building Design and Codes

Baseline energy loads are intrinsically related to building shape, orientation, structure, and materials, and are best addressed during the design and construction phases. Several considerable barriers impede energy efficient building design in the three cities, not least being the conflicting incentives among key stakeholders, the lack of available information and knowledge, and the issues relating to the visibility of energy cost signals. The enforcement of efficiency standards through building energy efficiency codes (BEECs) may overcome these barriers and offer the greatest impact on future energy consumption in the buildings sector.

BEECs are not currently implemented in the pilot cities. Both the Philippines and Vietnam have recently introduced national green building guidelines, but these are voluntary best practices rather than mandates for minimum standards for all buildings. In Cebu City, the Building for Ecologically Responsive Design Excellence code offers an initial and valuable guideline for BEEC development. In Da Nang, the Vietnam Green Building Council's LOTUS green building rating system has drawn inspiration from other building certification programs, including Leadership in Energy and Environmental Design in the United States and Building Research Establishment Environmental Assessment Method in the United Kingdom. These codes include specifications for building design beyond energy performance, such as landscaping and noise pollution. Although the remaining work on development and implementation of appropriate BEECs for warm-climate developing countries is considerable, Da Nang's adoption of building codes is a positive step toward enhancing energy efficiency in its buildings.

BEECs in Indonesia, the Philippines, and Vietnam are not compulsory, but it would be beneficial for Cebu City, Surabaya, and Da Nang to enforce BEECs to prevent themselves from being locked in to energy inefficient buildings that would contribute substantially to their cities' carbon emissions. To enforce BEECs successfully, cities will have to ensure the following:

- Government commitment to energy efficiency
- Effective government oversight of the construction sector
- Compliance capacity of domestic and local building supply chain
- Financing of energy efficiency measures (for example, retrofitting)

Major capacity building is necessary in the three cities; key stakeholders need to be trained and educated, and systematic approaches for enforcement with complementary incentives need to be prepared, implemented, and regularly updated.

## Solid Waste

The solid waste sector in each of the three cities is currently characterized by relatively low waste generation rates with varying waste recycling rates (see table 3.3; for comparison with other cities, see figure 3.3). Waste generation rates are projected to increase as the economies grow and lifestyles change, and city governments are expected to manage increased waste collection and treatment needs with well-functioning collection and transport systems.

Energy efficiency improvement opportunities are available primarily in the energy embodied in the waste and its associated GHG emissions. Energy used for waste transportation is of lesser significance. In particular, potential exists in the region to scale up organic waste management activities. About 60–70 percent of city solid waste is estimated to be high-moisture organic waste. Organic waste decomposes anaerobically in landfills and releases methane gas, which is 23 times more potent than carbon dioxide ($CO_2$) as a GHG and has other potential public safety risks emanating from the decomposition process.

### Segregating at the Source

A primary constraint to the effective management of organic waste, which comprises the bulk of solid waste, is weak performance in source segregation. Composting and biogas projects using mixed waste inputs often fail. Hence, mandatory organics separation at the household level, as practiced in both

**Table 3.3  Solid Waste Sector Performance for Three SUEEP Pilot Cities**

|                                            | Cebu City | Da Nang | Surabaya |
|--------------------------------------------|-----------|---------|----------|
| Waste generation (kg/capita/year)          | 183       | 248     | 256      |
| Capture of solid waste (%)                 | 80        | 90      | 70[a]    |
| Solid waste recycled (%)                   | 24        | 0       | 15       |
| Waste that goes to landfill (%)            | 26        | 90      | 64       |
| Contribution to citywide GHG emissions (%) | 0         | 7       | 1.4      |

*Source:* Phase I pilot study.
*Note:* GHG = greenhouse gas; kg = kilograms; SUEEP = Sustainable Urban Energy and Emissions Planning.
a. Because no data were available on the proportion of solid waste captured by volume in Surabaya, a proxy using the proportion of the population served by solid waste collection was used instead.

**Figure 3.3 Percentage of Solid Waste Recycled**

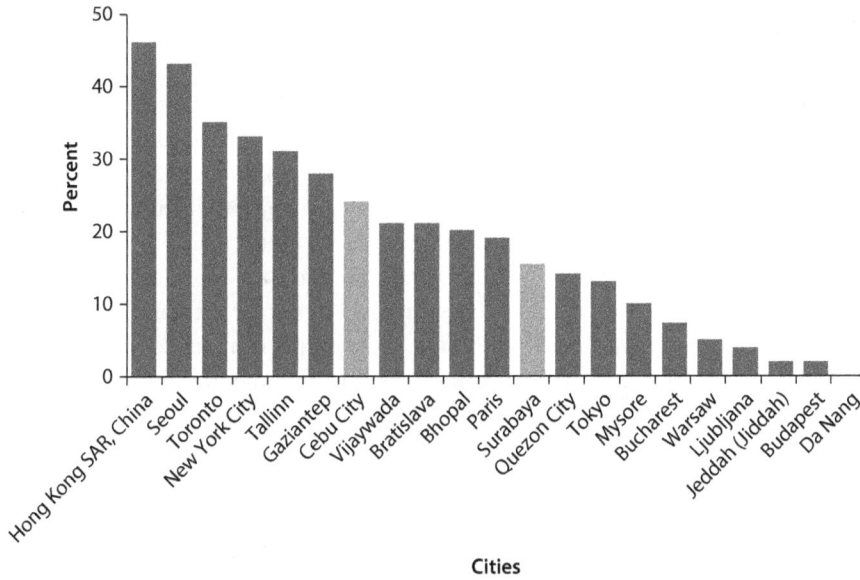

*Source:* Phase I pilot study.

Cebu City and Surabaya, is critical. Surabaya has a strong community-based waste self-management program based on the principles of "reduce, reuse, recycle." The community is involved in composting organic waste and recycling or selling inorganic waste to waste collectors. A similar scheme operates in Cebu City. Da Nang has no formal recycling program, but 28 percent of households report that they remove recyclable materials from their waste before discarding. The majority of these recovered resources are given to scavengers or sold to companies that recycle and reuse the materials.

Formal waste collection systems in the pilot cities have substantial capture rates (see table 3.3), and are organized through cooperatives, such as the daily collections at the barangay[2] level in Cebu City. In Da Nang, bins are collected daily and a flat rate is charged for residential customers. Businesses are charged a volume-based rate. Similarly, Surabaya's citizens are charged rates for collection that vary with distance from the main road.

### Reducing Volume to Landfill

Three principal treatment options can be applied to waste: landfilling, composting, and incinerating. The selection of treatment depends on a range of factors, including population size, waste volume, waste composition, availability of land, and existing policies. However, because the dense and moist waste in the region is characterized by low caloric values, incineration—which is ideal for dry waste with high caloric value—is not a suitable option, as was demonstrated in the failure of a waste incineration pilot project in Da Nang in 2003.

Energizing Green Cities in Southeast Asia • http://dx.doi.org/10.1596/978-0-8213-9837-1

The most prevalent current treatment option is landfilling because it remains the cheapest and simplest means of waste disposal. However, waste is dumped indiscriminately, and landfills in the three cities are poorly designed and are mostly beyond their design capacity. Waste pickers labor in all of the cities' landfills, and informal landfill recycling has substantial volume capture. In Cebu City, it is believed that a recycling rate of 16 percent is achieved through activities of the waste pickers. Only Da Nang has a landfill with a leachate collection system. An important benefit of scaling up organic waste management is that leachate management will be easier with less organic waste going to landfills.

Landfills require large areas of land for the disposal site itself and for the buffer that usually surrounds it. With expanding city limits, increasing land value, and increasing costs, authorities will find it increasingly difficult to locate affordable land reasonably close to the city. Reducing the volume of total waste treated at landfill therefore extends the operational life of such facilities.

### Capturing Landfill Gas

Landfill gas emissions present potential sources of energy generation that might receive financing through the Clean Development Mechanism. Studies have been undertaken in the pilot cities to determine the potential for such projects. Thus far, Surabaya has solicited alternative bids from potential partners for landfill gas capture projects on the Benowo landfill, but no contracts have been completed. Cebu City is considering capping the Inayawan landfill, and studies are being considered to reduce emissions by 5,000 tons of $CO_2$ per year through the project. Da Nang has identified a large project at the Khanh Son landfill that would generate 1 megawatt of electrical power and reduce GHG emissions by 140,000 tons of $CO_2$ per year or nearly 5 percent of total city emissions.

### Composting Organic Waste

Organic waste can be recycled into compost, and if managed aerobically, would not result in the production of methane. If managed properly, aerobic decomposition would result in carbon sequestration in soil solids. Because compost improves soil fertility, there is significant potential demand from the agricultural, residential, and landscape management sectors. Increased production of compost from city organic waste benefits urban agricultural productivity and urban food security, as experienced by Cebu City, which produces 25 percent of its own food and supplies urban farmers with compost from the city's composting facilities.

In Surabaya, thousands of residential composting bins have been distributed to residents by the city government, which also operates several composting houses that are supplied with organic waste from the park maintenance program and fruit and vegetable markets. These larger schemes use static pile methods with forced aeration. Compost is reused in the city's public parks. Cebu City has small-scale composting in the barangays, as well as larger-scale facilities at the landfill.

Second-generation biofuels, which are produced using nonfood sources (and therefore are not in competition with food production), are an expanding area of research and development. Of relevance to EAP coastal cities is the production of energy using fish waste. For example, Integrated Renewable Energy Solutions for Seafood Processing (Enerfish), a project developed by VTT Technical Research Centre of Finland in collaboration with the European Commission, produces biodiesel from waste generated by a fish-processing plant in Vietnam. EAP cities in coastal areas, especially those with fish-processing industries, could potentially harness such technologies, enabling them to produce local energy and reduce GHG emissions.

## Water and Wastewater

The primary concerns guiding city authorities in the water and wastewater sector are the need to reduce both costs and the negative impacts of treatment and distribution, while simultaneously expanding water and wastewater services to underserved populations. Many still lack access to safe drinking water and adequate sanitation facilities. In Surabaya, 70 percent of the population has access to clean drinking water; in Da Nang, the figure is 60 percent, and in Cebu City, above 50 percent. The lack of water supply, or its unreliability, can have detrimental health impacts, and result in significant social and economic costs.

Although only 2–3 percent of the world's energy consumption is used to pump and treat water for urban residents and industry, there is much scope to enhance energy efficiency in this sector. It is estimated that energy consumption in most water systems worldwide could be reduced by at least 25 percent by cost-effective efficiency actions alone.

### Enhancing Energy Efficiency in Pumping

The significant requirement for pumping means that sourcing water supplies from groundwater is a relatively energy-intensive process. Because Cebu City's potable water supply is sourced from groundwater, its energy intensity is higher than that of the surface water, gravity-fed systems that supply Da Nang and Surabaya (see figure 3.4). To reduce Cebu City's reliance on energy-intensive water sources, the city is considering the study of rainwater harvesting schemes in its buildings and ways to encourage the uptake of these schemes in private developments. Rainwater harvesting systems, which are not yet widely used in the region, are expected to receive much more attention in the future.

### Improving Distribution Networks

Improvements in distribution network performance, in particular, efficiency of pump operations throughout the network, can deliver large energy efficiency gains in the water sector. Considerable energy is used inefficiently when pump motors operate at suboptimal speeds. In addition, pumps that consistently maintain high rates of pressure even when demand is low can result in increased leakage across the system. Remedial measures to improve pump operation can

**Figure 3.4 Electricity Consumed per Cubic Meter of Potable Water Production**

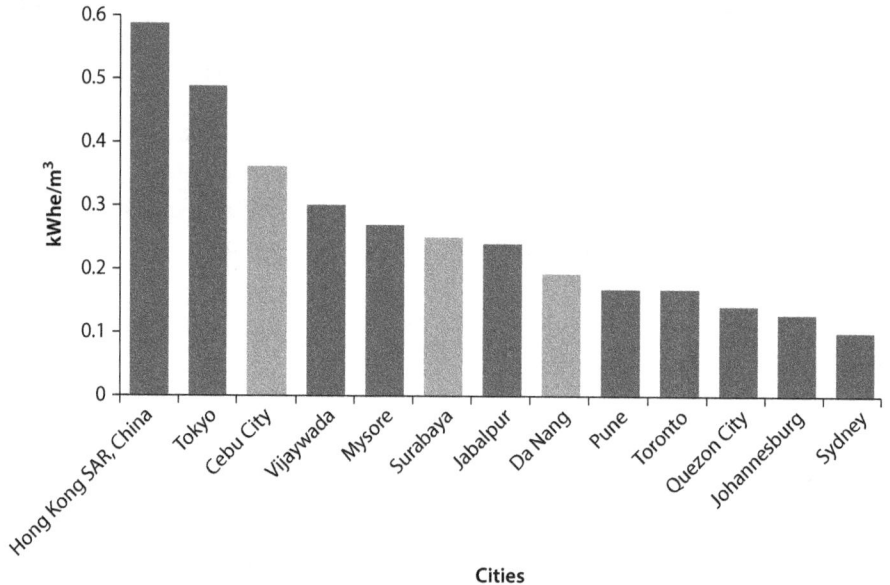

*Source:* Phase I pilot study.
*Note:* kWhe/m³ = kilowatt-hour equivalent per cubic meter.

provide attractive returns on investment. Low-cost measures, such as soft starters for motors and trimming impellers, as well as higher-cost measures such as the replacement of inefficient pumps and installation of variable speed drives that adjust pressure based on the demand on the system, are available. Da Nang has installed variable-speed drive pumps, and Cebu City has partnered with an external organization to implement a wide-scale pump efficiency program. Although no pump upgrade or maintenance is currently performed in Surabaya, a program to improve network pressure will introduce additional reservoirs and in-line lift stations across the network.

Enhancing the efficiency of the distribution system can also be accomplished by minimizing leakage—the cost of every liter of water delivered to consumers is increased by every liter lost to leakage. Wider water-loss management programs focus on theft as well as leaks, and water losses should be targeted to be brought to less than 10 percent. This recommended level is significantly lower than the levels of nonrevenue water experienced in the three cities' networks (see figure 3.5). Because old pipes are a primary cause of system losses, Da Nang and Cebu City have implemented audits and programs to replace them. In an effort to overcome the high initial investment cost of a leak-reduction program, the water utility company in Surabaya has considered performance-based contracting agreements to upgrade the network.

Pressure management is generally more cost effective than expensive repairs to leaks in buried pipes. This approach will likely be the preferred focus of distribution network improvement programs in the region, ahead of the replacement of existing networks.

**Figure 3.5  Nonrevenue Water as a Percentage of Total Potable Water Produced**

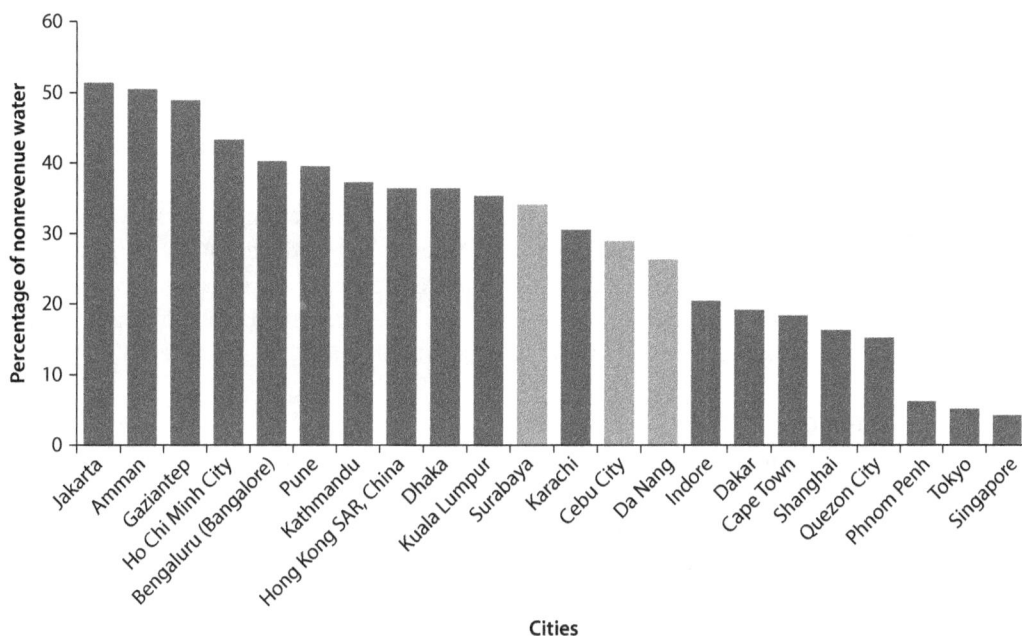

*Source:* Phase I pilot study.

## Scaling Up Treatment Infrastructure

Many cities in EAP have limited wastewater treatment infrastructure. Therefore, the impact on energy is minimal. However, more comprehensive systems are likely to be developed—thereby increasing overall energy use levels—as a result of increasing concerns about public health threats arising from the contamination of groundwater supply by septic tank seepage and septic waste dumping. Cities can take advantage of this opportunity to learn from global best practices and leapfrog inefficient systems design.

Cebu City has no citywide wastewater infrastructure. Its wastewater is currently managed through household septic tanks. The partially treated waste stored in the septic tank is removed by private septic waste haulers and disposed of, untreated, in the landfill. A similar process takes place in Surabaya, though several private wastewater treatment companies also operate treatment facilities for large buildings, industrial sites, and campuses. Da Nang is the only pilot city with wastewater treatment plants and a sewage pipe network that handles approximately 65 percent of wastewater. The remaining wastewater is stored in septic tanks. Da Nang's energy consumption for wastewater management is relatively low compared with other cities in the wider region, probably because the system has been designed as a "dry weather" system. During dry months, all wastewater flows through the system to treatment facilities where it is processed anaerobically. However, during the monsoon season, wastewater and rainwater are diverted directly into local waterways. This method increases pathogen levels in the receiving waters because the pumps are automatically switched off when rainfall floods the capacity of the system.

### Capturing Sludge Gas

In addition to treated effluent, wastewater treatment plants produce sludge composed of biomass and settled biological material. There is increasing awareness in the wastewater sectors in the three cities that sludge can be digested anaerobically or gasified to produce biogas for electricity generation. In Surabaya, smaller-scale composted sludge already contributes to soil fertilizer in city parks. A pilot project in Cebu City, aiming to treat septic tank waste using a biodigester, recently failed because of contamination of the sludge by industrial waste, which compromised the possibility of biological treatment.

### Managing Demand

Further energy improvements can be achieved by encouraging conservation and water-conscious behavior. Technological devices that promote water conservation, such as low-flow toilets, shower heads, and sink faucets, can be fitted in city government buildings and encouraged in private developments through outreach. In Cebu City and Da Nang, water utilities have implemented tariff structures and embarked on campaigns to educate consumers about the importance of water conservation. However, more can be done to improve customer knowledge about the importance of and means of reducing consumption at home because the initiatives are still not widespread.

## Power

With rapidly increasing energy demand, electric utilities in the region are challenged to establish and maintain generation and transmission capabilities and improve quality of service. All three cities are served by national electricity companies that provide subsidized power. Management of these companies is largely out of city government control, so the number of city interventions that can affect energy efficiency performance in this sector is limited. City governments, therefore, would need to work closely with national governments on energy policies to ensure the reliability and affordability of electricity supply to their cities. Despite this, as cities in the region increasingly localize and decentralize power generation in the future, city authorities will need to build capacity to understand the complexities and critical elements of power generation, distribution, and consumption.

### Managing Peak Loads

The daily load curves of the three cities are characterized by high peaks of short duration. Smoother load curves would enhance the efficiency of power plant operations, allowing generators to perform within their steady optimal performance ranges and reducing the need to add new generation capacity. City authorities are in a good position to influence energy demand to enhance the

**Table 3.4  Power Sector Performance for Three SUEEP Pilot Cities**

|  | Cebu City | Da Nang | Surabaya |
|---|---|---|---|
| Transmission losses (%) | 7.45 (combined) | 4.23 (combined) | 1.8 |
| Distribution losses (%) |  |  | 6.7 |
| Renewable generation (% of total) | 49 (geo and hydro) | 30 (hydro) | 0 |
| Peak load (MW) | — | 250 MW | — |
| Contribution to citywide GHG emissions (%) | 26 | 31 | 69 |

Source: Phase I pilot study.
Note: — = not available; GHG = greenhouse gas; MW = megawatt; SUEEP = Sustainable Urban Energy and Emissions Planning.

overall efficiency of the power sector. Authority can be wielded in various ways, such as the imposition of building codes (discussed earlier) and flexible tariffs to encourage energy use during off-peak hours, which is practiced in Da Nang.

### Generating Renewable Power

Both Cebu City and Da Nang rely on electricity generated outside the city using significant renewable sources (see table 3.4). No major renewable sources of energy, however, have been deployed within any of the city limits, despite significant potential for solar-powered technologies. The potential deployment of distributed and localized generation (smart metering) is not limited to renewable technologies such as solar sources on buildings—methane gas from the solid waste and wastewater sectors could also be captured.

To successfully encourage expansion of renewable and localized generation, each of the three cities would need to produce a renewable energy strategy. Steps should include the transformation of the transmission grid to enable two-way metering for localized generation as well as the development of smart-grid technology, which offers significant potential for improving electricity distribution and controlling demand. In addition, incentive schemes to encourage the uptake of renewable energy can be implemented. Such schemes include incentives such as feed-in tariff structures that can make renewable generation more financially attractive, given the failure of the market to account for environmental externalities. Although schemes like feed-in tariffs are usually implemented at the national level, city governments can encourage the use of renewables by favoring development that uses on-site renewables and by updating building codes to encourage the uptake of renewables in their cities.

### Optimizing Transmission and Distribution

As cities expand and the energy demand load increases, existing primary infrastructure, including transmission and distribution networks, experiences increased stress. Greater transmission inefficiencies and congestion occur, which, in turn, lead to reduced grid reliability and lower quality of electricity supply.

Transmission and distribution losses in the three cities are comparable (see figure 3.6). Several interventions, such as the installation of composite core conductors, high-efficiency transformers, energy storage, smart controls, and

**Figure 3.6  Transmission and Distribution Performance**

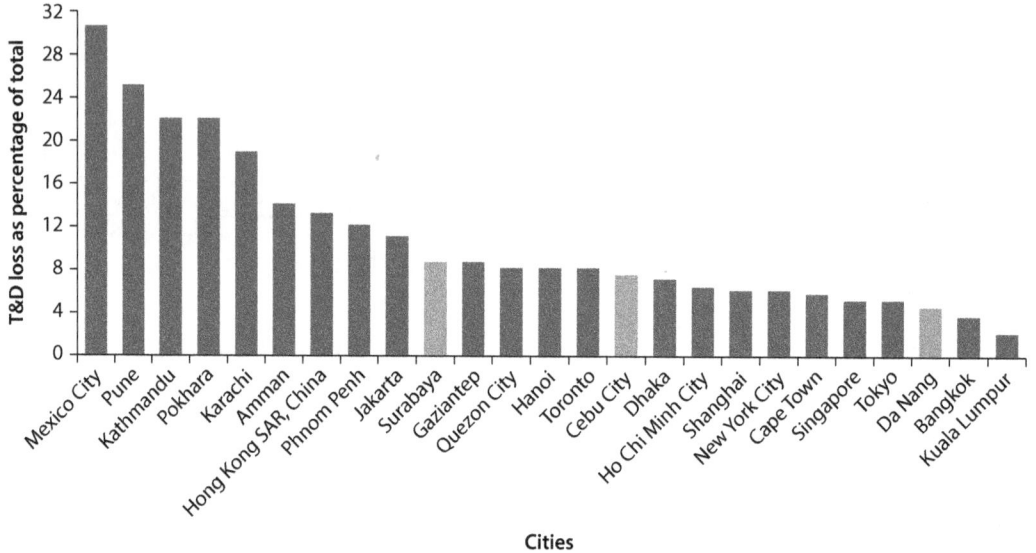

*Source:* Phase I pilot study.
*Note:* T&D = transmission and distribution.

localization of energy generation, can be made to reduce these losses. National energy suppliers to the three cities are aware of these engineering solutions, and have been performing ongoing maintenance on transformers and substations and making power factor corrections. In Da Nang, the power utilities are planning to ground distribution lines to minimize nontechnical losses from illegal connections, which is a persistent problem in the EAP region.

## Identifying City Government Priorities

The preceding sections of this chapter provide a review of sectoral energy use and issues across the three cities. Based on the opportunities and challenges posed, a range of potentially viable recommendations was derived for each sector that would enable emissions reductions and improvements in energy consumption efficiency. The challenge for city governments is to accurately determine which sectors and recommendations should be prioritized for action. This activity in the first phase of the Sustainable Urban Energy and Emissions Planning (SUEEP) program was undertaken using a prioritization process that forms the second module of the Tool for Rapid Assessment of City Energy (TRACE) system. The process is summarized below.

The following factors were taken into account in prioritizing sectors for action in the pilot cities:

- *Greatest potential for improvement.* Sectors that have the most potential for improvement in energy efficiency and GHG mitigation are identified. To this end, "improvement" was defined as a function of

- Current energy expenditures on a given sector (estimated citywide expenditures) and
- Energy savings potential (based on benchmarking work conducted as a part of the study).

- *Scope of control or influence of the city government.* Although city governments may be able to implement several measures that reduce energy consumption and emissions, the impact will be limited if the sector represents a small fraction of citywide energy and emissions. The degree to which the city government can affect energy consumption and emissions in sectors such as industry, wastewater treatment, building design, or power distribution varies substantially from city to city. Thus, in the sector prioritization process, knowing the level of influence a city ultimately has in improving energy efficiency in each sector is very important.

To prioritize specific actions and recommendations derived through the program, or additional recommendations that may be developed at a later date, the following factors should also be considered:

- *Resource constraints.* Within each sector, multiple fiscal and human resource constraints that will hinder the success of programs should be considered.
- *Compatibility with other development goals.* In addition to the improvement of energy efficiency and the reduction of a city's carbon footprint, a city government would also seek to improve the city's environmental, social, and economic conditions. Therefore, energy efficiency and low-carbon activities that would most effectively help the city achieve these parallel objectives would be given a higher priority.
- *Timing.* Given other ongoing initiatives in the city, it is useful to identify those activities that would be the most complementary in the short, medium, and long terms.

Based on the above, a methodology was formulated to prioritize programs using the following factors:

- *Energy expenditure information.* This information can be obtained either from city government budget offices or through the conversion of energy use across the city into a monetary value.

- *Energy efficiency opportunity.* Key performance indicators from the TRACE benchmarking process that are most indicative of energy use across a particular sector or subsector are chosen. To define opportunity, the mean value of sectoral energy use of the better-performing cities in the peer group is calculated; the difference between this value and the city's current performance provides an improvement target for the city. This is termed the "relative energy intensity" (REI) of the sector.

- *The control or influence of the city authority.* This factor ranges from minimum (national governments have full or greater control) to maximum influence (city has full budgetary and regulatory control). Each of the seven options in the range has a range of values; therefore, the TRACE user can determine the extent of the city authority's control or influence in each sector. This final component is used as a weighting factor, with values ranging from 0.01 to 1.

The prioritization calculation is based on simple multiplication of each factor's potential for energy improvement, spending, and degree of city authority influence. Thus the calculation is as follows:

$$P = (CGSe \times REI \times C),$$

in which
P = prioritization,
CGSe = city government expenditure on energy,
REI = relative energy intensity, and
C = level of city authority influence.

The prioritization table for Da Nang is illustrated in table 3.5.

Note that table 3.5 presents priorities for sectors in which the city authority has maximum influence, and for citywide issues over which the authority has limited influence. The ranking suggests that the city of Da Nang should prioritize programs as follows:

- Street lighting
- City buildings
- Solid waste
- Water treatment system

**Table 3.5  Summary Results of the TRACE Prioritization Process in Da Nang, 2011**

| Priority ranking | Sector | 2010 energy spending (US$) | Relative energy intensity (%) | Level of city authority control[a] | Savings potential (US$)[b] |
|---|---|---|---|---|---|
| *City authority sector ranking* | | | | | |
| 1 | Street lighting | 1,200,000 | 78.2 | 1.00 | 939,141 |
| 2 | City buildings | 2,069,047 | 15.1 | 1.00 | 312,426 |
| 3 | Solid waste | 452,380 | 48.8 | 0.97 | 214,277 |
| 4 | Potable water | 564,349 | 26.4 | 0.96 | 143,163 |
| 5 | Wastewater | 95,000 | 11.1 | 0.96 | 10,133 |
| *Citywide sector ranking* | | | | | |
| 1 | Power | 54,285,714 | 33.8 | 0.38 | 6,973,725 |
| 2 | Private vehicles | 44,665,149 | 10.0 | 0.14 | 669,977 |
| 3 | Public transportation | 361,773 | 65.8 | 0.90 | 214,416 |

*Source:* Phase I pilot study.
*Note:* TRACE = Tool for Rapid Assessment of City Energy.
a. 0 = no influence; 1 = maximum influence.
b. Based on TRACE benchmarking; these figures are indicative of the quantum of savings that may be possible, but not necessarily practicable.

On a citywide basis, the power supply system is a clear area of focus, as are public and private transportation.

This sector prioritization methodology relates only to energy efficiency. However, other factors that promote energy sustainability, such as increasing the share of renewable energy in the energy supply system, which will enhance the reliability of energy supply and reduce GHG emissions, can also be incorporated into the recommendations to be considered in due course without the need for further prioritization.

## Notes

1. Euro II and Euro IV are European vehicle emissions standards that define acceptable levels of vehicle exhaust emissions on new vehicles sold in Europe. Euro II was passed in 1996 and regulates passenger cars and motorcycles. Euro IV was passed in 2005 and regulates all vehicles.
2. A barangay is the smallest administrative division in a city in the Philippines, equivalent to a district or a ward.

## Reference

IPCC (Intergovernmental Panel on Climate Change). 2007. *Contribution of Working Group III to the Fourth Assessment Report of the Intergovernmental Panel on Climate Change*. Geneva, Switzerland: IPCC Secretariat.

# CHAPTER 4

# Governance

## The Importance of Governance

Energy governance refers to a city government organizational framework established specifically to deal with energy-related issues. It is a key mechanism for deploying effective energy efficiency projects on a citywide level. The city's approach to energy is undertaken in an integrated manner, and the city's projects are aligned with the national government's efforts to improve energy efficiency in the city.

Improved energy governance will benefit cities through better oversight and enhanced transparency and through the coordination of various energy-related initiatives across sectors. It will also enhance the potential for innovation by giving local agencies and stakeholders the opportunity to engage and exchange ideas.

Sustainable Urban Energy and Emissions Planning (SUEEP) cuts across sectors, and touches on most areas of public service provision and private enterprise. SUEEP also has a temporal dimension, especially with planning, designing, procuring, constructing, and operating new facilities and infrastructure. In this regard, SUEEP must also guide spatial and investment plans. For instance, suppliers of electrical power and potable water, among other stakeholders, should be actively engaged in the strategic and economic development planning of cities over the long term. This not only enables them to provide inputs but also allows them to identify where capital investment will be required for new or improved utility infrastructure. As a result of these various considerations, the active and structured management of a wide variety of stakeholder relationships is necessary to ensure successful and lasting outcomes.

## Governance of Energy-Related Issues

During SUEEP Phase I, an institutional mapping exercise established the principal agencies and actors involved in the delivery and management of services that directly or indirectly affect the energy and greenhouse gas (GHG) emissions profile of each pilot city. Institutional mapping is useful to understand the range of stakeholders involved in energy planning, delivery, management, and the SUEEP process and also serves to communicate to all parties the relevant responsibilities, ensuring that all understand why each stakeholder is involved. Institutional mapping is also a useful tool for understanding the interplay between agencies. In particular, being aware of which agency reports to whom, which agency has decision-making power, and which agency can influence the formulation and implementation of policy will be a crucial determinant of the success of SUEEP. Knowledge of such dynamics will also minimize the possibility that overlapping issues are left unaddressed, especially if more than one agency is responsible for the implementation of a policy.

Figure 4.1 illustrates the institutional arrangements in Da Nang. It shows clearly that planning and management of energy issues involved multiple agencies, underscoring the importance of cross-agency collaboration in the planning and management of sustainable urban energy and emissions issues. The institutional maps from Cebu City and Surabaya displayed a similarly

**Figure 4.1  Overview of Energy Planning System in Da Nang**

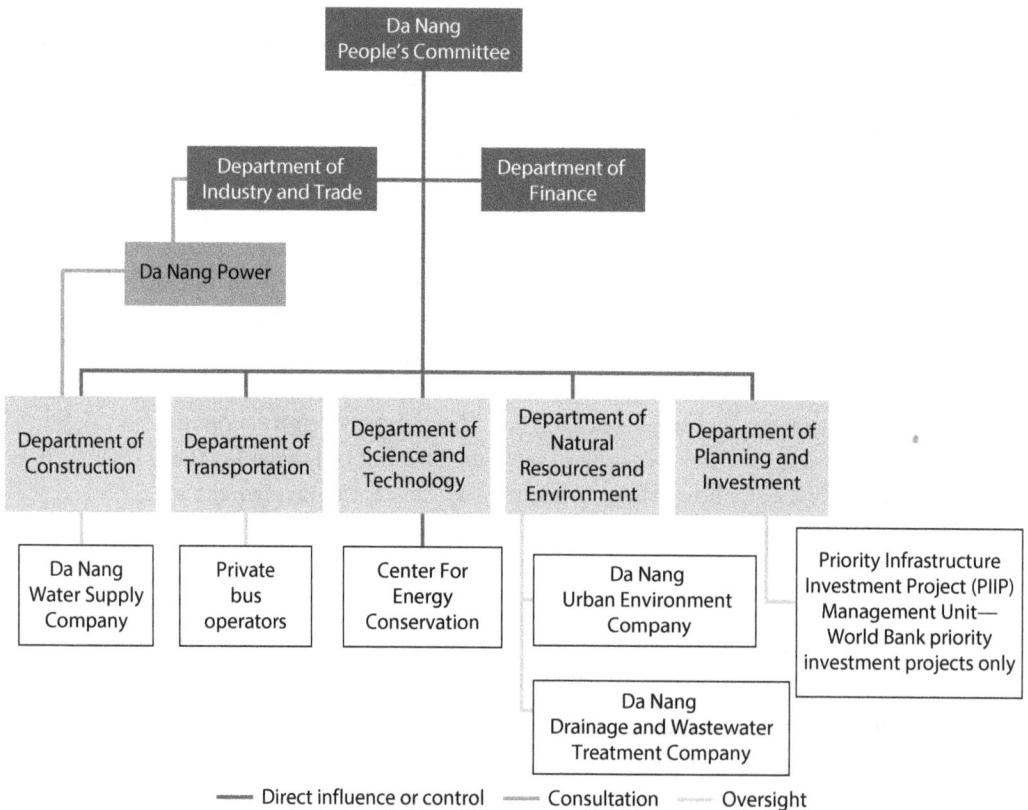

*Source:* Phase I pilot study.

complex pattern with respect to the numbers of organizations involved and the spread of control and influence over energy- and emissions-related matters. Despite this, differences in the governance structures in each of the pilot cities became evident, illustrating the need for city governments to fully understand city-specific relationships between stakeholders before embarking on an SUEEP exercise.

The degree of control or influence each city has was also mapped. The outcome, presented in figure 4.2, shows that each city generally has control of investment, budgets, and several local energy efficiency activities, as well as public lighting, public buildings, wastewater treatment, and waste tariffs. City authorities' influence was much more limited with respect to power supply and electricity tariffs. It is significant that in most other sectors, the pilot cities had varying degrees of influence, with coordination between local agencies more likely when facilitated by the city government. Thus, city governments need to be well aware of local circumstances so as to tailor plans to their unique needs and governance structures.

**Figure 4.2  Level of Influence of City Governments in Various Sectors**

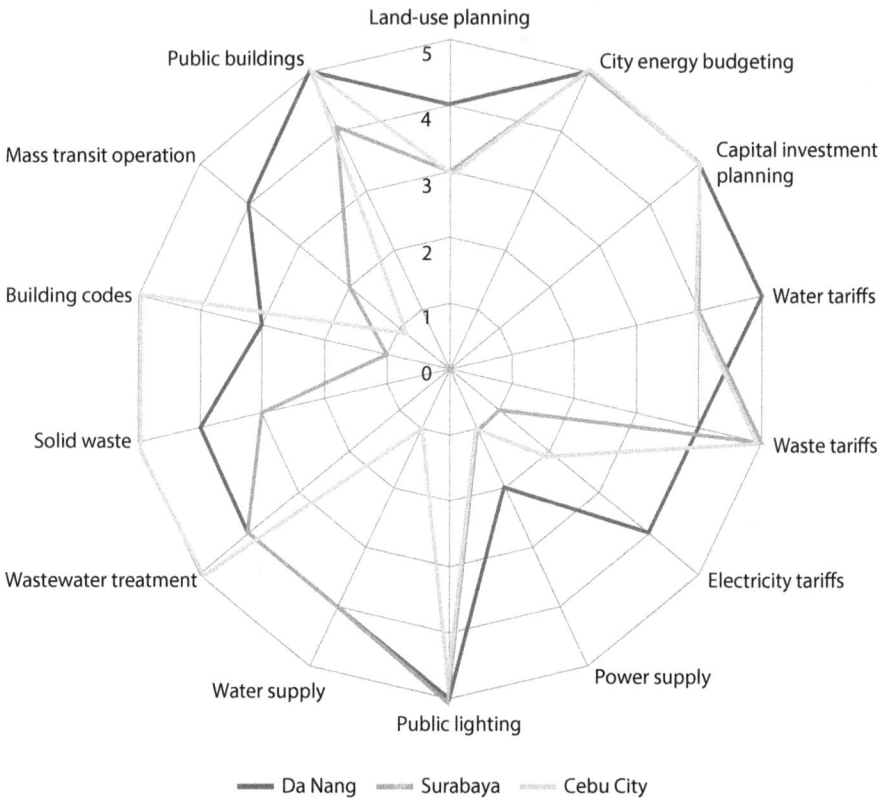

Da Nang        Surabaya        Cebu City

*Source:* Phase I pilot study.
*Note:* 0 = No influence; 5 = Maximum influence.

## City-Level and National Energy Programs

Although local governments have control or significant influence over policies at the city level, few energy policies, plans, and programs were implemented at the city level (see table 4.1). The limited number of policies enacted at the city level stems from the requirement to coordinate and institutionalize energy policies at a national level as part of the sustainable energy planning process. Implementing energy policies on a national level is more efficient and effective, and ensures consistency in application throughout the country. Examples of such measures include vehicle standards, energy labeling, and industrial energy efficiency programs. Likewise, programs that facilitate and promote low-carbon energy generation are also usually sponsored at a national level to enable and facilitate the country's economic development. Examples of national-level programs in each of the three pilot cities are listed in table 4.2.

**Table 4.1  Summary of Current Pilot City Energy Activities**

|  | Surabaya | Cebu City | Da Nang |
|---|---|---|---|
| City energy team |  | ✓ |  |
| City energy plan or program |  | ✓ | ✓ |
| Mayoral mandate for energy efficiency | ✓ |  | ✓ |
| Renewable energy research and development program |  |  | ✓ |
| Technical progams (for example, waste, buildings) |  | ✓ | ✓ |

*Source:* Phase I pilot study.

**Table 4.2  Summary of National Energy Efficiency Policies and Programs for Indonesia, the Philippines, and Vietnam**

| National policies, plans, and programs | Indonesia | Philippines | Vietnam |
|---|---|---|---|
| Government buildings energy efficiency management program | ✓ | ✓ | ✓ |
| State-owned energy services company (ESCO) | ✓ |  |  |
| Public-private partnership program on energy conservation | ✓ |  |  |
| Energy benchmark and best practice guide | ✓ |  |  |
| Building energy codes | ✓ | ✓ | ✓ |
| Energy labeling for appliances and products | ✓ | ✓ | ✓ |
| Energy efficient lighting program in the residential sector | ✓ | ✓ |  |
| Energy awards | ✓ | ✓ | ✓ |
| Lower rates on investments in energy conservation | ✓ | ✓ | ✓ |
| Government awareness raising for energy conservation in households and industries | ✓ | ✓ |  |
| Training and accreditation for energy managers and energy auditors | ✓ | ✓ | ✓ |
| Energy management services and energy audits |  | ✓ |  |
| Fuel conservation and efficiency in road transport program |  | ✓ | ✓ |
| Power conservation and demand management in all commercial, residential, and industrial sectors | ✓ | ✓ | ✓ |
| System loss reduction program |  | ✓ |  |
| Technical assistance to domestic producers on energy efficiency compliance |  | ✓ | ✓ |
| Support for industrial enterprises improving, upgrading, and optimizing technology | ✓ |  | ✓ |

*Source:* Derived from APERC 2010.

Given the close links between city and national energy policies, city and national governments must clearly understand the allocation of roles and responsibilities for bringing about a sustainable future. These roles are described below.

### The City Government's Roles and Responsibilities

Energy efficiency cuts across sectors and pertains in some way to most areas of public service provision and private enterprise. Cities often have the most direct line of public communication to residents, businesses, and industries, which means that education and incentives are most efficiently delivered through city governments. These factors, combined with the city government's intimate knowledge of city development and possession of the tools to influence or regulate sectors, mean that the city should be responsible for planning its development comprehensively and sustainably. To do this, the city government should take the lead in energy and emissions planning and implementation, and in advocating for changes that contribute to the advancement of the city's goals. One of its responsibilities would thus be to set up an appropriate organizational structure for energy planning that includes national, regional, and local governments as well as key stakeholders in the SUEEP process.

Achieving sustainable urban energy planning objectives through market transformation and integrated demand-side management will require the capability to better understand, characterize, and shape energy demand and use within urban built environments. In this regard, city governments will need to build their capacity to (a) understand the environmental, economic, and equity impacts of the embedded energy costs and operational energy needs of urban infrastructure systems and urbanization; (b) identify the local environmental, economic, and equity benefits of sustainable urban energy planning, especially with respect to the private sector; (c) develop information and materials that lead to a better understanding of planning options, particularly the costs and benefits of alternative technologies, practices, and development scenarios; and (d) develop effective decision-support tools and methods for community-based energy systems planning.

In planning its sustainable development, a city government must recognize its level of influence in implementing policies at the city level. To this end, the city government will have to be fully responsible for sectors over which it has significant influence (for example, street lighting, public buildings, and wastewater treatment). At the same time, city governments should determine those aspects of energy planning that can be aided by national programs and engage in a strong dialogue with national agencies from the outset of the process to determine the level of support that may be expected. In finding compatible interests between the city and national governments, cities can use the opportunity to secure support or financing from the national government to implement national policies in their cities. For example, there were programs in Da Nang that were implemented at the city level but were cofinanced by a national agency. The Department of Science and Technology, via its Center for Energy Efficiency and Conservation, coordinates the Da Nang City Project on Efficient Use of Energy

Energizing Green Cities in Southeast Asia • http://dx.doi.org/10.1596/978-0-8213-9837-1

and Application of Renewable Energy in accordance with the central government's energy efficiency mandates. The Center for Energy Efficiency and Conservation provides support to local businesses and government agencies on strategies to reduce energy consumption, conducts energy audits, and aids in implementation of audit findings at local businesses.

City governments should also work with national agencies and departments to coordinate activities with a clear understanding of individual responsibilities and expectations. Cities should also determine the kinds of policies and programs that must be undertaken at a national level (for example, specific aspects of energy generation and use) and those aspects of implementation for which national-level assistance is required (for example, capacity building and finance). Such feedback from city governments will be valuable to national government policy formulation and establish a starting point for future dialogue and collaboration.

### The National Government's Roles and Responsibilities

Many aspects of energy consumption, such as the sources of electricity and gas supply, gasoline subsidies, household appliance energy efficiency standards, and vehicle fuel efficiency standards, are influenced by national energy policies. Given the expected increase in energy consumption with rapid urbanization in the East Asia and Pacific (EAP) region, national governments will have to take the lead in implementing policies (in the form of regulations or incentives) that will promote the efficient use of energy. However, because the efforts of city governments will usually be essential to achievement of national policy goals or targets, it will be in the national government's best interests to work closely with individual cities on national-level policies and plans and to welcome interest by city governments in furthering efforts to implement national policies on the ground. For example, national energy, transportation, and environment departments or ministries should be included in city-level SUEEP discussions.

The national government should provide clear guidance to cities on the direction it plans to take on sustainable development so that cities can plan, and where possible, cooperate in areas in which national and city goals are aligned. In these areas, policies implemented by both the city and national governments can serve to reinforce each other, thus making efforts to develop the city sustainably more effective.

### City Planning

A strategic plan is usually guided by a "vision" and a set of objectives that city authorities seek to achieve. Thus, energy, GHG emissions, and the city's adaptation to climate change should be prominently featured as strategic objectives in this process. SUEEP, which is intrinsically linked to strategic spatial, transportation, and economic development planning, will determine the future location, scale, typology, infrastructure, and utility demands for development. With a long-term planning horizon, rapidly growing cities will benefit from early consideration of these issues while determining the future form and economic focus of their cities.

An integrated strategic planning process enables providers of public services, from mass transportation to wastewater treatment, to contribute and identify opportunities that would lead to greater reliability of energy supply, energy efficiency, and reduced GHG emissions, while at a minimum helping to define investment programs targeted to projections of future demand. Integrated planning is particularly important for transportation planning. Currently, some of the pilot cities suffer traffic congestion during peak hours, and growing car use will only exacerbate this problem. The energy balance studies and emissions inventories both point to the fact that transportation-related energy use already constitutes a significant share of each city's energy and emissions burden. Encouraging nonmotorized modes of transportation along with robust public transportation will be critical to mitigating the spread of congestion and improving the quality of transportation in all cities.

In addition to transportation management, local leaders may wish to consider preparing renewable energy master plans to improve the robustness of energy supply and harness the opportunities for renewables. Even if some renewable energy technologies are currently economically unattractive, "future proofing" the city for their widespread deployment (for example, structural strength requirements for flat-roofed buildings) will be hugely beneficial when the economics of renewables become more attractive. The establishment of new green building guidelines provides a particularly potent mechanism for encouraging renewables and future proofing cities.

These types of interventions are only possible if city governments have carefully considered energy issues during strategic planning cycles, and if collaboration across agencies is practiced to enable planning and implementation on a comprehensive basis. Transparency during collaboration is necessary to obtain support and buy-in for programs.

## Governance Mechanisms

This chapter thus far has focused on the need for city authorities to adopt an integrated and strategic approach to SUEEP. To maintain momentum and gain traction after plan, program, or initiative development, subsequent steps are required to formalize and govern activities.

### Energy and Emissions Task Force

One of the principal institutional recommendations is for each of the three pilot cities to establish a citywide energy and emissions task force to improve coordination and establish an integrated approach to energy planning and management. In all three pilot cities, committees have already been entrusted with the management of various aspects of energy and emissions. Where possible, existing committees can be used by extending their mandates through broader terms of reference and enhanced powers (if necessary) to take on the role of an energy and emissions task force.

Membership in the task force should not be limited to government—businesses and residential property managers, along with other members of industry and the public, can provide valuable input and offer a fresh perspective on the challenges. In addition, implementation of projects will require support from these stakeholders. The task force's roles would be threefold:

- Coordinate city-government and citywide energy efficiency projects and initiatives
- Develop long-term energy plans to identify and develop supply-side opportunities (for example, distributed energy generation and low-carbon generation projects)
- Act as a liaison between various departments and external agencies, especially with regard to energy infrastructure planning

In addition, the task force should establish the means to monitor energy-related issues over the long term, such that annual energy use and GHG emissions performance (both city government and citywide) can be monitored.

The effectiveness of the task force will depend on several factors, in particular,

- The body to which the task force reports
- The task force's decision-making powers
- The extent to which the task force can champion energy issues

### Formalizing Energy Plans

In cities in which SUEEP might take place outside a formal (that is, city or national government) urban planning process, the plan may require formal review and agreement by an executive authority. For example, in some EAP cities, the support of the mayor is insufficient for a program to be implemented. In this case, the program may need to be approved by the local parliament, which will also have to approve its budget.

### Monitoring and Reporting Systems

Monitoring and reporting systems are crucial for providing credibility in energy and emissions management. These systems need to be implemented carefully to ensure that they do not increase costs to the city unnecessarily and to ensure that they do fulfill the needs of the donor community that may, at some point, become involved in financing energy- or emissions-related projects. The establishment of a city-level platform for monitoring and managing energy and emissions is one of the first practical steps in implementing the vision of sustainable urban energy development and is a condition for mobilizing "green financing."

Regular collection of energy data is a prerequisite for the success of the SUEEP process. Of the pilot cities, Da Nang had the most readily available energy data from across a variety of government agencies, whereas collection of data was much less customary in Cebu City and Surabaya. Even when data were available, the three cities were characterized by limited sharing of data across departments. Sharing of information, experiences, and knowledge will be a major area of improvement opportunity for energy efficiency for all three cities. For example, analysis of trends in population, vehicle use, traffic, economic growth, and industry data would be useful for all city agencies to make better decisions with regard to energy consumption and energy policy.

Effective monitoring systems should be structured and formalized to demonstrate a high level of integrity and reliability, and should cover the following:

- *Institutional arrangements*—assign roles and responsibilities to individual agencies and personnel.
- *Boundaries*—define what has been included in and excluded from the energy and emissions inventories.
- *Sources of data*—include data from national agencies to city government collection arrangements and others; how data are collected and reported (for example, through surveys or meters); and the frequency with which data are collected.
- *Data collation methods*—provide transparency into how raw data are processed (particularly important if proxy data or extrapolations have been used or where data are incomplete).
- *Quality assurance processes*—demonstrate that the methods, processes, and sources used have been adequately audited or reviewed to identify gaps, omissions, and potential improvements to the monitoring and reporting process.

Monitoring and reporting systems should be continually improved for accuracy, reliability, consistency, transparency, and completeness.

## Reference

APERC (Asia Pacific Energy Research Center). 2010. *Compendium of Energy Efficiency Policies of APEC Economies.* Tokyo: APERC.

# Sustainable Urban Energy and Emissions Planning: The Way Forward

Chapters 1 through 4 of this report present the contextual background, citywide challenges, and sectoral issues faced by the pilot cities. These were illuminated with the use of the Energy Sector Management Assistance Program's (ESMAP) Tool for Rapid Assessment of City Energy (TRACE), which was designed to enable cities to assess how they use energy and to recommend energy efficiency initiatives to be implemented by the city government in a targeted manner. These chapters review and analyze the level of energy use and emissions in a small sample of East Asia and Pacific (EAP) region countries and identify areas in which reinforcement and capacity building may be required to enable these cities, as well as others in the region, to embed Sustainable Urban Energy and Emissions Planning (SUEEP) practices through city government leadership. However, these areas do not claim to be, and should not be, all-encompassing, given the need for cities to take into account their unique circumstances.

The SUEEP program has gone beyond the TRACE to explore ways for city governments to embark on citywide strategies to plan and manage the reduction of energy and emissions over the long term. Long-range energy planning will bring about economic growth and social benefits, such as the improvement of public health and safety, and will increase living standards. SUEEP's purpose is to provide a comprehensive approach to citywide energy planning to maximize energy efficiency across sectors. Its intent is to help cities to develop their own initiatives using different mechanisms and to define a governance system for implementation and monitoring and reporting. These outcomes will improve energy governance in the city and create a common platform for collaboration between the city and donors, civil society, and the private sector. The SUEEP process also provides a framework for city governments to plan a pipeline of investments in energy efficient infrastructure and to mobilize "green financing" support. To help cities establish and implement a road map for achieving a sustainable energy future, a SUEEP Guidebook, which recommends a set of steps based on the experiences in the pilot cities, has been prepared (see part III of this volume). It is envisaged that the Guidebook will be refined when experiences in implementing the SUEEP process from these pilot cities, as well as others using these guidelines, are incorporated.

Although a comprehensive planning approach is recommended, cities have varying levels of resources and capacity as well as differing priorities. For example, the pilot cities experienced difficulties in sustainable energy planning across multiple sectors. Thus, they had mainly undertaken project-oriented measures aimed at addressing issues on a sectoral basis, resulting in planning that was not geared toward a systemic strategy. Based on this reality, there is flexibility in working out different levels of engagement to suit cities' circumstances. The three levels of engagement that city governments could undertake follow:

- A high-level, rapid assessment of energy efficiency measures in a city (for example, the TRACE, an energy balance study, and a Greenhouse Gas [GHG] Emissions Inventory)
- Deeper sectoral engagements in selected areas (for example, public-private partnerships and sector-wide interventions)
- Implementation of the full SUEEP guidelines, subject to city governments' interest in and ownership of the process

Feedback from the pilot cities suggests that the SUEEP process would be useful, although the level of engagement may differ. Da Nang expressed interest in engaging at the first and most comprehensive level, whereas Cebu City and Surabaya found engagement at the second level, focusing on projects in specific sectors, to be more relevant.

This chapter defines the principal requirements for a full SUEEP process based on the observations from the study of the three pilot cities and experiences of other cities. The chapter is divided into six parts that lay out the key stages in the SUEEP process:

- City government leadership and commitment to "green growth"
- Energy and emissions diagnostics
- Goal setting and project prioritization
- Planning
- Implementation
- Monitoring and reporting of progress

A detailed description of these stages with examples of best practices in other cities is found in the SUEEP Guidebook (in part III of this volume).

## City Government Leadership and Commitment to "Green Growth"

City governments' multiple roles as regulator, enforcer, supplier, producer, and consumer provide great leverage in facilitating citywide, comprehensive, and strategic energy planning. As the sole body with the credibility, tenure, and strong linkages with the public and various stakeholders to put in place long-term measures to enable sustainable energy growth, city governments must exhibit commitment and leadership to enable "green growth" in their cities. Leadership is required at all levels, from the mayor's office to legislative bodies and city government departments, given their ability to understand the local situation and connect with the public and stakeholders. It is with strong leadership and commitment that sustainable solutions will gain traction and support to facilitate economic growth and social development.

In embarking on the journey toward sustainability, city governments must first create and clearly communicate the vision the city aspires to and that the public and stakeholders can identify with and thus support. To work toward this vision, a city government will have to engage with stakeholders in a transparent manner, identify the agencies and organizations involved in the planning and management of energy issues, and recognize its ability to control or influence specific issues and pertinent relationships with other agencies. With a clearer view of realizable solutions, city governments can establish energy and emissions task forces (or leverage existing ones) and hand them a strong mandate to enable change. Task forces must be empowered to oversee energy and emissions issues that cut across agencies and to establish plans and programs of energy- and emissions-related activities. City governments' roles will also include coordinating efforts with national agencies to facilitate sustainable practices. Finally, cities will be responsible for putting in place or strengthening monitoring and reporting systems to ensure that energy activities lead to the realization of the city's vision and goals.

Thus, city governments have an extensive role in enabling sustainable growth. Critical to effective energy planning and implementation will be governance, which will determine the success (or failure) of sustainable policies. Good governance improves oversight and enhances coordination and data sharing among agencies, which will facilitate the implementation of a comprehensive solution. Missed opportunities in comprehensive urban planning arising from insufficient coordination among government agencies were evident in some EAP cities where the Departments of Transportation, which were responsible for planning the cities' bus rapid transit systems, worked separately from the departments that were responsible for city planning. Because of this, the EAP cities could have missed the opportunity to reap new economic and energy-saving opportunities through density planning and zoning, which arise when new bus rapid transit systems are designed as an element of citywide planning strategies.

Institutional structures and clarity of responsibilities are also key to good governance. In Da Nang, for example, the institutional structures were modified to allow a comprehensive approach to energy governance to be taken.

However, overlaps in energy planning remained and involved a number of agencies, including the Department of Industry and Trade, Da Nang Power, the Department of Construction, the Department of Natural Resources and Environment, and the Department of Science and Technology. This meant that a segmented approach was embarked upon despite efforts to implement a cohesive strategy across agencies.

## Energy and Emissions Diagnostics

Having accurate baseline energy use and emissions data is important to enabling city governments to formulate policies in line with the SUEEP vision and overarching goals. The following are key diagnostics used to set energy and emissions targets.

- *An energy balance study.* This study maps primary and secondary energy supply and use in each city. Energy balance studies highlight the relative importance of different fuels through their contribution to the city's economy; the efficiency of different energy conversion technologies; and the areas where energy efficiency, reliability, and conservation efforts can be targeted. Benchmarks can also be set to enable a city to assess the effectiveness of different policies.
- *A GHG emissions inventory.* An inventory establishes the principal sources of GHG emissions, which helps to identify potential policy or project initiatives that would not be chosen based on a purely energy-based analysis. Just as important, this inventory provides the emissions information that may help local government agencies or utilities obtain carbon finance funding for renewable energy or energy efficiency projects.
- *TRACE.* TRACE is a decision support tool that compares one city's energy performance with that of other cities through a custom benchmarking mechanism. Benchmarking results are used to identify priority sectors to which a city should consider targeting its energy efficiency efforts. Policy recommendations built into TRACE are linked to each sector, triggering field interviews and further data gathering by the TRACE project consultant to assess which recommendations are most appropriate for the city context.

The application of these three processes will enable a city to establish a well-grounded overview of its current energy and emissions situation and, where data exist, build time series to provide trend and growth information.

In developing the energy balance study and GHG emissions inventory, the data- and information-gathering process must be tightly managed to ensure coordination between the various agencies that possess energy and emissions information and to ensure that data are evidence based rather than anecdotal. It is also important to ensure that sources of information are known and can be revisited, if necessary. For this purpose, a trained (but not necessarily large) team is needed to ensure the consistency, accuracy, and reliability of the data and information. Gaps in knowledge and information should also be accurately identified during and after data collection so that these gaps can be actively pursued as an area of future work.

## Goal Setting and Project Prioritization

### Set Goals

Goal setting is the process of linking the city's vision to its targets to reduce energy use and emissions through initiatives that do not compromise city development. Although the pilot cities had developed high-level visions, none had linked them to specific, measurable goals, indicating that robust connections between energy, emissions, and the city's aspirations had not yet been established.

The goal-setting process affects the SUEEP process in three principal ways:

- Linking citywide goals to energy and emissions goals by pairing them with city aspirations
- Building consensus, engaging stakeholders, and securing their buy-in
- Establishing key success factors to enable performance monitoring of the city's energy and emissions strategy

### Linking Citywide Goals to Energy and Emissions Goals

Linking wider economic, social, and environmental city goals to energy and emissions goals clarifies the role that energy and emissions play in economic development and builds the case for SUEEP to be a key contributor to the city's sustainable growth. Connecting energy goals to the city's priorities (for example, access to energy or potable water) and to its current and projected development path, and communicating the benefits and potential impacts of energy planning beyond the energy sector, will galvanize stakeholder support and strengthen the case for adoption of SUEEP. In the pilot cities, commitment to energy efficiency and formulation of the city vision were evident, but formalized links between citywide goals and energy planning have not been established. For example, the three cities had implemented energy conservation measures that included initiatives such as off-hours lighting and air conditioning energy savings, but none had an integrated strategy covering the short, medium, and long terms. This informal approach to energy efficiency and emissions limits cities' ability to link planning over varying time scales and sectors, build lasting internal capacity and partnerships, and secure financing for major projects as part of an integrated and comprehensive plan.

### Building Consensus, Engaging Stakeholders, and Securing Their Buy-In

Early engagement with potential stakeholders helps to gain support for energy and emissions planning. Seeking widespread participation from the general public, the private sector, and public agencies helps to ensure that all potential interests are represented and that no segment of society can claim exclusion from

---

**Box 5.1  Developing Da Nang: The Environmental City**

In 2008, the Da Nang People's Committee, through its Department of Natural Resources and Environment, produced a plan titled "Developing Da Nang: The Environmental City." This holistic environmental plan was prepared to redress the environmental damage caused by the recent urbanization and industrialization of Da Nang and covers air, water, soil, and urban environmental management over a 12-year period. The plan addresses energy use and planning via the "air environment" objective, with interventions tackling the public bus transportation network; encouraging the use of clean energy and the saving of energy and fuel; and developing models using renewable energy, in particular, wind and solar energy. The plan is particularly specific on governance and funding, as well as on how the plan will be monitored and reviewed.

---

the process. This becomes highly pertinent in the early stages of SUEEP, during development of a vision and setting of goals.

Effective buy-in from stakeholders can also be earned by demonstration of a city government's ability to manage energy issues through elements of city-government-controlled energy use. Typically, this would include activities such as the retrofitting of existing public buildings and public lighting and pursuit of renewable energy technologies. City authorities generally have good control over these issues and can use them to galvanize support within the city government and to show evidence of active management of energy to external stakeholder groups.

Clearly stated goals and expected outcomes engage stakeholders by formally setting out the benefits that can be expected from SUEEP (see box 5.1). Figure 5.1 shows an illustrative city's 2020 energy savings goal, indicating how the energy-growth trajectory can be "bent" over time in comparison with the business-as-usual scenario. More important, figure 5.1 shows clearly how individual sectors contribute to the process, starting with known trends during the first three years (2008–10). Figure 5.2 further illustrates a detailed plan by the city to achieve 25 percent energy savings against a business-as-usual scenario in the transportation sector through a targeted approach. Graphs such as these are powerful tools that help communicate the task at hand and the targets for each sector.

*Establishing Key Success Factors*

Goal setting identifies the means by which the success of the proposals can be measured. To set goals, two requirements must be satisfied:

- Quantitative indicators, either intensity (for example, "X" kWh per m$^2$ [kilowatt-hour per square meter]) or absolute (for example, "Y" tons of $CO_2e$ [carbon dioxide equivalent]), must be used.
- An accurate, complete, and replicable baseline must be established, from which changes will be measured.

**Figure 5.1 "Bending the Curve": Defining a 2020 Energy Savings Goal for an Illustrative City**

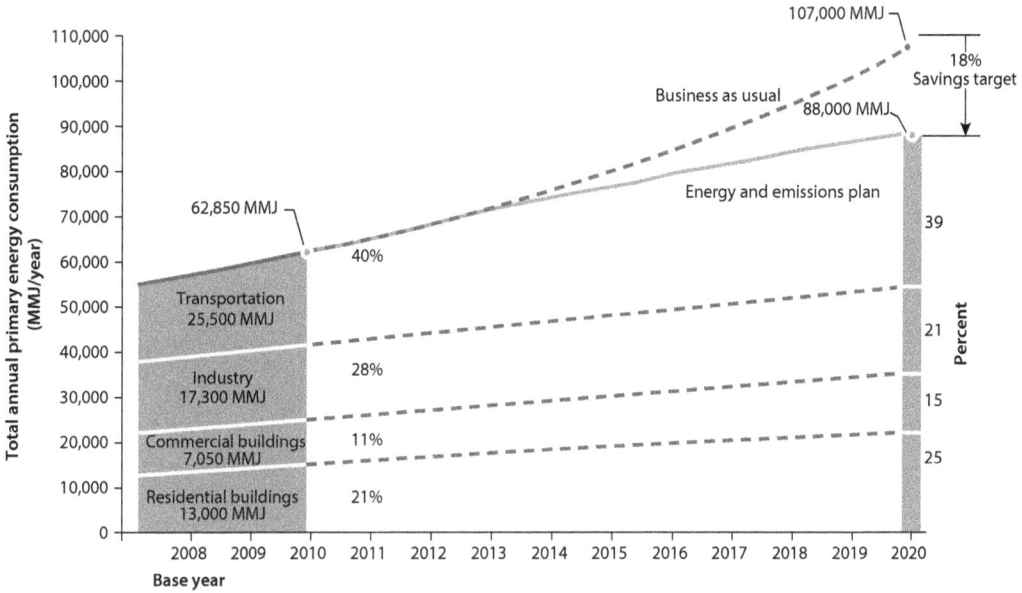

*Note:* MMJ = million megajoules.

The use of empirical and calculated evidence is vital to successfully creating and achieving goals. Using evidence-based numbers that are aggregated from several sources (that is, the sum of a number of technical interventions and application of other levers) also reduces the chances for goals to be overly influenced by political cycles or other extraneous factors.

In phased plans, the process can begin without a known baseline. However, creating the baseline would be one of the first enabling tasks, with goal setting taking place subsequently.

Ultimately, whether SUEEP succeeds or fails depends on the support and commitment of the stakeholders involved in implementing projects that enable the city to attain its goals. Energy goals that balance the political and technical aspects of energy use in cities provide the foundation for widespread acceptance.

In many instances, cities will be required to establish targets that address national energy or emissions goals. Ideally, cities should consider how they might work toward national goals while identifying practical measures that account for their unique conditions. For instance, a national target relating to energy consumption or efficiency in the manufacturing sector is of little relevance to a services-based city economy. Therefore, the city government must identify targets that navigate between national goals and city practicalities. National goals could be better addressed in the pilot cities.

### Prioritize Projects
Goal setting is not an abstract process. It should be undertaken with the full knowledge of the city's ability to bring to fruition a variety of operational, technical, and infrastructure-related proposals that collectively work to achieve a

**Figure 5.2 Energy-Demand Reduction Goals for the Transportation Sector in an Illustrative City**

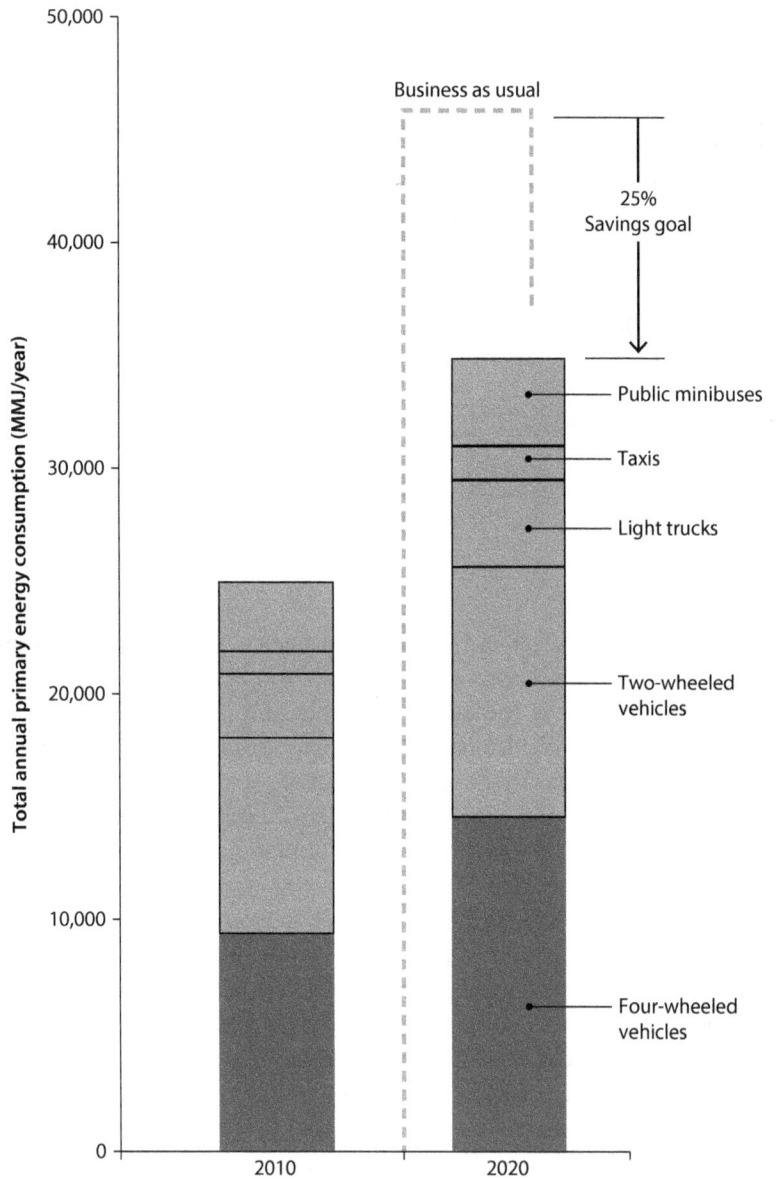

*Note:* MMJ = million megajoules.

specific goal. Given the number of barriers cities will face in implementing an SUEEP process, it is important that the city's goals are aspirational, but also realistic and achievable. Therefore, a focus on priorities is essential to successful deployment of the SUEEP process. Prioritization is part of a project management approach to SUEEP and is meant to break down initiatives into manageable activities. To this end, cities will have to prioritize projects to maximize their impact using available (but usually limited) resources. Factors such as the impact on energy use and emissions, funds available, extent of influence the city government has, existing institutional structures and culture of a city, as well as the relevance of the project to a city's overall sustainability and its relationship to national goals, all play a part in determining the success of a project. Therefore, prioritization of projects needs to combine common sense, quantifiable solutions aligned with the city's goals, and judgment by city leaders who understand the city's issues and politics.

## Planning

Armed with data on energy use and emissions and clarity about its goals, a city government can proceed to address its energy and emissions issues through a SUEEP process (see box 5.2) to help fulfill its vision. SUEEP enables cities to set out the case for sustainable energy, define actions they can take to improve energy and emissions performance, and garner support through the inclusion of stakeholders in the planning and implementation processes.

Despite the potential benefits that can be reaped using SUEEP, there are barriers that can hinder its implementation. These include unfamiliarity with the process; absence of formal institutional structures and capacity; and lack of political buy-in, cross-agency collaboration, and awareness of the importance of energy planning. These barriers were evident in the pilot cities, which generally approached energy and emissions planning on a fragmented basis that limited their capability to develop a plan. Despite this, the pilot cities have taken steps in the right direction. Da Nang, Surabaya, and Cebu City had begun instituting energy governance, indicating the potential for the cities' development in the future. Da Nang City People's Committee had identified the local Department of Industry and Trade as the lead agency on energy, supported by a number of other departments and local academic institutions. Similarly, Surabaya had initiated energy planning through Bapekko, which is a department in the mayor's office responsible for overseeing long-term planning initiatives.

---

**Box 5.2  Elements of an SUEEP Process**

- Identification of key energy and emissions issues faced by the city
- Establishment of goals that can be achieved by the city government and a plan for attaining those goals (with or without the aid of other agencies or external parties)
- Integration of energy and emissions issues into wider city planning processes
- Coordination across sectors and agencies and establishment of shared goals, in particular, engagement in a strong dialogue with the national government
- Establishment and improvement of governance, monitoring, and reporting processes essential to future management of the issues, as well as prerequisites for third-party involvement in project and carbon finance

---

The Philippine Department of Energy in Cebu City was responsible for a variety of energy efficiency initiatives fostered and supported by national policy, and the recently formed City of Cebu Government Committee on Energy Efficiency had been tasked with promoting energy efficiency across city government departments.

The scope of an energy and emissions plan can vary from citywide issues to internal city government energy management because of the city's plethora of energy-related functions—consumer, regulator, motivator, educator, and sometimes supplier. The approach to each of these functions should be an explicit part of the planning process to clarify the available levers for action and the potential impact of each measure, as well as the potential for conflicts of interest. For example, two key initiatives identified for Da Nang include the citywide deployment of green building codes and the development of a city procurement code for street light installations. Although the latter will positively impact energy efficiency in Da Nang, that impact is limited because street lighting represents only 2 percent of total city energy use. In contrast, the deployment of green building codes targets all potential development in the city and involves cross-sectoral stakeholders in city government and the private sector, expanding the potential impact of the intervention to all buildings in the city.

## Implementation

Successful implementation of the energy and emissions plan depends on many factors. In the pilot cities, the lack of capacity and financing were cited as the strongest impediments to the successful implementation of energy and emissions planning. In addition to these constraints, the pilot cities were faced with challenges that included the lack of personnel, low public awareness, poor communication with stakeholders, and lack of governance processes. This section discusses the most common barriers in greater detail.

### Policy and Regulatory Environment

Typically, sustainable energy and emissions issues are not considered in tandem with city policy development. Hence, cities may find that they face challenges related to misaligned policy (or even cases in which policies work against each other) or policy gaps that will need to be addressed when they embark upon an SUEEP process. For example, street lights in Cebu City were being upgraded, but energy efficiency was not a key consideration and carbon or other financing options had not been sought. In the transportation sector, the three pilot cities have yet to develop integrated and effective policy approaches for tackling rising congestion and private vehicle ownership despite the large contribution of the transportation sector to energy use and emissions. Regardless of these challenges, cities should persevere in formulating and implementing policies and projects on a comprehensive basis because the benefits that can be reaped are substantial.

In some instances, city governments may feel that their ability to influence or control issues is limited if national policies take precedence. For sustainability issues for which national and local governments' interests are aligned, city governments should review existing policies, projects, and initiatives to determine whether the policies can be tweaked or strengthened so that the city can build on or contribute to national policies. If so, the SUEEP process will gain national-level support.

Policies are vulnerable to political cycles and are subject to change by incoming mayoral administrations. Because energy and emissions plans require consistent application spanning a number of political cycles, codifying SUEEP policies into law could minimize political uncertainty. EAP cities will have to review this option carefully, taking into consideration their respective governance structures.

### Financing

The manner in which energy and emissions projects are financed is particularly important. Financing tends to be subject to a range of factors such as knowledge and experience with financing similar projects, access to capital, perceived risk, and the need for delivery partners that possess implementation skills.

A fundamental challenge facing cities wishing to implement energy efficiency and emissions projects is achieving a return on investment acceptable to investors while mitigating perceived risks. Many such projects tend to face higher investment hurdle rates than more conventional projects because investors are not familiar with the calculation methodologies used to determine investment return, and energy efficient technology tends to be relatively new to the market. This problem is amplified in developing economies in which the level of perceived risk is heightened, especially for newer technologies in the environmental sector, thereby resulting in higher hurdle rates. Another barrier to financing is the improbability of a private financial institution accepting a return on investment lower than that provided by a government bond, as opposed to a city government that might consider investments with very low or even negative returns if the socioeconomic benefits are significant.

Still, city governments have a number of tools at their disposal to encourage investment, including the following:

- Reducing regulatory risks stemming from uncertainty about the long-term policy environment (which can be mitigated if the city is willing to make policy guarantees)
- Pursuing a range of partnering options (for example, public-private partnership initiatives and partnering with energy services companies)
- Providing incentives and tax breaks (for example, by providing loan guarantees for initial seed funding of investments or fast-tracking regulatory permits)

City governments can also actively seek out alternative forms of funding that offer concessional financing for climate-positive projects. The potential to attract financing from international institutions and bilateral and multilateral agencies should be reviewed during the evaluation of potential projects that the city plans to undertake. Some financing schemes are region specific or will only fund certain types of projects or programs. For example, if an analysis of return on investment demonstrates that the proposed project cannot compete with projects in nonenvironmental sectors, tailored finance can be sought from investors that are experienced in environmental project risk and have obligations to direct capital into certain market segments. Socially responsible investment funds, for instance, deploy capital according to refined criteria that include environmental and social performance indicators.

Various financing mechanisms can be leveraged to support projects. They differ not so much by sector as much as they do by applicable project stage. The following sections describe the different subsidized international mechanisms applicable to different stages of development: (a) design and piloting, (b) construction and implementation, and (c) operations and maintenance. The common thread among them is that they are made possible through the estimation and ongoing monitoring of GHG emissions associated with specific investments.

### Project Design and Piloting

Significant barriers can obstruct the implementation of new energy efficiency measures, including the lack of experience or technical capacity, challenging regulatory environments, investor concerns about risks associated with untested projects, and most important, uncertainty about the potential payback. To overcome these barriers, local governments can leverage an array of international grant programs.

Grants are made available at various times by multilateral financial institutions for specific project objectives, such as GHG mitigation. Grants are often provided by international agencies, including the World Bank Group, the Asian Development Bank, and the United Nations, along with many multilateral agencies. Some large nongovernmental organizations and research agencies may also provide funding and technical assistance for financing certain types of projects.

The most widely known program that supports project development is the Global Environment Facility (GEF), which was established to provide incremental cost financing for projects that support energy efficiency, renewable energy, new clean energy technology, and sustainable transportation projects. Its approach focuses on removing barriers to "win-win" mitigation projects by providing support for technical assistance, policy reform, capacity building, piloting, and partial risk guarantees (PRGs). GEF grants through the World Bank Group average between US$8 million and US$10 million per project.

These grant programs can be very successful in getting new ideas and technologies off the ground and establishing greater certainty on payback periods for some initiatives, but grant funding is generally too small to support large-scale implementation of projects and policies or to support construction of new infrastructure facilities.

*Project Construction*

After project design and piloting hurdles are overcome, a new, more robust source of financing is needed to support construction or wide-scale implementation. Financial products such as soft loans, which are becoming more common than grants, are rising to meet this need. Soft loans are also provided by international agencies and multilateral financial institutions as well as by some larger commercial banks. These loans make funds available at reduced interest rates, and often with longer repayment times than a commercial loan. Terms can be negotiated directly between the funder and the city. Commercial banks may provide these loans only with an external guarantee or partial loan guarantee, or an external agency may provide funding to reduce the interest rate.

Another type of financing that can be used to facilitate construction is guarantees, such as the PRG. These guarantees are particularly helpful in supporting energy efficiency projects that do not fall into traditional lending categories. PRGs protect private lenders or investors against the risk of a government (or government-owned) entity failing to perform its obligations with respect to a private project. In the case of default resulting from the nonperformance of contractual obligations undertaken by governments or their agencies in private sector projects, PRGs ensure payment and thereby significantly reduce the risk assumed by investors. PRGs are available for projects with private participation depending on certain government contractual undertakings, such as build-operate-transfer and concession projects, public-private partnership projects, and privatizations. PRGs are available for both greenfield and existing projects.

Combined with grants for removing initial barriers and developing pilots, these financing mechanisms can be strong incentives—but more assistance may be needed to ensure long-term sustainability during the operations and maintenance phase of GHG mitigation projects.

*Project Operations and Maintenance*

A final hurdle in supporting low-carbon investments is ensuring that sufficient funds are available for ongoing operations and maintenance, especially in the short term. Very often, green financing and funding mechanisms that support project design and implementation or construction do not provide sufficient resources for ongoing monitoring to ensure that the project is meeting expected environmental objectives—a critical step to ensuring that future investments are suitably allocated.

This is an area in which early engagement of carbon financing mechanisms, such as the Clean Development Mechanism (CDM) or the voluntary carbon markets, can improve the feasibility of GHG mitigation projects. The CDM, which can be leveraged exclusively by developing countries, is a performance-based funding mechanism, through which annual payments are made to project sponsors based on realized emissions reductions (for example, energy efficiency performance). If the potential benefits of carbon financing are recognized and pursued early in the project's life (before commission), these mechanisms can ensure optimal and continuous project operation and maintenance. The three pilot cities have identified opportunities to leverage CDM financing across most sectors.

City governments can also actively seek alternative forms of funding. For example, tailored finance may be sought from investors experienced in environmental project risk who have obligations to direct capital into certain market segments.

### Procurement and Implementation Supervision

Procurement and implementation supervision are aspects of energy efficiency and emissions reductions programs that require strong technical capacity and experience, which often is not available in city governments. Technical and resource capacity may not be an issue for SUEEP initiatives that modify governance structures or for specific small-scale projects. However, successful delivery of the majority of technical and infrastructure projects requires a robust procurement process and stringent contract management, neither of which are often realistically available. For example, in Cebu City, a landfill gas capture project had been proposed on a number of occasions, but progress was limited because of a lack of institutional capacity to implement it. To preempt procurement and implementation problems, city authorities may consider a number of strategies:

- Develop a favorable policy environment, for example, enact regulations enabling public-private cooperation, contract law enforcement, and so forth.
- Establish a procurement policy to enable SUEEP projects.
- Nurture a collaboration network to gain access to technical assistance from nongovernmental organizations.
- Share resources and knowledge with neighboring cities or areas.
- Develop pilot projects to increase the skill level of the city workforce.
- Prioritize projects to ensure achievable workloads are maintained for city employees.

All of these strategies can enable procurement and implementation that reduce the burden on the city, thereby increasing the probability of effective project rollout.

## Monitoring and Reporting of Progress

EAP cities will need to establish citywide monitoring and reporting systems as well as systems specific to individual technical projects, given that performance monitoring is a requisite component of both goal setting and performance measurement. The pilot cities all have scope for improvement in this aspect of SUEEP, based on the scarcity of city-specific energy performance data. In addition, energy and emissions plans require regular review, based on measured outcomes, to realign efforts to achieve stated goals, if necessary, during the course of the plan's operation. SUEEP Phase I found that performance monitoring in the pilot cities had not been adequately implemented except in some cases as required by the national government.

### Identification of Performance Measures

The first requirement for performance monitoring is the selection and deployment of performance indicators. Typically, performance indicators will be either "gross" or "intensity" related. Gross indicators evaluate the totality of a given measure, for example, tons of carbon dioxide per year, and are useful for measuring urban issues at a city scale. Intensity indicators are more usually associated with efficiency (for example, kilowatt-hours per square meter of city buildings per year), and measure the intensity of energy use or emissions on the basis of indicative units (for example, per person, unit of area, or unit of length). Both types of indicators may be used in benchmarking performance. Table 5.1 presents the range of indicators deployed in the TRACE that enable benchmarking of city energy use against other cities; these indicators are also useful for long-term performance measurement.

### Data Acquisition and Management

City governments and external agencies make long-term policy and investment decisions based on data. Hence, it is imperative that city governments give appropriate consideration to the requirements of reliable data management systems and processes, as well as the risks inherent in data acquisition and management, to ensure that data are reliable, accurate, consistent, and complete.

Figure 5.3 provides an overview of the principal components of a data production chain. The first two bubbles relate to the means of measurement and are generally defined through international standards and best practices, or by using a generic energy and emissions tool such as the SUEEP Guidebook (part III of this volume). The remaining bubbles relate to the systemization of data acquisition and management.

Data management systems arc used to establish, document, implement, and maintain an effective process for data management. The typical characteristics of increasingly sophisticated data management systems are presented in figure 5.4.

In practice, most systems start at the informal level and develop following a management system approach such as the International Organization for Standardization's recently released energy management systems standard (ISO 2011), which specifies requirements applicable to energy use and consumption, including measurement, documentation and reporting, design, and procurement practices for equipment, systems, processes, and personnel that contribute to energy performance.

**Table 5.1  City-Specific KPIs Used in the TRACE**

*Citywide KPIs*

| | |
|---|---|
| CW-1 | Electricity consumption (kWhe/capita) |
| CW-2 | Electricity consumption (kWhe/GDP) |
| CW-3 | Primary energy consumption (MJ/capita) |
| CW-4 | Primary energy consumption (MJ/GDP) |

*Transportation KPIs*

| | |
|---|---|
| T-1 | Total transportation (MJ/capita) |
| T-2 | Public transportation (MJ/passenger-km) |
| T-3 | Private transportation (MJ/passenger-km) |
| T-4 | Transportation nonmotorized mode split (%) |
| T-5 | Public transportation mode split (%) |
| T-6 | Kilometers of high capacity transit per 1,000 people |

*Buildings KPIs*

| | |
|---|---|
| B-1 | Municipal buildings (kWhe/m$^2$) |
| B-2 | Municipal buildings heat consumption (kWhth/m$^2$) |
| B-3 | Municipal buildings energy spent as percentage of municipal budget |

*Street lighting KPIs*

| | |
|---|---|
| SL-1 | Electricity consumed per km of lit roads (kWhe/km) |
| SL-2 | Percentage of city roads lit |
| SL-3 | Electricity consumed per light pole (kWh/pole) |

*Power and heat KPIs*

| | |
|---|---|
| PH-1 | Percentage of heat loss from network |
| PH-2 | Percentage of total T&D losses |
| PH-3 | Percentage of T&D loss due to nontechnical losses |

*Water and wastewater KPIs*

| | |
|---|---|
| WW-1 | Water consumption (liter/capita/day) |
| WW-2 | Energy density of potable water production (kWhe/m$^3$) |
| WW-3 | Energy density of wastewater treatment (kWhe/m$^3$) |
| WW-4 | Percentage of nonrevenue water |
| WW-5 | Electricity cost for water treatment (potable and wastewater) as a percentage of total water utility expenditures |

*Waste KPIs*

| | |
|---|---|
| W-1 | Waste per capita (kg/capita) |
| W-2 | Percentage capture rate of solid waste |
| W-3 | Percentage of solid waste recycled |
| W-4 | Percentage of solid waste that goes to landfill |

*Source:* TRACE model.
*Note:* GDP = gross domestic product; kg = kilogram; KPI = key performance indicator; kWhe = kilowatt-hour equivalent; kWhth = kilowatt-hours thermal; m$^2$ = square meter; m$^3$ = cubic meter; MJ = megajoule; T&D = transmission and distribution; TRACE = Tool for Rapid Assessment of City Energy.

Data management systems can take many years to perfect, especially if data are only collected annually. Data management systems need to be regularly tested and reviewed to ensure that accurate, reliable, and consistent data are being reported. In many instances, the data provider's role is not a priority (especially if it is an external organization), which results in inaccurate and inconsistent assumptions that affect data. The review process should be designed to analyze the data production chain critically, identify gaps or inaccuracies, and deploy corrective actions to resolve these issues.

**Figure 5.3  The Principal Components of a Data Production Chain**

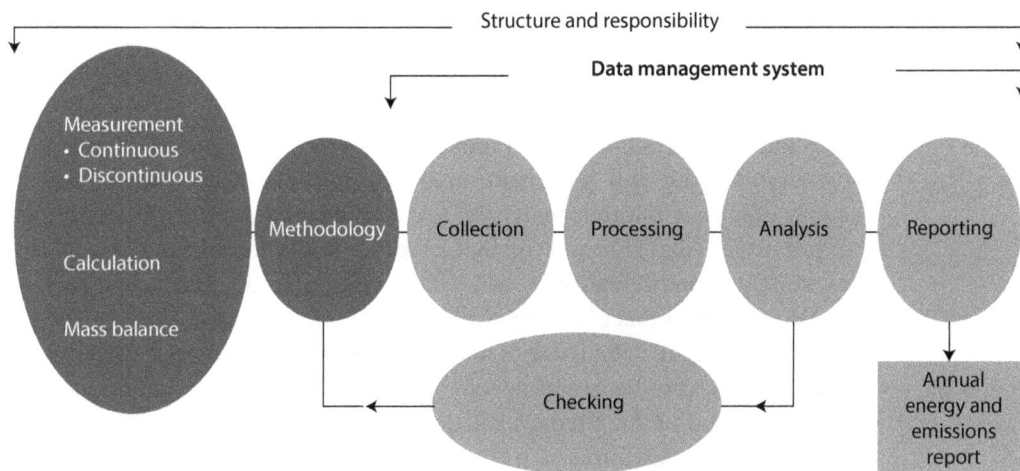

*Source:* Robert Carr, 2005, for British Standards Institute.

**Figure 5.4  Typical Characteristics of Data Management Systems**

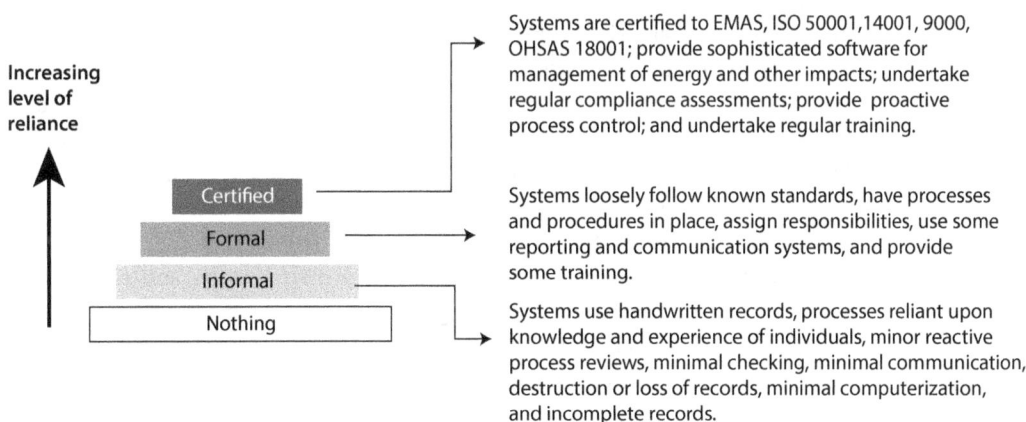

*Source:* Robert Carr, 2005, for British Standards Institute.
*Note:* EMAS = Eco-Management and Audit Scheme; ISO = International Organization for Standardization; OHSAS = Occupational Health and Safety Assessment Series.

## The Way Forward

The attainment of long-term, sustainable urban energy and emissions development is not a goal easily defined or achieved. However, the SUEEP process can facilitate cities' efforts toward a more sustainable future through the following contributions:

- **Institutional Development and Capacity Building.** SUEEP introduces a number of key foundation-building activities required to support long-term urban green growth strategies. The Guidebook in part III brings clarity and international best practices to the institutional reform, policy development, and stakeholder outreach processes necessary to achieve targets. Also, the SUEEP process includes accounting tools cities can use to quantify their energy consumption and GHG emissions for use in target setting, as well as for ongoing monitoring and reporting of results and implementation progress.

- **Creation of a High-Quality Pipeline of Green Investments.** Policy and institutions alone will not create green growth outcomes—investments in energy efficiency improvements and GHG mitigation activities will also play an important role. Through the SUEEP process, city leadership can evaluate investments comprehensively, based not only on fiscal return, but also on relative green impact and contribution to other social and economic development goals. The result is a well-defined pipeline of green investment projects that can be communicated to local stakeholders and financing institutions and to the international donor community and potential partners, including private investors.

- **Mobilization of Financing.** The international donor community has substantial interest in supporting sustainable infrastructure for green growth in rapidly developing EAP cities. However, there have been many challenges: defining the green city goals, identifying those activities that would optimally support green growth goals, ensuring local governments have the capacity and institutional structures needed to support both the construction and maintenance of green investments, and identifying means to measure success. The SUEEP process brings the donor and investor communities and city governments one step closer by (a) building an institutional and policy foundation for supporting green investments; (b) setting up a system of quantitative indicators for identifying green growth targets and monitoring and reporting progress over time; and (c) creating a long-term green growth plan and a well-defined, thoroughly evaluated pipeline of bankable investments that can be easily communicated to potential investors and financiers.

## Reference

ISO (International Organization for Standardization). 2011. "ISO 50001: 2011 Energy Management Systems—Requirements with Guidance for Use." ISO Secretariat, Geneva. http://www.iso.org/iso/catalogue_detail?csnumber=51297.

# Sustainable Urban Energy and Emissions Planning in Three Pilot Cities

# CHAPTER 6

# Cebu City, the Philippines

Cebu City is the second-biggest growth area in the Philippines after Manila. Cebu City is a midsize coastal city with a population of 799,763. Metro Cebu includes several other cities, Lapu-Lapu, Mandaue, and Talisay, which cumulatively have a population of about 2 million. Cebu City occupies mountainous terrain and has a land area of 291 square kilometers ($km^2$). The highest point in Cebu City reaches an elevation of 900 meters. Cebu City's plateau area occupies 23 $km^2$, about 8 percent of its total area, but is home to two-thirds of the city's population. Cebu Island occupies a strategic location, easily accessible by air and water, and is an important port. Cebu City is less than an hour away from Manila by plane and within a few hours of any city in Southeast Asia (map 6.1). It is served by an international airport and busy seaport.

The city has a tropical climate and an average temperature of 25.6 degrees C (centigrade) with an average relative humidity of 75 percent. Rainfall is at its lowest levels from February to April and gradually increases from May to July.

Cebu City has a thriving commercial seaport, and a majority of the city's labor force (73 percent) is employed in trade and other related services such as banking, real estate, insurance, and community and personal services. About 19 percent of the population is employed in industry, and 8 percent in agriculture and related services. The services sector is growing and is expected to maintain its economic dominance. Cebu City's strategic location and seaport help support trade and services. The majority of establishments in Cebu City are still considered micro or small enterprises with an average capitalization of 1.5 million Philippine pesos (Php) (US$34,445) or less. Some 67 percent of these business establishments are situated in the central part of the city and collectively account for about 77 percent of the city's economy.

The central Philippine government and Cebu City government are the focal points for policy action, but there is a strong base of grassroots action in Cebu City. Under a new local government code, which aims to devolve certain powers away from the central government, the city has the power and authority to establish an organization responsible for the efficient and effective implementation of its development plans, programs, and priorities. The main

**Map 6.1  Cebu City, the Philippines**

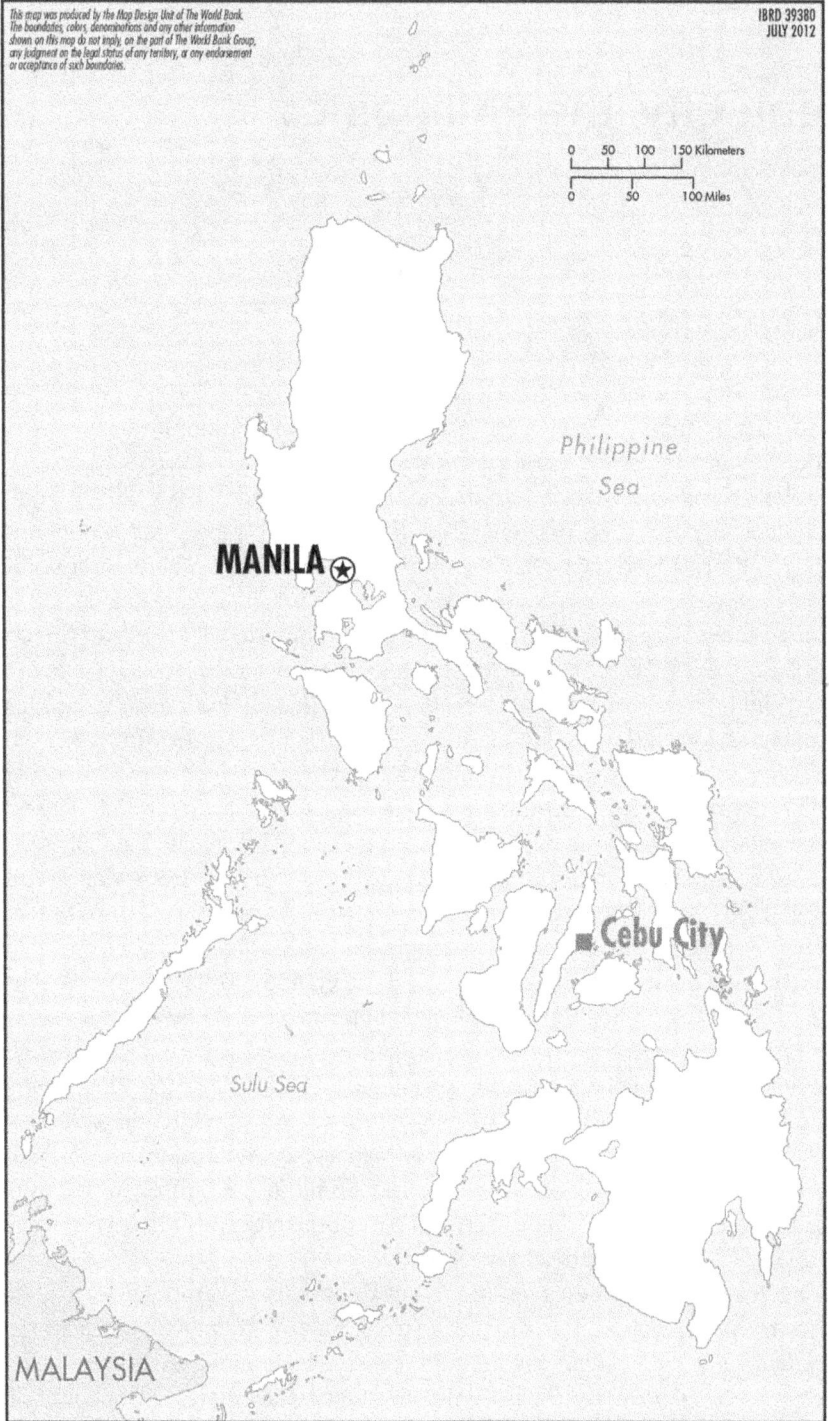

*Source:* World Bank.

departments, committees, and external agencies responsible for the planning, development, and operation of energy-consuming sectors in Cebu City are represented in the flow chart in figure 6.1.

The city government fully controls solid waste management and its government buildings. It has partial control over transportation through traffic planning and regulation. Other agencies exercising control over the transportation sector are the Land Transportation Office and the Land Transportation Franchising and Regulatory Board, which enforce vehicle emissions and registrations, respectively. Public lighting is largely within the jurisdiction of the city, although there is a mix of responsibilities, with the national government responsible for lighting of trunk roads and Visayan Electric Company (VECo, the electricity utility) owning all street lighting poles and responsible for installing luminaires. The city does not generate electricity, which is provided by VECo. The majority of water is supplied by Metropolitan Cebu Water District (MCWD), a public water utility, with a number of private water companies engaged in smaller-scale potable water supply. Cebu City has no formal sewerage system, although septic tank waste (septage) is currently disposed of at the city landfill.

## National Energy Efficiency Strategy

At the national level, President Benigno Aquino III announced his intention to encourage more public-private partnerships to improve Cebu City's economy. Cebu City officials and businessmen have identified seven areas of concern: the need for more infrastructure, improvement of the tourism industry, reduction of power costs, modernization of Mactan-Cebu International Airport, realization of Panglao International Airport, implementation of a mass transit system, and building of the Cebu-Bohol Bridge. The priority attached to these sectors illustrates the importance to the Philippines of energy efficiency and infrastructure projects, which are also relevant to Cebu City.

The Philippine government has initiated a number of energy efficiency programs to achieve the targeted annual savings of 23.4 million barrels of fuel oil equivalent in 10 years through the National Energy Efficiency and Conservation Program. Programs and initiatives include the following:

- The Road Transport Patrol program targets a 10 percent reduction in fuel consumption by providing consumers with information on the efficient use of fuel through proper vehicle maintenance and efficient driving through seminars and workshops and the use of all types of media.
- Voluntary agreements between government and the private sector encourage industrial economic zones in which businesses voluntarily monitor energy consumption and implement energy efficiency and conservation programs.
- An energy audit service is provided by the Philippine Department of Energy to manufacturing plants, commercial buildings, and other energy-intensive companies to help them understand their energy use patterns and identify initiatives to improve their energy performance.

**Figure 6.1 Cebu City Government Structure for Energy-Consuming Agencies**

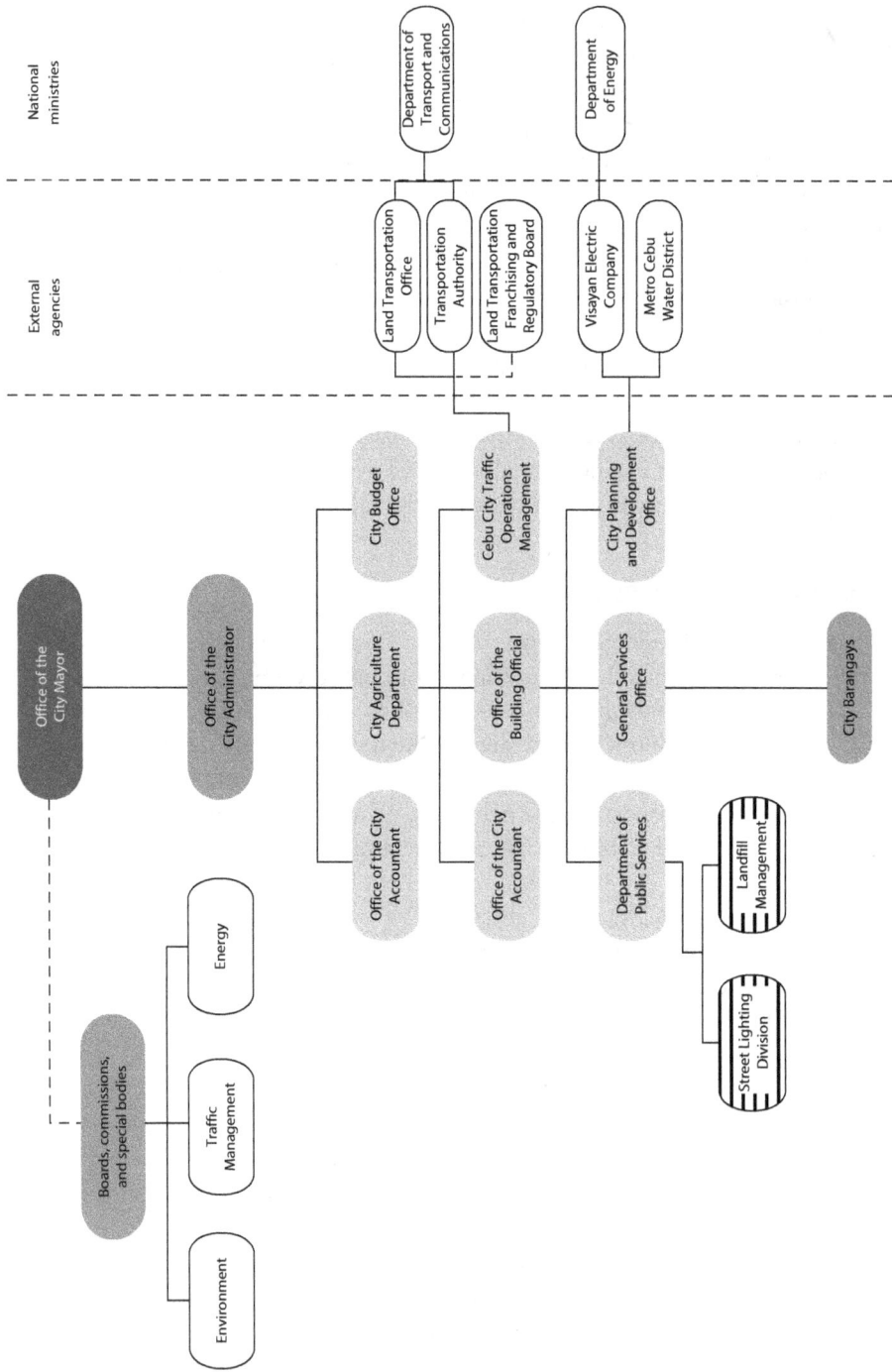

Source: Phase I pilot study.

- Infomercials and publications encourage energy efficiency.
- The Partnership for Energy Responsive Companies encourages industrial and commercial establishments to voluntarily monitor their energy consumption and implement energy efficiency and conservation programs.
- The Partnership for Energy Responsive Ecozones is a partnership of private sector companies formed in conjunction with the Power Patrol Project (http://projects.wri.org/sd-pams-database/phillipines/power-patrol-program) to encourage energy efficiency and conservation in economic zones.
- Energy standards and labeling of appliances and equipment aim to improve the efficiency of appliances by requiring energy-rating labeling of products.
- The Enercon program is an initiative to incorporate energy efficiency into the procurement guidelines of government agencies, bureaus, and offices.
- Energy use standards for buildings have been developed, but are not yet mandated, to give guidance on energy efficient building design.
- Initiatives for heat rate improvement in power plants will bring power plant generating units to their optimum performance levels by improving their operational capability.
- The systems' loss-reduction program requires energy utilities to implement measures to cut nontechnical losses.

## City-Level Energy Efficiency Strategy

The city government has enacted a number of energy efficiency initiatives:

- The Cities for Climate Protection Campaign in 1999 is administered by Local Governments for Sustainability.
- The International Resource Cities Program is a technical cooperation program between Cebu City, the Philippines; Larimer County, Colorado; and Fort Collins, Colorado. The program was funded by the US Agency for International Development (USAID) and implemented by the International City/County Management Association. In March 2001, an action plan was prepared that establishes the objectives of the partnership and serves as the framework for the relationship between the communities. It also identifies how Cebu City intends to target the financial assistance from USAID.
- A climate change forum was initiated and conducted by the Cebu City government in 2008.
- Cebu City formed action plans on energy.

The city government already has a focus on energy issues covering a wide range of departments and services. Considerable progress has been made in addressing energy consumption, and the city has proposed and engaged in a variety of energy efficiency activities, including the following:

- *Building initiatives*—policies for light fixture replacements and air conditioning operation hours

- *Waste sector initiatives*—barangay-level composting and the "No Segregation, No Collection" policy, which requires residents to sort their trash into three containers: recyclables, organic waste, and others
- *Efforts to encourage sustainable transportation*—for example, bicycle lane pilot areas and sidewalk repair schemes

Building on these initiatives to develop a strategic approach to demand management will be beneficial to Cebu City's future development, contributing to the city's energy security, improving its citizens' quality of life, and attracting businesses.

## Energy Use and Carbon Emissions Profile

An overview of Cebu City's energy supply profile provides the city government with valuable insights for strategic planning with respect to energy security and economic growth. Cebu City imports almost all of its energy; other than the 11 percent of electricity consumed that is generated within city boundaries, all energy sources (electricity, oil products, and natural gas) are imported. Although Cebu City has significant potential to harness solar power, it does not do so because no renewable power technology is installed in the city. Cebu City's energy flows and profile are summarized in a Sankey diagram (figure 6.2) to illustrate citywide energy supply and demand characteristics of its different sectors.

The total primary energy supplied to Cebu City is 21.8 billion megajoules (MJ; or 6.5 gigajoules per capita), or 6 billion kilowatt-hours (kWh; or 6 megawatt-hours per capita). Some 20.4 percent of this is in the form of electricity and 79.6 percent is in the form of petroleum products. A very small amount of natural gas is also imported into the city. The reliability of energy supply is consequently an important consideration for future energy planning. Transportation is the biggest energy user in Cebu City, consuming 51 percent of the city's primary energy and accounting for a substantial share of fuel combustion losses (see figure 6.3). Industry is the second-biggest energy user in Cebu City, consuming 36 percent of energy. The residential sector accounts for 7 percent of consumption and the commercial sector 5 percent. Industrial, residential, and commercial users are the biggest consumers of electricity. The residential sector was the dominant electricity user in 2008 but saw a 26 percent decline in consumption in 2009 and only a moderate 4 percent increase in 2010. The commercial sector has held relatively steady. In contrast, industrial sector electricity use has grown explosively, with a 50 percent increase in 2009 and a more moderate 8.2 percent increase in 2010.

On the whole, citywide usage increases steadily year-on-year, with electricity demand growing 4.3 percent from 2008 to 2009 and 4.5 percent from 2009 to 2010. If Cebu City maintains this trend, electrical energy demand will double in 16 years. From a policy-making perspective, untempered electricity consumption will negatively affect energy security, economic growth, and quality of life and should be addressed by Cebu City officials.

**Figure 6.2  Cebu City Energy Flows, 2010**

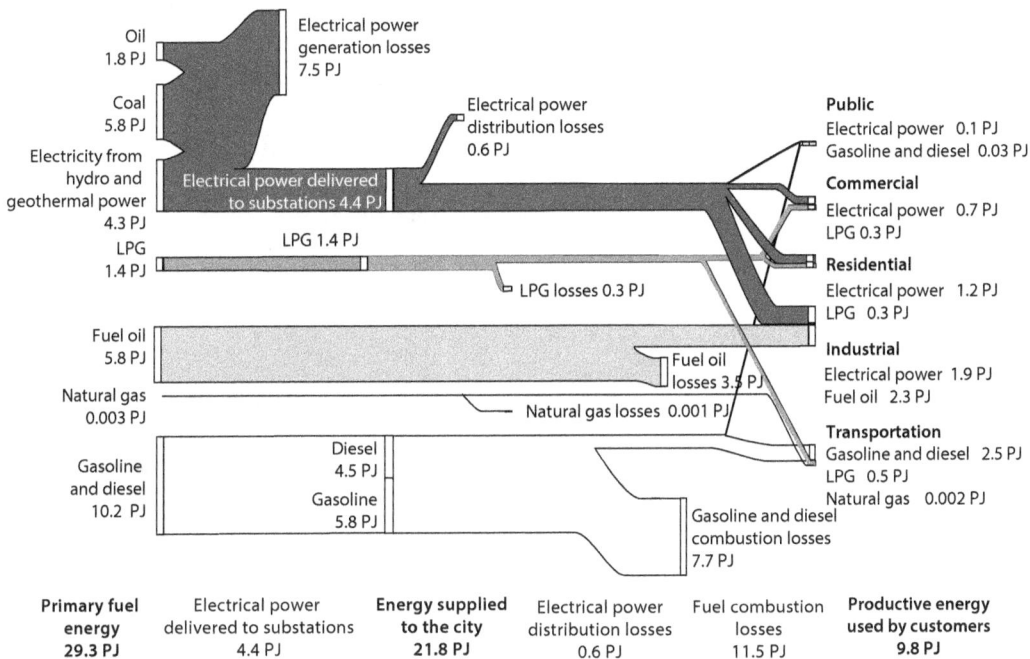

| Primary fuel energy 29.3 PJ | Electrical power delivered to substations 4.4 PJ | Energy supplied to the city 21.8 PJ | Electrical power distribution losses 0.6 PJ | Fuel combustion losses 11.5 PJ | Productive energy used by customers 9.8 PJ |

*Source:* Phase I pilot study.
*Note:* LPG = liquefied petroleum gas; PJ = petajoule. "Public" includes the end-use energy of city buildings, street lighting, city vehicles, water, wastewater, and solid waste management.

**Figure 6.3  Cebu City Energy Consumption by End Use**
*Percent*

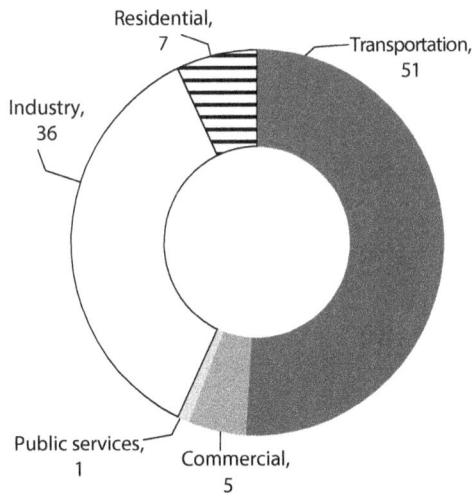

*Source:* Phase I pilot study.

Greenhouse gas (GHG) emissions present a slightly different picture from the energy balance results (see figure 6.4). Close to 1.8 million tons of carbon dioxide equivalent ($MtCO_2e$) were emitted by all sectors in Cebu City in 2010. Residual fuel oil and electricity are the biggest contributors by source, at 31 percent and 26 percent, respectively. Both of these energy sources are used almost entirely by the industrial, commercial, and residential sectors. Gasoline and

**Figure 6.4  Cebu City GHG Emissions by End Use and Fuel Source**

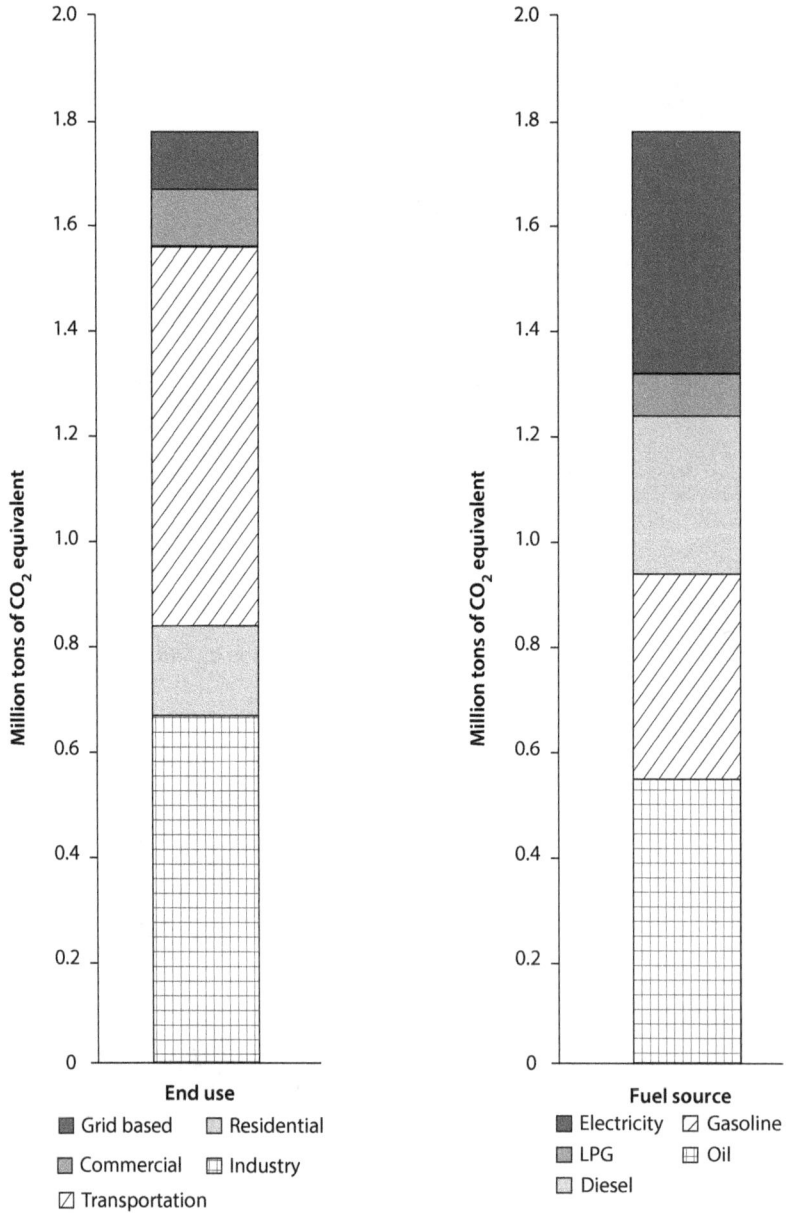

Legend (End use): Grid based, Residential, Commercial, Industry, Transportation

Legend (Fuel source): Electricity, Gasoline, LPG, Oil, Diesel

*Source:* Phase I pilot study.

diesel, which make up 22 percent and 17 percent of emissions, respectively, are used exclusively by the transportation sector.

The transportation and industrial sectors are the largest contributors to Cebu City's emissions with 40 percent and 38 percent shares, respectively. The residential sector accounts for 9 percent of emissions and the commercial sector accounts for 6 percent.

Transportation and industry are the most significant sectors for both the energy balance and the GHG inventory results. It is worth noting, however, that transportation uses just over half of the city's energy, but its share of GHG emissions is 40 percent, whereas industry consumes 36 percent of Cebu City's energy but produces a disproportionate 38 percent of emissions. This points to the importance of fuel choice and systems efficiencies in linking energy and GHG emissions.

## Sector Review and Prioritization

Cebu City's interest in pursuing the TRACE (Tool for Rapid Assessment of City Energy) diagnostic underscores its commitment to achieving optimal energy efficiency. The analysis was carried out across six city sectors: passenger transportation, city buildings, water and wastewater, public lighting, solid waste, and power and heat. These were, in turn, assessed against the performance of a range of peer cities through a benchmarking process. This review provided a number of significant findings that helped focus activity during the early part of the study and contributed to the definition of priority sectors for further study.

Key findings of the Cebu City diagnostics in comparison with the cities in the TRACE database follow:

- Low electricity use per capita and low energy use per unit of gross domestic product (GDP)
- High transportation energy use per capita, despite very high dependence on public transportation, arising from the dominance of inefficient jeepneys, as well as from congestion and low fuel efficiency (Many jeepneys are aging, with older, inefficient engines, and also typically carry fewer passengers than they are designed to carry. These factors, as well as the common practice of operating the vehicles for many years with occasional second-hand engine replacements, point to significant potential for improvement in the energy and operational efficiencies of this public transportation mode.)
- Low per capita water use but high energy density of potable water production
- Mid-range electricity consumption per kilometer of lit roads with room for improvement in public lighting
- Low but increasing electricity consumption in city buildings
- Low levels of recycling due to the absence of a formal solid waste recycling program
- Low levels of transmission and distribution losses in the electricity network

**Table 6.1  Cebu City Sector Prioritization Results**

| Priority ranking | Sector | 2010 energy spending (US$) | Relative energy intensity (%) | Level of city authority control[a] | Savings potential (US$)[b] |
|---|---|---|---|---|---|
| *City authority sector ranking* | | | | | |
| 1 | Street lighting | 2,238,761 | 20.0 | 1.00 | 447,752 |
| 2 | City buildings | 1,363,920 | 19.9 | 1.00 | 271,420 |
| 3 | City vehicles | 2,187,659 | 10.0 | 1.00 | 218,766 |
| 4 | Solid waste | 622,922 | 15.0 | 1.00 | 93,438 |
| *Citywide sector ranking* | | | | | |
| 1 | Public transportation | 89,696,607 | 10.0 | 0.21 | 1,883,629 |
| 2 | Potable water | 1,410,895 | 30.0 | 0.15 | 63,490 |
| 3 | Power | Unknown | 15.0 | 0.11 | Potentially large |

*Source:* Phase I pilot study.
a. 0 = no influence; 1 = maximum influence.
b. Based on TRACE (Tool for Rapid Assessment of City Energy) benchmarking; these figures are indicative of the savings that may be possible, but not necessarily practicable.

The TRACE analysis identifies priority areas in which significant energy savings are possible. Table 6.1 indicates the amount of energy spending in each of these sectors, the relative energy intensity (the percentage of energy that can be saved in each sector, based on TRACE benchmarking), and the level of influence the city government has over these sectors. The savings potential is calculated by multiplying these three factors. The TRACE contains a playbook of 58 energy efficiency recommendations applicable across all sectors.[1] The recommendations themselves are not meant to be either exhaustive or normative. They simply outline a number of policies and investments that could help local authorities in Cebu City achieve higher energy efficiency standards.

The table shows priorities with respect both to sectors over which the city authority has maximum influence, and to citywide issues for which the authority has limited influence. The ranking suggests that the city government should prioritize street lighting, followed by city buildings and city vehicles. On a citywide basis, public transportation is clearly a focal area, followed by potable water supply.

Following the sector-by-sector analysis, each recommendation was reviewed to establish its applicability to the Cebu City context. This filtering helped focus the process on those recommendations that are both viable and practical.

## Recommendations

### Transportation
Transportation in Cebu City is dominated by jeepneys, with private automobiles, taxis, and motorcycles also playing increasingly important roles. A high proportion of Cebu City's taxis run on liquefied petroleum gas as a response to the incentives in place for drivers to upgrade their engines to combat urban air pollution. With regard to fuel use minimization, the city's General Services Office (GSO) implemented a biofuel project in which ethanol was used as an additive in city

government automobiles. This activity is expected to result in a 15 percent reduction in fuel use in the transportation sector.

Cebu City experiences significant congestion during rush hour. Parking restrictions are in place and are regulated by zoning ordinances. However, these restrictions have not been rationalized by any strategic plan for the city. The last transportation plan was produced in 1972 and a new plan is currently being prepared. The high-profile proposal to introduce a bus rapid transit (BRT) system in combination with the city's efforts to reduce traffic congestion indicate that a comprehensive transportation plan that is aligned with overall city planning will be essential.

Nonmotorized modes of transportation such as bicycles are not common, which may be due to safety concerns associated with using crowded and often chaotic streets. In an effort to encourage sustainable transportation, a local nongovernmental organization has spearheaded the Road Revolution initiative. The Road Revolution involves closing Osmena Boulevard to vehicular traffic for a day to enable the public to enjoy entertainment activities and to experience the street's improved environment when it is occupied by pedestrians rather than automobiles, all the while reducing GHG emissions. This program has been implemented twice and received mixed reactions. Some members of the public enjoyed the initiative while some, especially businesses located on Osmena Boulevard, were negatively affected by the lack of access.

### Public Transportation Development

The public transportation sector in Cebu is privately run and is dominated by jeepneys. The public transportation mode share is 80 percent, which is particularly high, but jeepneys tend to work inefficiently, with regard to both fuel consumption and route optimization. One of the most notable observations made by the Sustainable Urban Energy and Emissions Planning (SUEEP) team is the need for an integrated planning approach to ensure that plans for BRT, land use, street signals, parking policies, vehicle registration pricing, and sidewalk policies are all adequately integrated and that there is an effective means for turning plan into practice. Integrated planning is especially important for transportation planning. Currently, traffic flow in Cebu City is congested during peak hours and growing private car use will exacerbate this problem. Encouraging nonmotorized transportation alongside the development of robust public transportation will be critical to mitigating congestion and improving the quality of transportation. The Cebu City Planning and Development Office and the City Traffic Operations Management are responsible for ensuring that nonmotorized modes are encouraged through city planning.

Coupled with nonmotorized transportation, the promotion of public transportation should be part of an energy efficient strategy for Cebu City. A formal public transportation system has been in the works for a number of years. A BRT pre-feasibility study has been completed and further studies are under way. The city plans to improve public transportation, and authorities have placed an emphasis on the need for capacity building to institutionalize transportation data gathering and management. The city could take advantage of a few carbon financing resources, such as the Clean Development Mechanism (CDM), for its BRT program.

### Vehicle Emissions Standards Testing

Vehicle emissions standards testing infrastructure could be more effectively applied to encourage and enforce better vehicle emissions. Lower emissions will lead to better air quality and reduced energy consumption. The current testing and enforcement system is fragmented because different city and national government departments regulate (license) test centers, and enforcement is either weak or ineffective because cars can be back on the road after a few simple measures are applied to get the vehicle to pass inspection. New testing equipment that meets standards for measuring an engine's efficiency would be required. Enforcement activities and resulting sanctions should also be revised to ensure that poorly performing vehicles would be identified and removed from service. The implementation of such a measure is potentially administratively difficult because of the fragmented nature of Cebu City's existing testing system and the magnitude of enforcement required. One way to address this issue would be to establish a government-run emissions testing center as a model for private testing centers.

Inefficient two-stroke motorcycle engines are widespread in Cebu City. Fifteen two-stroke to four-stroke engine replacement programs have been successful in a number of other cities, and this is an area in which Cebu City could make significant gains with minimal investment.

Cebu City's fleet vehicles do not follow a formal maintenance regime beyond the first few years following purchase. Initially, cars are well maintained to keep up their warranties. After the warranties expire, a city-operated vehicle maintenance center repairs faulty vehicles as necessary. The replacement parts are usually the cheapest available, leading to early failure of replacements in many cases. Life-cycle costing is not factored into parts procurement.

It is highly recommended that the GSO spearhead a city vehicle fleet efficiency program. Procurement and maintenance policies should be implemented to maximize the efficiency of the city's fleet. This includes incorporating life-cycle costing in replacement parts procurement for vehicles.

### Solid Waste

Residential and commercial waste is collected at the barangay level and by city garbage trucks through the Department of Public Services. Barangay-collected waste is brought to the landfill in barangay-owned vehicles using fuel provided by the city government. Waste collection and processing in Cebu City is currently characterized by grassroots initiatives at the barangay level. Barangay Luz, in particular, has spearheaded a number of actions, including the Kwarta sa Basura (Money from Trash) program, which organizes a women's group to create marketable products from waste.

Composting is widespread in Cebu City and operational at the city, barangay, and household levels. At the city level, the Inayawan landfill includes composting facilities for collected compost while barangays engage in small-scale composting, which benefits the city in two ways. First, it minimizes waste sent to the landfill, and second, it responds to the high demand for compost in Cebu City, which produces 25 percent of its own food. Households use the Takakura home method of composting, which was introduced to Cebu City in 2008 by Koji Takakura of Japan. Takakura's research team monitors the progress of composting in Cebu City as part of an ongoing project to establish a pilot area for successful home composting. All these initiatives have collectively contributed to Cebu City's recent reduction in the percentage of domestic solid waste that is disposed of at the landfill.

Disposal of hazardous and medical wastes is an ongoing challenge because no appropriate treatment facilities are available for these waste streams. Currently, they are disposed of at the landfill, exacerbating the risks posed by its current operation and ultimately, its restoration. To address this problem, the city government, through the office of Councilor Nida Cabrera, chairperson of the Committee on Environment, and in conjunction with a local nongovernmental organization and educational institutions, has initiated a household survey in the urban barangays to determine the types of household hazardous wastes that are generated and how they are disposed of.

Until the end of 2011, Cebu City had one landfill in Barangay Inayawan that accepted hazardous, septic, and noncompostable domestic solid waste. The landfill was constructed in 1998, with a design life of seven years. Dumping activities are no longer allowed in the area. Barangays located in the North District have started delivering their waste to a privately owned sanitary landfill facility in Consolacion, about 27 kilometers (km) from the Cebu City Hall Building. Barangays in the South District will use a transfer station that is currently being developed by the city in a private lot of nearly 8,000 square meters about 1 km from the old Inayawan landfill.

Landfill leachate at Inayawan currently remains untreated, posing further land and water contamination issues. The city has earmarked Php 11 million (US$250,000) for funding of a facility to treat the landfill's leachate, but no action has been taken to date.

As of 2011, recycling activities were largely carried out by the approximately 300 waste pickers operating as part of a collective at Inayawan landfill. It is believed that a recycling rate of 16 percent was achieved through the activities of the waste pickers. There are plans to implement waste segregation at source and potentially to install a centralized materials recovery facility, but no conclusive actions have been taken to date.

Efforts to minimize the amount of solid waste disposed of at landfill have been highly effective. Some 60 percent of domestic solid waste is biological or compostable waste. Some of this is composted at the barangay level, and some is brought to the landfill where there are larger-scale composting and wormery facilities. This policy has been in operation since April 2011.

Although the Department of Public Works is responsible for waste infrastructure, the barangays do much of their own waste haulage and management. A separate nonprofit organization at the landfill manages waste treatment. Because of the slightly fragmented institutional arrangements for waste management in Cebu City, the SUEEP team recommends that the Department of City Planning, with the support of the mayor, identify the appropriate party to undertake waste studies or to review previous recommendations made in such studies for their relevance.

### Landfill Gas Capture

The city is developing plans to close the Inayawan landfill, creating an opportunity to capture landfill gas and produce energy. It is recommended that a study be undertaken to investigate the feasibility of a landfill gas capture project. A successful landfill gas capture project would potentially remove up to 4,900 $tCO_2e$ annually from Cebu City's GHG emissions profile, and if used to generate electricity, could contribute to satisfying the city's electricity requirements.

### Energy from Waste

The city has approved an allocation for waste from energy projects to be established at four sites for implementation in 2012: three at barangay material recovery facility (MRF) cluster sites in Luz, Quiot, and Talamban, and one upland site. The three MRF sites will use organic waste as feed and the upland site will use animal waste. The energy generated will be accessible to barangay residents for electrification and cooking purposes.

### Collection Route Optimization

Following the cessation order issued by the mayor, no more dumping will occur at the Inayawan landfill. As a consequence, garbage trucks operated by the barangays and the city will deliver their waste to a private landfill located 27 km north of the city. Barangays in the north deliver their waste directly to the landfill and barangays in the south deliver waste to a transfer station 1 km away from the Inayawan landfill to reduce the distance waste must be transported by trucks originating in the south.

The SUEEP team notes that the city has already established a new route scheme for the diversion of waste from the old landfill site at Inayawan to the new Consolacion site and recommends that Cebu City continue to assess ways to optimize waste collection to minimize traffic and enhance fuel efficiency.

### Septage Treatment

There is potential for the development of a septage waste treatment facility in Cebu City. The Spanish government funded an initiative to treat septage waste using a biodigester, but this project failed because the septage waste was often contaminated by industrial wastes before treatment, compromising the biological treatment process. Although technical knowledge in Cebu City for managing septage waste is limited, it is understood that the landfill will be closed to septage by early 2012. Thus, the need to address treatment of landfill leachate and septage is pressing. It is suggested that the city government consider the possibility of a combined septage and leachate treatment facility located at the Inayawan landfill site. A feasibility study should first be undertaken to assess the technical viability of such a proposal and how the issues with contamination of septage can be overcome.

### Solid Waste Treatment

With the closure of the Inayawan landfill and the pressing need to identify alternative means for the disposal of domestic solid waste, it is suggested that the city government undertake a feasibility study for development of a centralized MRF and, possibly, a new composting facility. The MRF can be staffed by former waste pickers (who are skilled in identifying waste types). Recovered materials could be bulked and shipped to local markets. Alternatively the city government may wish to consider providing incentives for the establishment of materials reprocessing businesses colocated with the MRF, thereby creating a positive economic impact.

Because the city pays for garbage truck fuel and many of the vehicles are old and inefficient, the city would benefit from vehicle upgrades and improved procurement processes. Maintenance regimes and driver education to encourage energy efficient driving practices are also recommended to improve vehicle performance.

### Water

Cebu City's potable water supply is sourced from groundwater. MCWD is the principal water utility, serving 50 percent of Cebu City's population. The rest of Cebu City's water is provided by independent suppliers that sell piped potable water from private wells. There is strong competition between these private suppliers and MCWD. Although MCWD's competitors are technically subject to the same regulations as MCWD, there is reportedly minimal enforcement of regulations relating to the siting, abstraction rates, and water quality from private wells.

Groundwater resources are a relatively energy-intense source of water supply because of the requirement for pumping, in comparison with surface water gravity-fed systems. Despite this, MCWD's water system is generally well maintained and uses energy efficient equipment, resulting in energy costs constituting about 30 percent of MCWD's operating costs.

MCWD has been the beneficiary of a number of energy efficiency projects funded by USAID. In 2006, the Alliance to Save Energy assessed MCWD's system and helped the agency to make a 0.5 kWh per cubic meter ($m^3$) reduction in overall energy use. MCWD is currently working with the Las Vegas Water District to improve energy efficiency by an additional 5 percent. MCWD has also taken strong action to improve leakage rates, but has run into some difficulties distinguishing its pipes from those of private suppliers, and has limited abilities to fix leaks even when they are known to exist. It is of interest that MCWD has undertaken a thorough audit of the system to identify and remedy leaks, but losses have not been reduced significantly. The company does not have a complete understanding of the reasons behind this and is conducting further research into the issue.

MCWD has started an awareness-raising campaign to educate consumers about the importance of water conservation. However, this initiative is not yet widespread and could be more widely promoted.

### Energy Efficient Water Resources

Prioritizing energy efficient water sources by deploying rainwater harvesting in city buildings and providing incentives for its uptake in private developments would reduce reliance on energy-intensive groundwater and reduce the currently unsustainable water withdrawal rate.

### Water Efficient Fixtures and Fittings

Water efficient fixtures and fittings should be installed in government buildings and encouraged in private developments through educational measures. Types of energy and water efficient fittings applicable to both retrofit and new-build developments include the following:

- Low-flow taps and showers
- Water efficient household appliances
- Dual, very low, or siphon flush toilets
- Low-flush or waterless urinals
- Rainwater harvesting tanks

Water efficient fixtures and fittings help to reduce water consumption by reducing the volume of water used in each application, which also reduces the associated energy needed to treat and convey the required flows.

Efficient fittings can help to raise consumer awareness of the link between water use and energy consumption and generally leads to the consumer installing additional energy efficient products.

### Wastewater

Cebu City has no citywide wastewater treatment facility. Wastewater is currently managed through household, or clusters of household, septic tanks. The septage is removed by tankers and disposed of, untreated, at the landfill. For this reason, Cebu City's energy use for wastewater is technically minimal (fuel use in haulage), but in the interests of public health, it is expected that a more comprehensive system will be developed. Wastewater management in the city will encounter significant challenges, including contamination of the groundwater supply from septic tank seepage and landfill leachate, and illegal dumping of septage waste.

Cebu City government can do little to influence the activities of the private septage haulers. Technical interventions may only be leveraged by the city government's use of the planning system, which requires citywide technological measures the city can undertake.

It is recommended that the potential for sludge reuse be investigated. Cebu City does not currently have a sustainable sludge management strategy, but CDM funding could increase the feasibility of developing a sludge reuse project. Such a project would include treating the sludge under aerobic conditions (for example, dewatering and land application), or installing a new anaerobic digester that treats wastewater or sludge (or both), from which the biogas extracted is flared or used to generate electricity, and the residual after treatment is directed to open lagoons or is treated under clearly aerobic conditions.

### City Buildings

The Department of Engineering and Public Works (DEPW) and Office of the Building Official (OBO) have complete control over the design of all city buildings with the exception of schools. There is no formalized refurbishment cycle for government buildings, which poses a significant challenge to achieving energy efficient performance in the existing building stock. In addition, energy efficiency is not currently a consideration in capital investment planning or life-cycle costing. This omission poses a barrier to the uptake of energy efficiency initiatives that require higher upfront investment but result in long-term savings.

As an initial step, the city government is using the landmark city hall building as a pilot project for achieving energy savings in buildings. The building is being retrofitted for a central air conditioning system as one of the project's major efficiency measures.

VECo has provided energy-saving tips to city departments and an ad hoc energy conservation committee was established to address energy use in the city.

Additionally, the GSO has been intensifying efforts to save energy. The GSO has begun collating data on energy use, issuing advice on energy efficiency tips, and collecting information on energy saving commitments from all departments across the city. Although this is a positive start, there is still a considerable range of opportunities to improve energy efficiency throughout the life cycle of city buildings.

Little focus has been given to energy efficiency across the city estate, although recent initiatives are starting to gain momentum. Audits of public buildings and energy efficiency calculations show that energy efficiency retrofits have good payback potential (less than three years), with continued savings thereafter over the lifetime of the project.

### City Building Energy Efficiency Task Force

Currently, some measures are being implemented at the barangay level to improve energy efficiency in buildings. For example, Barangay Luz has collaborated with a paint company to provide homeowners with materials to create "cool roofs" for their homes. However, these initiatives are not coordinated at a city level and are therefore difficult to track.

In light of the above, as well as the need to coordinate energy efficiency measures in the building sector, it is recommended that a city building energy efficiency task force, comprising representatives from GSO, OBO, DEPW, VECo, and the national government's Department of Energy, be set up. The task force would design, implement, and manage energy efficiency initiatives in city buildings. The establishment of a task force will put in place a formal structure for these agencies to share ideas and approaches and streamline the city's efforts to enhance energy efficiency in buildings.

A number of efforts have been made in the city buildings sector to improve energy performance, such as lighting replacement programs and the implementation of air conditioning schedules. Although the SUEEP team endorses these efforts, opportunity remains for improvement by replacing old air conditioning units and other inefficient appliances and by improving the design and construction of building envelopes. It is recommended that the existing efforts be continued and a city building audit and retrofit program be implemented through a city building energy efficiency task force.

It is noted that the Philippine Department of Energy supports energy efficiency through its Advisory Services for Major Industries and Commercial Buildings program. An energy audit is provided for a fee by the Department of Energy to manufacturing plants, commercial buildings, and other energy-intensive companies. A team of engineers from the department evaluates the energy utilization efficiencies of equipment, processes, and operations of these companies and recommends energy efficiency and conservation measures to attain energy savings. These advisory services may be a further avenue for the city government to pursue.

### Green Building Guidelines

A consistent, citywide approach to procurement and green building guidelines is needed. Formalizing sustainability standards in the local design and construction industries is a huge challenge. The city should carefully consider adopting the Building for Ecologically Responsive Design Excellence (BERDE) code, which was recently introduced by the Philippine Green Building Council, in city building ordinances and future new build or renovation work. Both the DEPW and OBO have expressed interest in formalizing the design and construction process to encourage energy efficient green building design. The city has the power to enact ordinances for building codes, which will be advantageous should authorities wish to mandate compliance with green building codes.

If a green building code is adopted, rigorous implementation and monitoring will be big challenges for the construction industry. Typically, construction materials procurement and contractor tenders are the biggest impediments to the implementation of green building codes in industries that have not evolved to incorporate sustainability requirements. Thus, these areas will require particular attention and focus from the parties responsible for overseeing the implementation of a Cebu City green building code. Before deployment of the BERDE system, city government personnel with knowledge of Cebu City building practices and contractor tenders will need to review the system's requirements. Identification from the outset of the aspects of the BERDE system that may not be practical for Cebu City, or components that are better applied to select developments, is imperative. It may also be appropriate to initially target high-end developments or develop a phased rollout that prioritizes some elements of BERDE above others.

The city can support its building industry by drawing on the experiences of Quezon City, including the identification of key challenges and opportunities associated with complying with such codes, and providing the resources needed to address them. It is strongly recommended that Cebu City establish contacts with the relevant agencies to capture the knowledge that Quezon City will have gained from its experience.

## Public Lighting

Cebu City has good street lighting coverage, with the majority of areas being lit by low pressure sodium fixtures (mainly used in smaller streets), as well as significant numbers of fluorescent bulbs, mercury halides, and compact fluorescents. Along the coastline, high pressure sodium fixtures are used.

The city's previous efforts to upgrade street lighting encountered problems with fixture quality, which has resulted in undesirable or unacceptable lighting in some cases. General Electric, Phillips, Sylvania, and American Electric fixtures have been used in Cebu City, but it was unclear to the SUEEP team which fixtures had delivered the best performance. The city currently operates a program to upgrade low pressure sodium fixtures to metal halides along main thoroughfares, which does not take advantage of the more energy efficient technologies available.

However, the city government has put aside a budget of Php 5 million (US$114,800) for the installation of light-emitting diode (LED) fixtures along Osmena Boulevard in 2012, a pilot area for this project. The purpose of the project is to test the benefits of LEDs for power consumption and GHG emissions to inform decision making about a future rollout of LEDs. The city has concerns about potential theft of the expensive LED fixtures, indicating the importance of site selection to avoid this problem.

Public lighting in Cebu City is in the remit of the Department of Public Services: Street Lighting Division. Responsibility for street lights on national roads lies with the national government, and individual barangays maintain the lights on smaller streets. The imprecise ownership and accountability for proper operation of the system has resulted in several maintenance problems. Given this, definitive responsibilities of each agency should be provided to ensure clear ownership and accountability.

### Street Light Audit and Retrofit Program

Despite the high cost of LED luminaires when compared with other types of fixture, the low energy and maintenance costs of efficient LED lamps makes them financially and technically viable. A simple cost-benefit analysis undertaken by the SUEEP team suggests that an investment of Php 312 million (US$7.1 million) would have a payback period of 4.95 years, and would save an additional Php 540 million (US$12.4 million) over the 10-year lifespan of the project. This would also lead to an annual reduction in GHG emissions of 4,043 tons of carbon dioxide.

The SUEEP team notes that Cebu City had piloted the use of LED luminaires in parts of the city previously and is aware that the beam pattern of the luminaires did not meet current national street lighting standards. Although this issue could be addressed by altering the height or spacing of light poles, all of which adds cost to the program, using LED luminaires from alternative manufacturers that meet national street lighting standards could also be considered.

The team recommends that the city tries a range of LED luminaires in pilot areas to test their efficacy as a precursor to rolling out a more comprehensive street lighting replacement program. The city can apply CDM methodology AMS-II.C, which comprises activities that encourage the adoption of energy-efficient equipment and appliances (for example, lamps, ballasts, refrigerators, motors, fans, air conditioners, pumping systems).

### Lighting Timing Program

The current use of photosensors to determine lighting on and off times may not be optimal. Because safety is the primary purpose of public lighting, the team has avoided drawing specific conclusions about the deployment of a lighting timing program but recommends that the Department of Public Services work in conjunction with other city authorities to determine if a lighting timing program would be effective for the city.

## Conclusion

Cebu City government has recognized the need for a sound energy strategy as evidenced by its numerous excellent existing energy initiatives. The steps Cebu City has taken to improve energy use provide a sound starting point and clearly show the city's commitment to improving energy efficiency. In particular, better energy governance in government operations; on a citywide scale, better coordination between city planning and agencies; and enhanced procurement and green building codes are all key steps Cebu City officials must initiate to work toward a genuinely coherent energy strategy and policy for the city.

## Note

1. For further details on TRACE, see chapter 3.

# CHAPTER 7

# Surabaya, Indonesia

Surabaya is a large city with a population of 2,765,908 people. The city occupies coastal terrain and has a land area of 327 square kilometers ($km^2$). The highest point in Surabaya reaches an elevation of about 30 meters. Surabaya is located in the Brantas River Delta, an area that has a high hazard of flooding.

Surabaya's population density is 8,458 inhabitants per $km^2$. The city is highly urbanized, and the numerous industries located in the city have attracted migrants, contributing to growth of slum areas. The city's annual population growth rate is 0.65 percent, and much of the city's center is densely populated.

Surabaya is a tropical city characterized by distinct wet and dry seasons. The city enjoys plenty of sunny weather, with temperatures regularly peaking at more than 30 degrees C (centigrade, or 86 degrees Fahrenheit). The city's wet season runs from November through May, and the dry season covers the remaining five months. Surabaya on average sees approximately 1,500 millimeters (more than 59 inches) of precipitation annually.

Surabaya is located in the northeastern corner of Java (see map 7.1), and is a key node in various national and international air, water, and land transportation networks. Surabaya is less than two hours away from Jakarta (the country's capital) by plane and within a few hours of any city in Southeast Asia. It is served by Juanda International Airport and Perak Port, one of Asia's largest and busiest seaports. These two international nodes serve as important gateways to the province of East Java not only for passengers, but also for the transport of goods. Surabaya has a large shipyard and numerous specialized naval schools.

As the provincial capital, Surabaya is also home to many offices and business centers and is an educational hub for Indonesian students. Surabaya's economy is also influenced by the recent growth in foreign industries and the completion of the Suramadu Bridge. Surabaya is currently building high-rise apartments, condominiums, and hotels as a way of attracting foreigners to the city.

The city has a gross domestic product (GDP) of US$22,850 million, which grew at the rate of 6.3 percent in 2008 (compared with national GDP growth of 6.1 percent). The primary industries contributing to the city's GDP are the trade, hotel, and restaurant (together accounting for 36 percent) and

**Map 7.1  Surabaya, Indonesia**

*Source:* World Bank.

manufacturing (32 percent) sectors, with smaller contributions by the transport and communication, construction, financial services, and services sectors.

The formal sector accounts for 44.1 percent of employment. A significant factor in Surabaya's economic profile is the large contribution of the informal sector to employment (22 percent). The main employment sectors (both formal and informal) are trade, hotel, and restaurant (41.5 percent); community and personal services (21.2 percent); and industry (15 percent). Historical employment data were only available for the manufacturing industry, which saw a decline of 2.8 percent in the 2000–07 period.

Surabaya city governance (Kota Surabaya) comprises the city government and the city's parliamentary body. A decentralization policy implemented in Indonesia in 1998 has devolved public services provision to district and city levels; details of the structure and institutional relationships between central and local government are articulated in Law No. 32/2004. In Surabaya, it is primarily the departments and agencies, under the guidance of the main coordinating planning body, Bappeko, that implement policy changes and actions that could affect energy supply and consumption. Figure 7.1 provides an overview of Surabaya's institutional structure and the relationships involved in the city's energy management.

## Energy Efficiency Initiatives

### National Level

At the national level, energy policies are formulated by the National Energy Council. Energy efficiency and conservation programs are implemented by the Ministry of Energy and Mineral Resources. The Indonesian government has initiated a number of energy efficiency programs, including the following:

- *National Energy Conservation Master Plan (RIKEN) (2005).* RIKEN stipulates a target for Indonesia to decrease its energy intensity by about 1 percent per year on average. It identifies sectoral energy savings potential: 15–30 percent

**Figure 7.1  Surabaya Government Structure for Energy-Consuming Agencies**

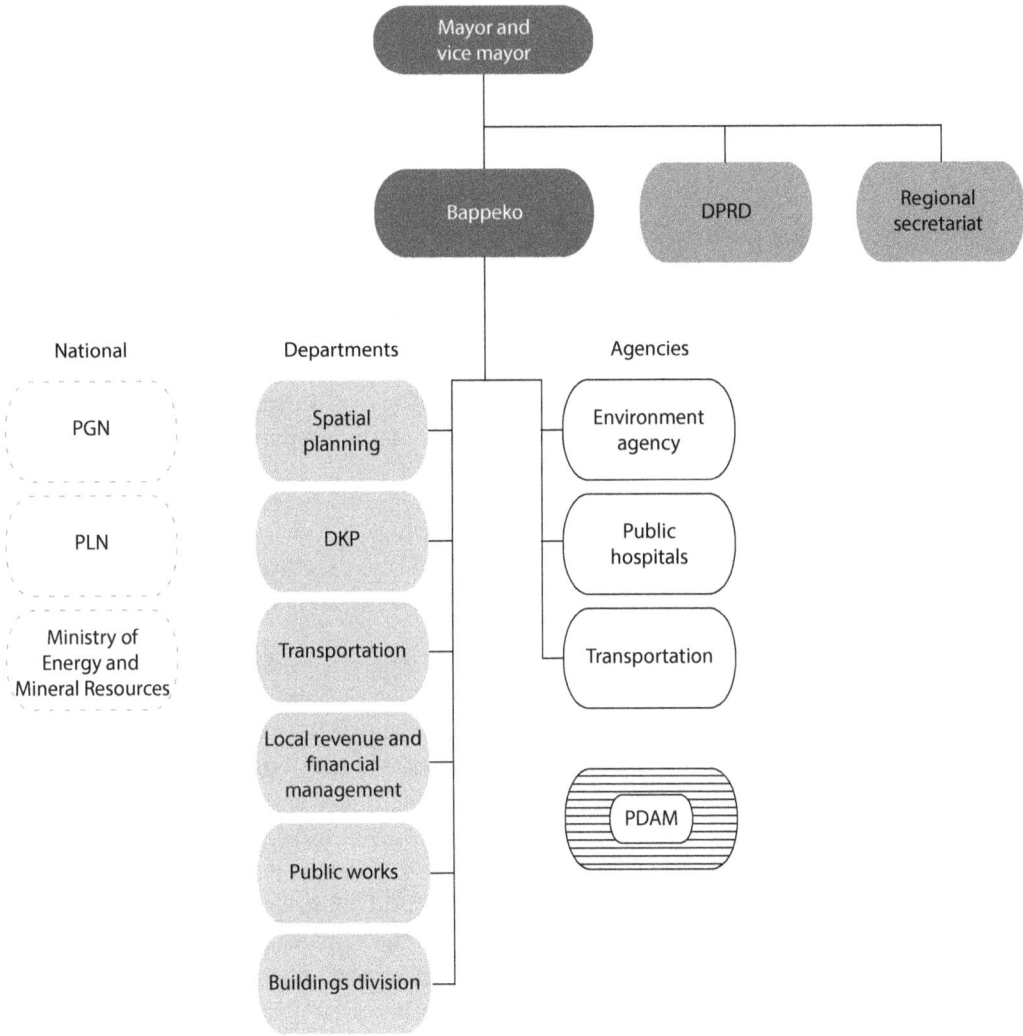

Source: Phase I pilot study.
Note: Bappeko = city spatial and development agency; DKP = Cleansing and Park Department; DPRD = city parliament; PDAM = Regional Drinking Water Company; PGN = state-owned gas company; PLN = state-owned electricity company.

in industry, 25 percent in commercial buildings for electricity, and 10–30 percent in the households sector. Without this plan, energy use is projected to increase rapidly under a business-as-usual scenario by 41 percent in 2025.

- *National Energy Management Blueprint (PEN) (2006).* PEN supports the RIKEN through the implementation of energy efficiency and conservation measures. It provides development road maps for various sectors that involve the implementation of supply- and demand-side management, intensification of efforts to search for and use renewable energy sources, implementation of fiscal measures such as tax allowances, development of energy infrastructure, community participation in commercial energy, and the restructuring of energy institutions.

- *National Energy Policy (2006).* This policy stipulates national targets for an optimal energy mix in 2025: less than 20 percent from oil, more than 30 percent from gas, more than 33 percent from coal, more than 5 percent from biofuel, more than 5 percent from geothermal, and more than 5 percent from other renewables. It further stipulates a national energy elasticity target of less than 1 by 2025.

- *Presidential Decree No. 2/2008 on Energy and Water Efficiency.* This decree mandates energy conservation practices in government office buildings. Government departments and agencies and regional governments are required to implement best-practice energy-saving measures outlined in the government's guidelines and directives and are mandated to report their monthly energy use in buildings to the National Team on Energy and Water Efficiency every six months.

- *Building energy codes.* Indonesia has four energy standards for buildings that cover the building envelope, air conditioning, lighting, and building energy auditing.

- *Fuel and electricity subsidies.* Fuel and electricity subsidies were scheduled to be phased out by 2014, mainly prompted by an increasing deficit in the state budget—electricity subsidies peaked at 83.9 trillion Indonesian rupiah (Rp) (US$11.05 billion) in 2008 and were estimated to be Rp 65.6 trillion (US$8.6 billion) in 2011. Currently, all categories of customers pay for electricity at rates far below market price—the average electricity tariff is about Rp 655 per kilowatt-hour (kWh) (US$8.62 per kWh), whereas the market price is about Rp 1,030 per kWh (US$13.5 per kWh). In May 2011, the government announced an increase in the base tariff of 10–15 percent to reduce the swelling subsidy. Funds formerly used for subsidies will be used to fund energy investments, including geothermal electricity generation, energy efficiency, and other low-carbon energy generation projects. These will be implemented via the Clean Technology Fund, which has amassed more than US$4 billion since its establishment in 2008. The change in policy and approach from energy subsidy to investment in low-carbon, highly efficient technologies is a major component to the background in which energy policy decisions will be made in the future in Surabaya. Despite the government's aspirations and efforts to phase in the reduction of subsidies, the threat of faster inflation amid gains in oil prices have delayed the national government's plans, which will likely impact the timeline for the proposed phasing out of subsidies.

The primary national focus is to transition public transportation to using gas and the household sector from kerosene to liquefied petroleum gas (LPG). In June 2010, the government planned to improve fuel efficiency for private cars by imposing limits on engine capacity to no greater than 2,000 cubic centimeters.

However, this plan was delayed for reconsideration. Although the national government has long promoted LPG and compressed natural gas for transportation, uptake will continue to be limited as long as fuel and power continue to be underpriced.

### City Level

Currently, the city has neither an energy plan nor a policy directed at issues of energy efficiency. However, a number of relevant energy efficiency initiatives have been enacted by the city government, including the following:

- *Surabaya Development Plan 2010–14 (RPJMD).* This is the city's urban development plan, which is renewed every five years. It addresses several issues, including the development of clean water networks for the city, utilities development, development of the transportation system, and spatial planning in the city. (There is a separate Surabaya Spatial Plan 2009–29; however, it has yet to be codified.) The development plan does not explicitly deal with issues of energy management.

- *Bus rapid transit (BRT) studies.* The city has prepared several transportation studies for the development of a BRT system in the next few years.

- *Mayor's letter.* Following Presidential Decree No. 2/2008 on Energy and Water Efficiency, the mayor issued a letter to city departments mandating the implementation of energy efficiency measures such as energy efficient light bulbs.

- *Eco² Cities.* Surabaya is hosting a pilot World Bank Group Eco² Cities program, which will focus on strengthening its core urban planning, management, and finance capacities while investing in a catalyst waterfront redevelopment project. The waterfront redevelopment will enhance environmental and quality of life aspects of the city while increasing accessibility and social inclusiveness, and will revitalize the urban economy.

## Energy Use and Carbon Emissions Profile

Surabaya's 2010 energy flows and profile are summarized in a Sankey diagram, shown in figure 7.2, to illustrate the citywide energy supply and demand characteristics of its different sectors.

Although 62.4 billion megajoules of primary fuel energy is supplied to Surabaya, a substantial portion is lost through thermodynamic conversion processes in vehicles (about 77 percent) and in electrical power generation. Similarly, low-efficiency combustion motors in the transportation sector result in only 6.3 out of 27.4 petajoules of gasoline and diesel energy supplied to the sector being used effectively. In comparison, LPG use clearly shows its efficiency advantage because no conversion is involved until its final use. Thus, it would be beneficial both to increase the use of LPG and to locate gas-fired distributed

**Figure 7.2 Surabaya Energy Flows, 2010**

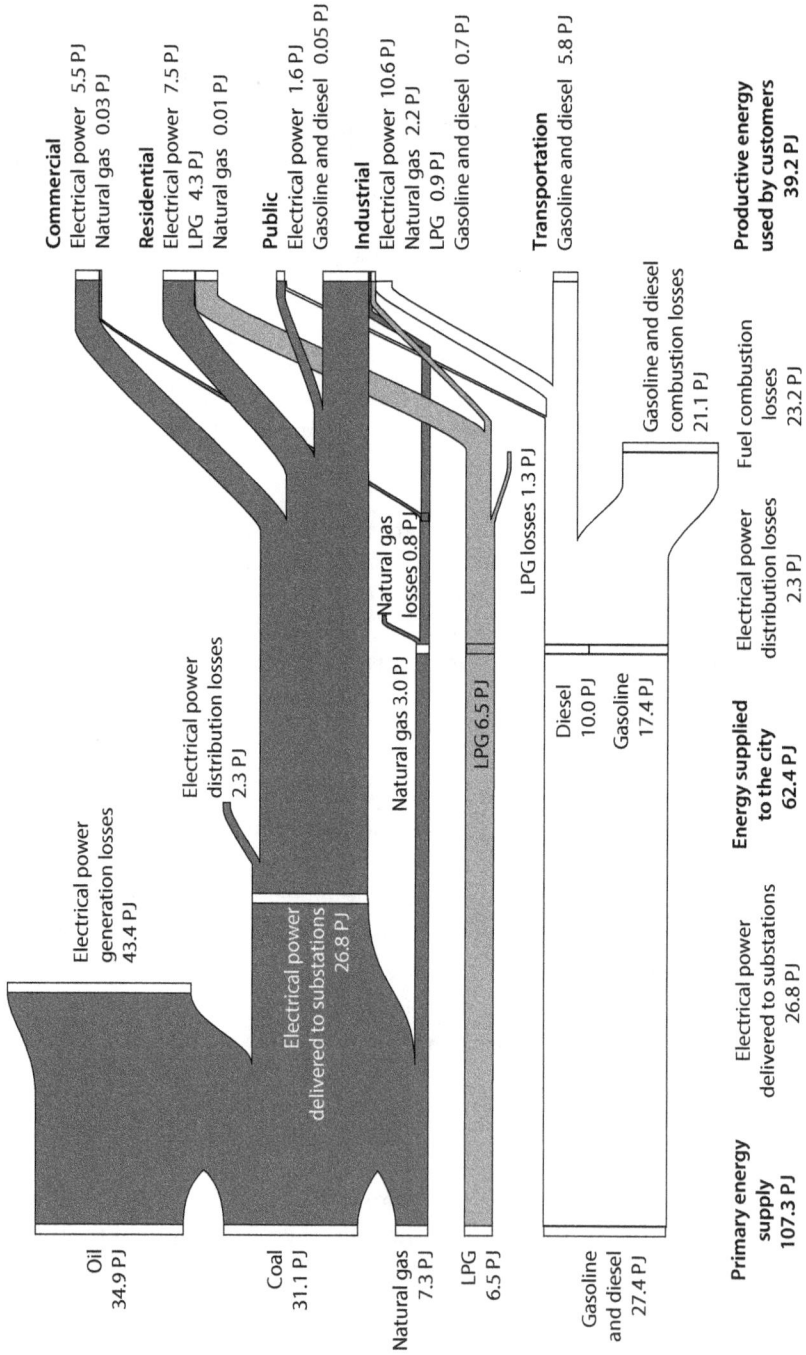

Commercial
Electrical power  5.5 PJ
Natural gas  0.03 PJ

Residential
Electrical power  7.5 PJ
LPG  4.3 PJ
Natural gas  0.01 PJ

Public
Electrical power  1.6 PJ
Gasoline and diesel  0.05 PJ

Industrial
Electrical power  10.6 PJ
Natural gas  2.2 PJ
LPG  0.9 PJ
Gasoline and diesel  0.7 PJ

Transportation
Gasoline and diesel  5.8 PJ

Productive energy
used by customers
39.2 PJ

Electrical power
generation losses
43.4 PJ

Electrical power
distribution losses
2.3 PJ

Electrical power
delivered to substations
26.8 PJ

Natural gas
losses 0.8 PJ

Natural gas 3.0 PJ

LPG 6.5 PJ

LPG losses 1.3 PJ

Gasoline and diesel
combustion losses
21.1 PJ

Diesel
10.0 PJ

Gasoline
17.4 PJ

Fuel combustion
losses
23.2 PJ

Electrical power
distribution losses
2.3 PJ

Energy supplied
to the city
62.4 PJ

Electrical power
delivered to substations
26.8 PJ

Primary energy
supply
107.3 PJ

Oil
34.9 PJ

Coal
31.1 PJ

Natural gas
7.3 PJ

LPG
6.5 PJ

Gasoline
and diesel
27.4 PJ

*Source:* Phase I pilot study.

*Note:* LPG = liquefied petroleum gas; PJ = petajoule. "Public" includes the end-use energy of city buildings, street lighting, city vehicles, water, wastewater, and solid waste management.

generation closer to consumers. In particular, the industrial sector can make further efficiency gains by using heat (through cogeneration), which is otherwise wasted in coal-fired plants.

Although the transportation sector accounts for the highest proportion of Surabaya's primary energy consumption, the proportions of energy consumed in the commercial, industry, and residential sectors are also significant (see figure 7.3). In contrast, energy use is insignificant for city public services (solid waste, public lighting, water supply).

Apart from analyzing the energy end-use profile, an overview of Surabaya's energy supply profile can provide the city government with valuable insights for strategic planning with respect to issues of energy security and economic growth. A negligible amount of energy is currently generated from renewable energy sources or primary energy fuels sourced from within the city boundary, but this could be a major potential source of energy production (through thermal solar) for residential, industrial, and hotel uses.

Of the energy supplied to the city, 58 percent is in the form of petroleum products. The majority (68 percent) of the petroleum products consumed in Surabaya is used in the transportation sector; another 18 percent of the city's petroleum use occurs in the industrial sector. The remaining 42 percent of energy imported is electricity. Currently, only 2 percent of energy supplied to Surabaya is generated within the city boundaries. For electricity generation both within and outside its city boundaries, Surabaya relies on oil, natural gas, and coal.

Greenhouse gases (GHG) totaling 8.6 million tons of carbon dioxide ($CO_2$) equivalent were emitted by all end-use sectors in Surabaya in 2010 (see figure 7.4). Industrial energy use represents 35 percent of GHG emissions. Commercial and residential energy use represents 43 percent of GHG emissions. Transportation fuel results in 20 percent of the city's GHG emissions, with the remaining emissions emanating from methane released from the city's wastewater treatment operations and local landfills. On a fuels basis,

**Figure 7.3  Surabaya Energy Consumption by End Use**
*Percent*

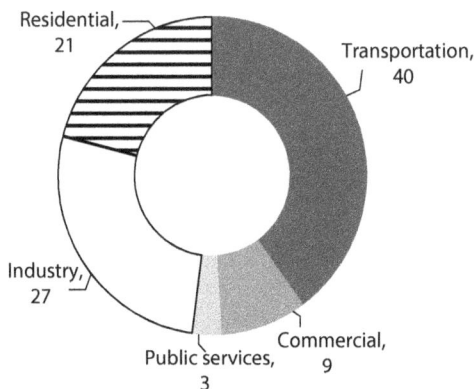

Residential, 21
Transportation, 40
Industry, 27
Public services, 3
Commercial, 9

*Source:* Phase I pilot study.

**Figure 7.4  Surabaya GHG Emissions by End Use and Fuel Source**

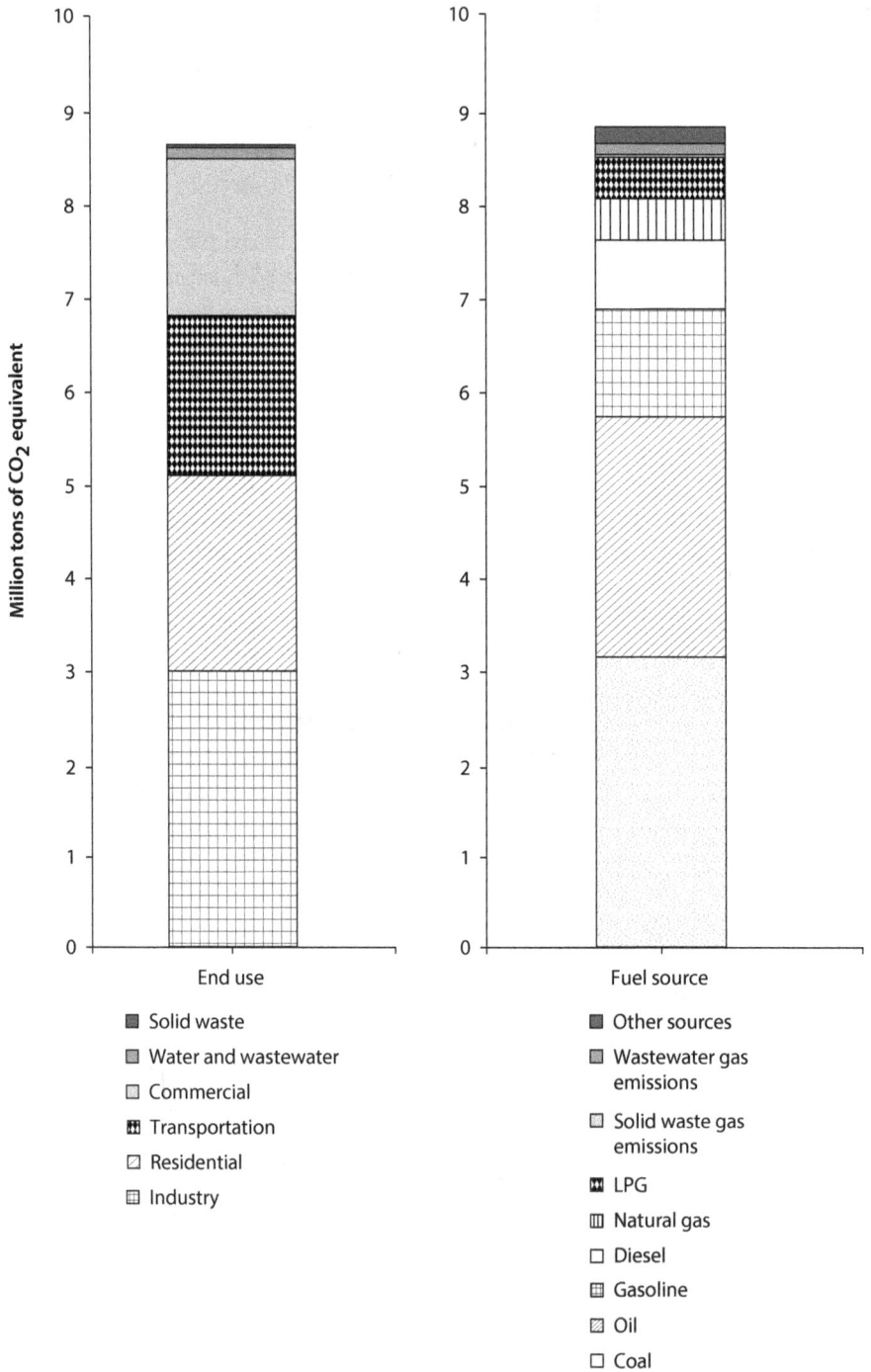

Source: Phase I pilot study.
Note: $CO_2$ = carbon dioxide; GHG = greenhouse gas; LPG = liquefied petroleum gas.

coal is the dominant contributor to GHG emissions in Surabaya (36 percent) and oil the second-largest contributor (29 percent). Gasoline, diesel, LPG, and natural gas collectively account for another 31 percent of emissions from fuel use.

## Sector Review and Prioritization

Surabaya's interest in pursuing the Tool for Rapid Assessment of City Energy (TRACE) underscores its commitment to achieving optimal energy efficiency. The analysis was carried out across six city sectors: passenger transportation, city buildings, water and wastewater, public lighting, solid waste, and power and heat. These were, in turn, assessed against the performance of a range of peer cities through a benchmarking process. This review provided a number of significant findings that helped to focus activities during the early part of the study and contributed to the definition of priority sectors for further analysis.

Key findings of the Surabaya diagnostics in comparison with the cities in the TRACE database are the following:

- High electricity use per capita and high energy use per unit of GDP
- Relatively low energy consumed by transportation due to the low level of automobile use and high usage of relatively fuel-efficient motorcycles
- Low use of public transportation coupled with growing private vehicle ownership and widespread use of private motor scooters, resulting in relatively low operating energy intensity for mobility
- High per capita water consumption and relatively high water losses from the distribution system, but midrange energy density for potable water production
- Low electricity consumption per light pole but room for improvement in public lighting
- Low energy consumption in city buildings but electricity use is on the rise
- Low level of recycling and high amount of solid waste that goes into one landfill
- Low levels of transmission and distribution losses in the electricity network

The TRACE analysis identified priority areas for which significant energy savings are possible. Table 7.1 indicates the energy spending in each of these sectors, the relative energy intensity (the percentage of energy that can be saved in each sector, based on the TRACE benchmarking), and the level of influence the city has over these sectors. The savings potential is calculated by multiplying the three factors. The TRACE contains a playbook of 58 energy efficiency recommendations applicable across all sectors analyzed.[1] The recommendations are not meant to be either exhaustive or normative. The recommendations only outline a number of policies and investments that could help local authorities achieve higher energy efficiency standards. Following the sector-by-sector analysis, each recommendation was reviewed to establish its applicability to Surabaya. This filtering process helped concentrate the process on those recommendations that are both viable and practical.

Energizing Green Cities in Southeast Asia • http://dx.doi.org/10.1596/978-0-8213-9837-1

**Table 7.1  Surabaya Sector Prioritization Results**

| Priority ranking | Sector | 2010 energy spending (US$) | Relative energy intensity (%) | Level of city authority control[a] | Savings potential (US$)[b] |
|---|---|---|---|---|---|
| *City authority sector ranking* | | | | | |
| 1 | Street lighting | 6,089,000 | 20 | 1.00 | 1,217,000 |
| 2 | City buildings | 2,237,000 | 10 | 0.95 | 212,515 |
| 3 | Public vehicles | 1,617,840 | 10 | 0.05 | 6,235 |
| *Citywide sector ranking* | | | | | |
| 1 | Potable water | 6,528,000 | 36 | 0.96 | 2,256,000 |
| 2 | Public transportation | 68,889,000 | 5 | 0.38 | 1,309,000 |
| 3 | Power | Unknown | 12 | 0.05 | Potentially large |
| 4 | Solid waste | 1,306,000 | 15 | 0.75 | 146,000 |

*Source:* Phase I pilot study.
a. 0 = no influence; 1 = maximum influence.
b. Based on TRACE (Tool for Rapid Assessment of City Energy) benchmarking; these figures are indicative of the savings that may be possible, but not necessarily practicable.

Table 7.1 shows priorities with respect both to sectors over which the city authority has maximum influence and to citywide issues over which the authority has limited influence.

The rankings suggest that the city government should prioritize street lighting, followed by city buildings, and then public vehicles. On a citywide basis, potable water supply is clearly deserving of attention, followed by public transportation, power, and solid waste.

## Recommendations

The recommendations to improve the city's energy efficiency concentrate on areas over which the city has direct influence and are, for the most part, derived from TRACE. Many of the recommendations are targeted at reducing energy use to lower the city's energy expenditure. The recommendations will help city officials identify how the initiatives can be implemented in Surabaya. The recommendations are embedded in TRACE and their details can be made available. The recommendations are expected to be refined by the city and the Sustainable Urban Energy and Emissions Planning (SUEEP) team as further analysis and discussions occur. Although the energy balance and GHG emissions inventory provide an overview of the energy and emissions profile of Surabaya, additional analysis beyond the public sector is required to develop a sustainable urban energy and emissions plan that includes sectors outside the direct influence of the city government.

### Transportation

More than 1.3 million motorcycles and motor scooters dominate private vehicle transportation in Surabaya as they do in many cities in Southeast Asia. The other important modes of transportation in Surabaya are private automobiles, taxis, and angkot buses (minibuses). Surabaya currently experiences a high-volume rush hour, during which traffic flow is severely impeded.

Although motor scooters in Surabaya are predominantly new and fuel efficient, a shift toward private cars—influenced by higher wages and increasing standards of living, as well as the lack of restrictions on new vehicle registration (although progressive taxes apply to second, third, and fourth vehicles)—is driving up fuel usage. Fuel usage is rising despite the fact that new vehicles tend to be smaller and more fuel efficient than cars in other developed cities.

No high-capacity public transit system serves Surabaya aside from regional commuter trains that run only three or four times each morning and account for a very small share of mode split, estimated at less than 1 percent. Angkot minibuses, another mode of public transportation in Surabaya, generally have older and inefficient engines, and typically use kerosene fuel that may be blended with gasoline, which can damage the engine and is highly polluting. Despite the low usage of public transportation in the city, transportation energy intensity per capita remains fairly low because of the widespread use of fuel-efficient motor scooters.

These factors point to significant potential for improvement in the energy and operational efficiencies of the public transportation system. Furthermore, the lack of access to nonmotorized modes as well as safety concerns regarding the use of crowded streets have made nonmotorized modes of transportation such as bicycling unattractive. Bicycle lanes could potentially be established on a few main thoroughfares wide enough to accommodate them, though the vast majority of streets in Surabaya are too narrow to safely make room for bicycle lanes.

With respect to Surabaya's city operations, its fleet vehicles are maintained at a single facility so projects to improve maintenance and energy efficiency could be easily implemented at this location. The city government is responsible for more than 500 vehicles that support the following services: garbage collection and transfer, official vehicles, street sweeping, and street light maintenance.

### Public Transportation Development

The traffic congestion in Surabaya, which is exacerbated by growth in private vehicle ownership, has led city officials to enhance planning efforts to increase public transportation networks. Irregular public transportation schedules and a declining perception of the attractiveness of road-based transit have been identified as the main challenges. Therefore, the city is focusing its planning efforts on the implementation of multimodal transportation systems based on a light rail transit line, the development of an Intelligent Transportation System, and park-and-ride facilities. Campaigns to increase public awareness include programs such as car-free days and license plate restrictions. A recent increase in the use of bicycles has also prompted plans for the development of bicycle and pedestrian path networks.

One of the most notable observations was the need for an integrated planning approach to ensure that plans for public transportation systems, land use, street signals, parking policies, vehicle registration pricing, and sidewalk policies are adequately integrated and that there is an effective means for turning plan into practice. Integrated planning is especially important for transportation planning. Encouraging nonmotorized modes alongside the development of robust public transportation will be critical to mitigating congestion and improving the quality of transportation in Surabaya.

Coupled with nonmotorized transportation, the promotion of public transit should be part of an energy efficient strategy for Surabaya. Various proposals for enhancing the public transportation system have been in development for a number of years. A BRT pre-feasibility study was completed with the support of Japan International Cooperation Agency (JICA), and the World Bank Group is involved in helping Surabaya conduct more detailed studies. Because of the complexity involved in the development of public transportation and the level of documentation available for the proposed BRT, this report will not go into detail on this issue. Surabaya is also developing plans for a mass transportation system comprising tramway and monorail systems to serve heavy traffic corridors into the city center and is initiating studies to integrate the rest of its public transportation network with this system.

### Vehicle Emissions Standards

Surabaya's transportation system is dominated by motorcycles and cars, which are currently fairly new and fuel efficient. However, vehicle emissions standards testing could be more effectively applied to encourage and enforce better vehicle emissions, which will be important as vehicles begin to age. This, in turn, will lead to better air quality and reduced energy consumption. The current system of testing and enforcement of vehicle emissions is fragmented because different city and national government departments regulate (license) test centers, and enforcement is either weak or ineffective because cars can be back on the road after a few simple measures are applied to get the vehicle to pass. New testing equipment that meets standards for measuring an engine's efficiency would be required. Enforcement activities and resulting sanctions should also be revised to ensure that poorly performing vehicles are identified and removed from service. The implementation of such a measure is potentially difficult because of the fragmented nature of Surabaya's existing testing system and the magnitude of enforcement required.

It is highly recommended that the Bappeko lead a city vehicle fleet efficiency program. Procurement and maintenance policies should be implemented to maximize the efficiency of the city's fleet.

In response to the limited transportation data available to support the Department of Transportation's decision making for energy efficiency issues, a data collection program is also recommended. A variety of informal and inexpensive data collection systems are now becoming available with software applications that collect data from cell phones or city vehicles and do not rely on manual counts to collect transportation data.

Finally, the establishment of a Surabaya regional transportation planning authority is highly recommended to study all types of regional transportation issues and to allocate funds to the lowest-cost solutions for the most pressing problems.

### Solid Waste

Surabaya has one landfill, which accepts hazardous, septic, and noncompostable waste. The landfill is fairly new and has capacity for approximately 10 years, with expansion sites nearby. Residential waste is collected by the kampungs[2] and brought to transfer stations. Dinas Kebersihan dan Pertamanan (DKP), or Cleansing and Park Department waste transfer trucks take waste from the transfer stations to the landfill.

Sludge from wastewater treatment is collected from private wastewater treatment companies and processed by DKP. Composted sludge is used as soil fertilizer for city parks.

Landfill leachate currently remains untreated, posing contamination and disposal issues. Surabaya produces nearly 707,000 tons of domestic solid waste annually, giving rise to approximately 37,800 tons of carbon dioxide equivalent ($CO_2e$) from landfill gas emissions.

Programs have started in four of the transfer stations to extract compostable waste to be turned into soil for use as fertilizer in parks throughout the city. Kampungs are also engaged in small-scale composting with thousands of residential-scale composting bins distributed to households. Surabaya has reviewed numerous energy-from-waste proposals for the landfill, and proposals for a number of different technologies by a variety of bidders are being evaluated; however, no projects have yet been implemented. The landfill in Surabaya is currently set up to capture methane gas, but the gas is not being used as a resource.

### Vehicle Maintenance Program and Vehicle Operations Program

The first easy win for energy efficiency in the waste sector is a vehicle maintenance program for the collection and transfer vehicles under control of the city government. These vehicles, if tuned up and operated with the correct tire pressure and with clean fuel, could run at least 10 percent more efficiently.

An effective vehicle operations program would improve the efficiency of routes for the trucks by running them through less-congested areas and making the routes shorter and more direct.

### Composting Program

The existing composting programs should be expanded and be run at all of the waste transfer facilities, not just the four sites where composting is already happening.

### Landfill Gas Capture Program

Finally, the proposal for a landfill gas capture program should be executed. The program would provide quick and easy energy and reduce GHG emissions at a very low cost to the city.

### Water

The regional drinking water company (PDAM) produces and distributes Surabaya's drinking water and manages the treatment and distribution infrastructure. PDAM is owned and operated by the city government. About 70 percent of the population has access to clean drinking water. Domestic wastewater is mostly treated with septic tanks and absorption technology.

Energizing Green Cities in Southeast Asia • http://dx.doi.org/10.1596/978-0-8213-9837-1

Industrial wastewater is often discharged directly into the Surabaya River without treatment. However, a small number of companies install and manage their own facilities to treat their wastewater discharge. The Surabaya River is a primary source for the city's drinking water, so coordination and action on upstream industrial sites is particularly necessary.

Raw water is gravity fed to the two potable water treatment facilities by the Surabaya River; thus, the energy intensity of potable water is fairly low because no energy is required for transmission pumping or groundwater pumping. Because the water consumption rate in Surabaya is relatively high at 290 liters per person per day, reflecting both actual consumption and loss, water leakage reduction programs and water conservation will be effective.

Surabaya has no citywide wastewater infrastructure and wastewater is currently managed through household, or clusters of household, septic tanks. Seventeen private wastewater treatment companies run treatment facilities for large buildings or campuses. For this reason, Surabaya's energy use for wastewater is very low. However, in the interest of public health, it is expected that a more comprehensive system will eventually be developed.

Surabaya has a large number of storm water ejector pumps in the canals to move water through and out to the sea during rain events. Although many of these facilities are quite old and use large, inefficient pumps, the limited time that these pumps are operated each year means that total energy savings potential is unlikely to be significant.

The city is facing significant challenges in wastewater management, including the contamination of water supply from septic tank seepage, contaminated landfill leachate, and the lack of treatment of wastewater sludge. However, the Surabaya city government has the opportunity to overcome these issues because it owns PDAM (as is common in Indonesia). The large leakage rates and the low water pressure in the east and north sectors of the network also provide opportunities to greatly reduce nonrevenue water losses.

### Pump Replacement Program
Considering the age of Surabaya's water network and water treatment facilities, a pump replacement program is strongly recommended for Surabaya.

### Leak Reduction Program
The city should implement a leak reduction program and hire a long-term partner to deliver a performance-based contract.

### Water Awareness Program
A public water awareness program would also be a productive complement to the water use reduction efforts in Surabaya.

### Power
Surabaya is primarily an importer of electricity, with only one 57 megawatt power plant in Perak (in northern Surabaya), which is fueled by natural gas.

PLN (Perusahaan Listrik Negara) Distribution East Java is the state-owned enterprise that provides electricity to meet Surabaya's needs. Hence, although Surabaya can make recommendations and ask for improvements to the electrical network, the city government cannot make direct decisions or even allocate funding for improvements in this sector.

With respect to transmission and distribution, Surabaya has relatively low losses (1.8 percent for transmission and 6.7 percent for distribution). PLN has implemented energy efficiency programs and performs ongoing maintenance on transformers in substations and seems to have a good program for addressing transmission and distribution losses. Because the PLN network is performing relatively well, any improvements would be incremental.

JICA recently provided technical assistance to PLN for the study of "smart metering" in Jakarta. This concept would be beneficial to Surabaya as well.

### Distributed Generation Program

In light of the city government's limited influence in the power generation sector, the SUEEP team recommends that a distributed generation program, which would use the capacity of the natural gas network and the nearby natural gas fields to generate low-cost, local electricity within the city, be developed. A distributed generation program would produce additional benefits, such as the use of waste heat for hot water heating, generation of chilled water with waste heat, and reductions of distribution losses in the electrical grid.

### Public Lighting

Public lighting in Surabaya is in the remit of the DKP. Responsibility for street lights on most city streets is under DKP, whereas small roads in residential areas are maintained by the kampung or by the local developer. Surabaya has good street lighting coverage at 79 percent, but its implementation of a program to achieve 100 percent coverage does not suggest that it is a top priority. The SUEEP team noted that the program did not appear to have significant funding.

Surabaya has 40,000 street lights, 95 percent of which use high pressure sodium lamps, which is good practice today, although not the most energy efficient lamp on the market. The lighting levels do not meet international standards, hence electricity consumed per km of road lit is low. A bulb replacement program to ensure all street lights use high pressure sodium lamps was nearly complete.

Only 12 maintenance vehicles were identified for street lighting repair and lamp replacement for the entire city. The team of people assigned to maintenance was also quite small for a city of close to 3 million people. Construction costs for new street lamps can be high, and fitting new wiring and poles into already crowded streets filled with old existing pipes and wires can be difficult. Despite the above, there is scope for improving public lighting in Surabaya (see below).

Energizing Green Cities in Southeast Asia • http://dx.doi.org/10.1596/978-0-8213-9837-1

### Public Lighting Assessment Program

The first recommendation for public lighting is to improve data collection processes and data availability through a public lighting assessment program so that future programs and funding allocations can be adequately informed with accurate data.

### Lighting Timing and Dimming Program

There is an opportunity to create a lighting timing and dimming program that would test and install new technologies for new street light installations and replacement lamps.

### Public Lighting Research and Development Program

Because DKP does not have the capacity or funding to test new technologies, the most important recommendation in this sector is the public lighting research and development program. This program could accept demonstration poles and lamps from manufacturers and measure their performance and test their lighting output to satisfy decision makers that new low-energy technologies can perform satisfactorily.

### City Buildings

The city buildings category covers all buildings owned by the city, including government offices, city schools, and city hospitals. The city building stock is generally characterized by the use of natural daylight, fans and natural ventilation, compact fluorescent lighting fittings, and limited meeting rooms and offices augmented by air conditioning on a timed basis. Hallways, lobbies, and open-plan offices are generally open air and naturally ventilated. Very few buildings in Surabaya have central air conditioning, and the SUEEP team did not identify any city buildings with chillers or central ventilation. These are excellent examples of sustainable design—on the sea coast, where natural ventilation is quite good, not every facility should be fully enclosed with sealed windows and air conditioning. The city buildings in Surabaya consume very little energy, and the opportunities to further enhance energy efficiency seem to be very limited.

No new construction projects were under way for city buildings during the time of the SUEEP mission, and no new building construction projects were identified. Only minor renovation projects were under way to improve city buildings.

No green building codes or ordinances were identified, nor were any building energy code revisions. National-level green building guidelines have not been set up in Indonesia.

Surabaya's city buildings are low energy users, so maintaining this low consumption while increasing the quality of buildings and services will be a challenge. For example, the SUEEP team's walk-throughs of the hospitals in Surabaya showed that the old hospital was open to the air in all hallways and had minimal air conditioning and lighting, whereas the new hospital was fully air conditioned and overlit through all lobbies, hallways, and spaces. This trend presents both a challenge and an opportunity.

Each department has control over the buildings it uses. There is no central facilities management group within the Surabaya government. Neither is there a formalized refurbishment cycle for government buildings, which poses a significant challenge for achieving energy efficient performance in the existing building stock. In addition, energy-efficiency-oriented capital investment planning and life-cycle costing are not used, hindering the uptake of those energy efficiency initiatives that require higher upfront investment but yield long-term savings.

Any new city building design and construction projects would be good opportunities for the city to show leadership in energy efficient building design practices. The city has the power to pass ordinances for building codes, which will be advantageous for any agency that wishes to mandate compliance with green building codes. The current mayor has been a vocal proponent of energy efficiency and it has been through her leadership that the existing energy efficiency projects have been implemented.

Because of the low energy use by city buildings in Surabaya, only two recommendations are put forward for this sector.

### Computer Power Save Program
First, based on the walk-throughs of more than a dozen city buildings, the computer power save program was deemed to have the most potential because of the large number of desktop computers in evidence.

### Energy Efficient Building Code
Second, because of the growing stock of higher-quality and higher-energy-using buildings, the SUEEP team recommends development of an energy efficient building code to appropriately address the construction industry in Surabaya and mitigate the trend of increasing energy consumption in new buildings. The city government can use future new building projects to demonstrate the techniques and benefits of energy efficient buildings.


## Conclusion

Robust energy planning and management will help shape Surabaya's future. Without a formal energy and emissions strategy, Surabaya's economy and quality of life will not reach their full potential. A comprehensive and strategic approach to energy now will pay dividends by future proofing the city's infrastructure against increases in energy use, a growing population, and the pitfalls of energy-intensive development that have hindered so many other cities.

Surabaya's government has recognized the need for a sound energy and emissions strategy through the many existing excellent energy initiatives reviewed above.

Energizing Green Cities in Southeast Asia • http://dx.doi.org/10.1596/978-0-8213-9837-1

In the future, energy governance should be prioritized by the city government because it will help strengthen the city's internal energy management practices as well as engage other key stakeholders who have not played a significant role in the city's energy planning efforts to date. Better governance practices include not just enhanced oversight and data tracking but also improved procurement practices and a willingness to "lead by example" by showcasing best practices for the benefit of local businesses and households.

## Notes

1. For further details on TRACE, see chapter 3.
2. Kampungs are the local neighborhood level of city governance.

**CHAPTER 8**

# Da Nang, Vietnam

Da Nang is a major harbor city and the largest urban center in central Vietnam. The city is spread over 1,283 square kilometers, with 911,890 people and a population density of 711 inhabitants per square kilometer. With the fourth largest seaport in the country, Da Nang (map 8.1) is an important gateway city to the Central Highlands of Vietnam, the Lao People's Democratic Republic, Cambodia, Thailand, and the Republic of the Union of Myanmar. After relatively slow population growth (1.7 percent annually) between 2000 and 2007, Da Nang appears poised for a significant increase in the next 10 years, primarily due to migration from rural areas. By 2020, Da Nang hopes to become one of the country's major urban centers with a population of about 1.65 million.

Da Nang is in a tropical monsoon zone with high temperatures and a stable climate. There are two seasons, with the wet season lasting from August through December and the dry season from January through July. Winter cold spells tend to be short and not severe. The annual average temperature is 25.9 degrees C (centigrade), with average humidity of 83.4 percent. On average, the city receives 2.5 millimeters of rainfall per year, and enjoys 2,156 hours of sunshine annually.

Da Nang has recorded remarkable changes in economic development. Its gross domestic product (GDP) growth rate has been higher than the country's average rate. Between 2000 and 2007, Da Nang's regional GDP grew 12.3 percent annually, totaling US$1.48 billion in 2009. The production of industrial, agricultural, and aquatic products has increased, as has export value. Promising growth in tourism, commerce, and services has also occurred. Da Nang's economy has historically been dominated by the industry and construction sectors, but this is slowly changing. In 2006, the services sector became the largest economic sector in the city as measured by gross output. This shift is in keeping with local policy targets, which seek to develop the city as a rail, road, and seaport hub, in addition to other services-oriented industries (financing, banking, insurance, telecommunications, and consulting). The tourism sector is also expected to grow, as the city strives to become a major national tourist sector that capitalizes on the city's beaches and proximity to the old capital, Hue; Hoi An Ancient Town; and the ruins at My Son.

**Map 8.1  Da Nang, Vietnam**

*Source:* World Bank.

Energizing Green Cities in Southeast Asia  •  http://dx.doi.org/10.1596/978-0-8213-9837-1

**Figure 8.1 Da Nang Government Structure for Energy-Consuming Agencies**

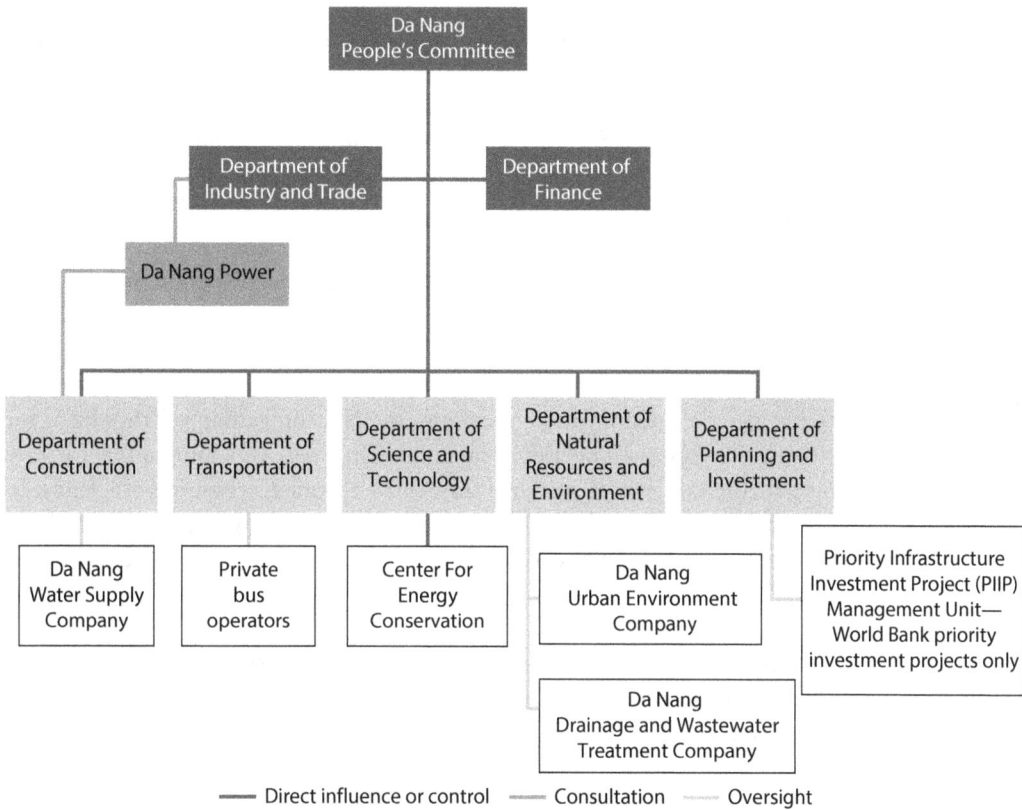

*Source:* Phase I pilot study.

Da Nang became a centrally governed city in 1997. The central Vietnamese government and Da Nang City People's Committee are the focal points for policy making, but a strong base of local entities is involved in energy planning matters and energy systems operations in Da Nang. The main departments, committees, and external agencies responsible for the planning, development, and operation of energy-consuming sectors are represented in the city government structure as shown in figure 8.1.

## Energy Efficiency Initiatives

### National Level

The Vietnamese government has acknowledged that increasing energy efficiency is a national priority. The National Energy Efficiency Program (2006–15), which is run by the Department of Industry and Trade, comprises a set of activities to encourage, promote, and propagate energy efficiency and conservation to the public. Elements of the program include the following:

- A targeted reduction in national energy usage of 3–5 percent between 2006 and 2010, and of 5–8 percent between 2011 and 2015

- Coverage across sectors including organizations, households, and individuals using energy
- Prescription of energy efficiency and conservation measures, for example, product labeling to encourage energy efficiency technology and phasing out of inefficient equipment
- Promotion of energy efficiency and conservation through incentives, scientific and technological development, education, and engagement of consultants by the city government
- Engagement of consultancy support for energy efficiency measures in enterprises and buildings in Da Nang in 2010, for example, provision of training to companies on energy auditing

### City Level

The City People's Committee Decision on "Promulgation of the Plan for Developing Da Nang—The Environmental City" lays the foundation for city planning in the context of sustainability and encourages resource efficiency. In addition, a number of relevant energy efficiency initiatives and studies have already been enacted by the local Da Nang city government, including the following:

- The Wastewater Management Strategy in Da Nang City with Addendum produced for the Priority Infrastructure Investment Project, 2009
- The Integrated Development Strategy for Da Nang City and Its Neighboring Area
- A city energy efficiency and conservation program covering six energy efficiency projects to be cofinanced by the National Target Program on Energy Efficiency and Conservation, the city budget, private investors, and other sources
- A city priority infrastructure investment project
- The Da Nang Master Plan for Water Supply Systems for 2020, which lays out the city water supply company's strategy for expanding Da Nang's current potable water system
- A program for public lighting
- A program for public agencies and utilities
- A school lighting program
- Renewable energy research and development and renewable energy application in the city

Da Nang has already deployed and assigned different agencies, districts, and enterprises to employ and follow the Government Guidelines on the National Target Program on Energy Efficiency. The city has also set targets for public agencies to save energy in office buildings, and encouraged energy conservation through lighting turn-off times and air conditioning use guidelines. The city has also considered such initiatives as using nanotechnologies for lighting on bridges and solar energy for traffic signal lighting. Figure 8.1 provides an overview of

Da Nang's institutional structure and the relationships involved in the city's energy management.

## Energy Use and Carbon Emissions Profile

Da Nang's energy flows and profile are summarized in the Sankey diagram in figure 8.2, illustrating citywide energy supply and demand characteristics by sector. Da Nang currently has no significant energy resource base of its own (for example, indigenous coal, natural gas deposits, or hydropower facilities), so virtually 100 percent of the city's energy supply is imported. The city does enjoy significant solar energy potential and may be able to benefit from local wind resources, but because no renewable power technology is installed in Da Nang, no power is harvested from these resources.

Two sectors (transportation and industry) dominate energy use and emissions, and energy demand is increasing rapidly, with electricity demand alone likely to double between 2011 and 2015. In 2010, the city used roughly 17.9 petajoules of energy in various forms. Transportation was responsible for 45 percent of the city's energy use, followed by 21 percent by the industrial sector. The residential sector consumes 13 percent of the city's energy (figure 8.3). Commercial uses (3 percent) and public services (that is, the government sector, at 2 percent) lag far behind these other sectors. Some 16 percent of energy is consumed in other sectors.

Of the energy imported, 73 percent is in the form of petroleum products, out of which 61 percent is used in the transportation sector. This 61 percent comprises mainly diesel fuel and gasoline, in equal shares. Some 11 percent of the city's petroleum use is attributable to the industrial sector, of which 83 percent is fuel oil, which presumably is used to create process heat in boilers. Electricity accounts for 27 percent of energy imported into the city. In the residential sector, electricity is the primary form of energy used, although a sizable percentage of liquid petroleum gas is used for cooking and hot water. Current peak electricity demand in Da Nang is approximately 250 megawatts, which is a significant increase from 2007, when peak demand totaled just 176 megawatts. Electricity usage in Da Nang steadily increased between 2007 and 2010 (an increase of 44 percent in that period), and is expected to double in six years. The industrial and residential sectors dominate electricity use. Although the commercial sector accounts for a small share of electricity consumption, it has seen the most rapid growth from 2007 to 2010, with demand increasing 44 percent. Industrial demand has kept pace with the overall increase, at 41 percent for that time period, whereas residential demand has grown at a slightly slower pace of 39 percent.

Greenhouse gas (GHG) emissions tell a similar story. A total of 1.54 million tons of carbon dioxide equivalent ($CO_2$e) were emitted by all sectors in Da Nang in 2010 (figure 8.4), and the city's carbon intensity (0.89 kilograms of $CO_2$ per unit of GDP) is more than twice the national carbon-intensity estimate for Vietnam. Da Nang's carbon intensity is quite high even in comparison with other developing or highly industrial countries. Transportation fuels are responsible for

# Figure 8.2  Da Nang Energy Flows, 2010

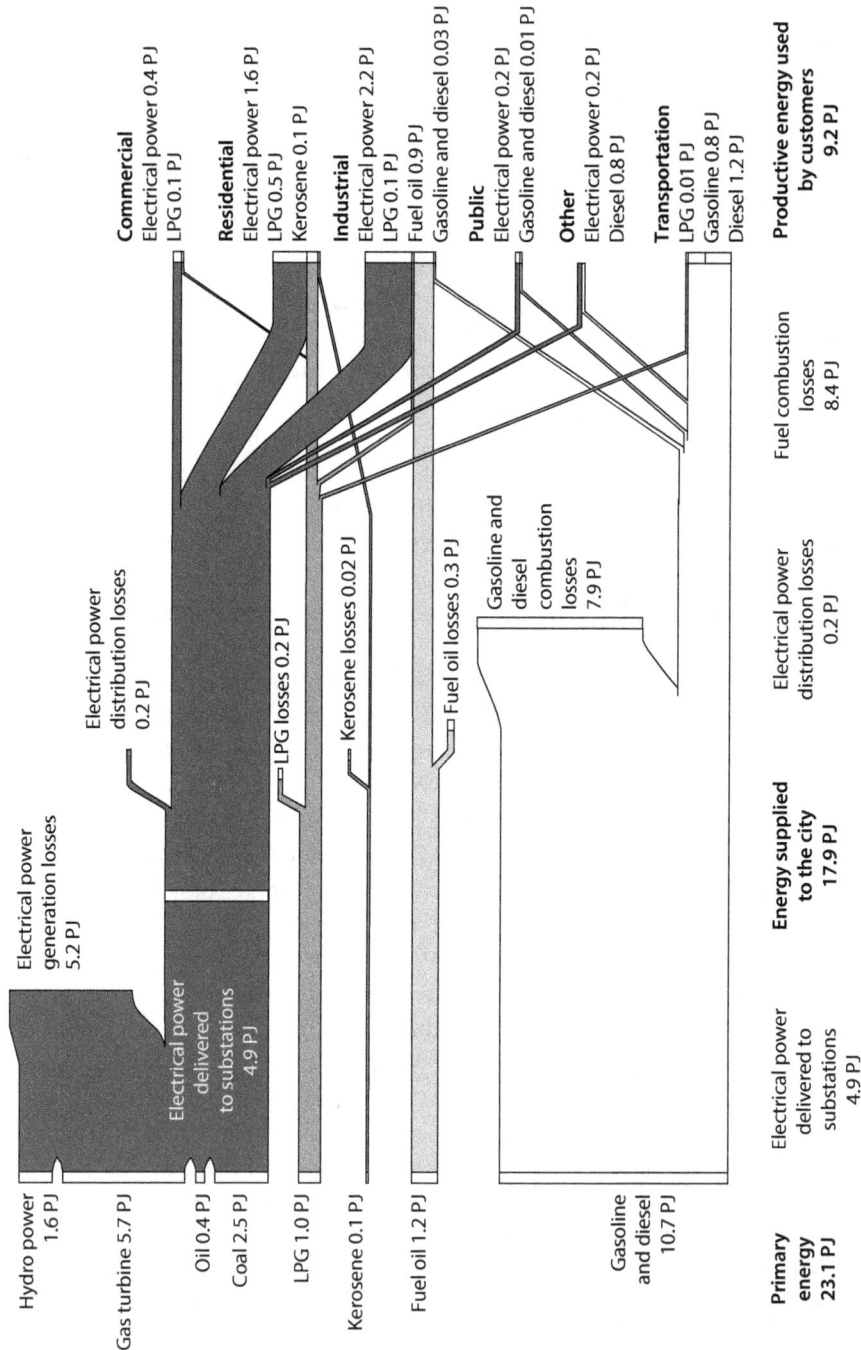

**Primary energy 23.1 PJ:**
- Hydro power 1.6 PJ
- Gas turbine 5.7 PJ
- Oil 0.4 PJ
- Coal 2.5 PJ
- LPG 1.0 PJ
- Kerosene 0.1 PJ
- Fuel oil 1.2 PJ
- Gasoline and diesel 10.7 PJ

Electrical power generation losses 5.2 PJ

Electrical power delivered to substations 4.9 PJ

Electrical power delivered to substations 4.9 PJ

Energy supplied to the city 17.9 PJ

Electrical power distribution losses 0.2 PJ

LPG losses 0.2 PJ

Kerosene losses 0.02 PJ

Fuel oil losses 0.3 PJ

Gasoline and diesel combustion losses 7.9 PJ

Electrical power distribution losses 0.2 PJ

Fuel combustion losses 8.4 PJ

**Commercial**
- Electrical power 0.4 PJ
- LPG 0.1 PJ

**Residential**
- Electrical power 1.6 PJ
- LPG 0.5 PJ
- Kerosene 0.1 PJ

**Industrial**
- Electrical power 2.2 PJ
- LPG 0.1 PJ
- Fuel oil 0.9 PJ
- Gasoline and diesel 0.03 PJ

**Public**
- Electrical power 0.2 PJ
- Gasoline and diesel 0.01 PJ

**Other**
- Electrical power 0.2 PJ
- Diesel 0.8 PJ

**Transportation**
- LPG 0.01 PJ
- Gasoline 0.8 PJ
- Diesel 1.2 PJ

**Productive energy used by customers 9.2 PJ**

*Source:* Phase I pilot study.

*Note:* LPG = liquefied petroleum gas; PJ = petajoule. "Public" includes the end-use energy of city buildings, street lighting, city vehicles, water, wastewater, and solid waste management.

**Figure 8.3  Da Nang Energy Consumption by End Use**
*Percent*

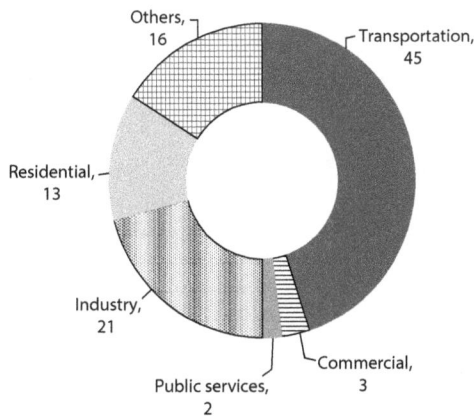

*Source:* Phase I pilot study.

46 percent of citywide GHG emissions. Of the total citywide $CO_2$ emissions that originate from the transportation sector, two-thirds are derived from the use of diesel fuel vehicles. Gasoline use (primarily from local motorbikes) accounts for roughly 16 percent of total citywide emissions. The industrial sector is the second largest contributor to GHG emissions in Da Nang. The city's wastewater treatment and water supply operations account for 6.8 percent of the city's emissions, and the solid waste system accounts for 6.7 percent, adding another 13.5 percent to the total. The residential sector is responsible for roughly 14.6 percent of total emissions, far outpacing the commercial sector's share of the emissions load (3.4 percent).

## Sector Review and Prioritization

Da Nang's interest in pursuing the Tool for Rapid Assessment of City Energy (TRACE) underscores its commitment to achieving optimal energy efficiency. The analysis was carried out across six city sectors: passenger transportation, city buildings, water and wastewater, public lighting, solid waste, and power. These were, in turn, assessed against the performance of a range of peer cities through a benchmarking process. This review provided a number of significant findings contributing to the definition of priority sectors.

Key findings of the Da Nang diagnostics in comparison with other cities in the TRACE database are the following:

- Low electricity use per capita but high level of energy use per unit of GDP
- Relatively low energy use in transportation due to a low level of automobile use and high usage of relatively fuel-efficient motorcycles
- Very low use of public transportation coupled with growing ownership and use of private motor vehicles, which is increasing energy intensity for the transportation sector

**Figure 8.4 Da Nang GHG Emissions by End Use and Fuel Source**

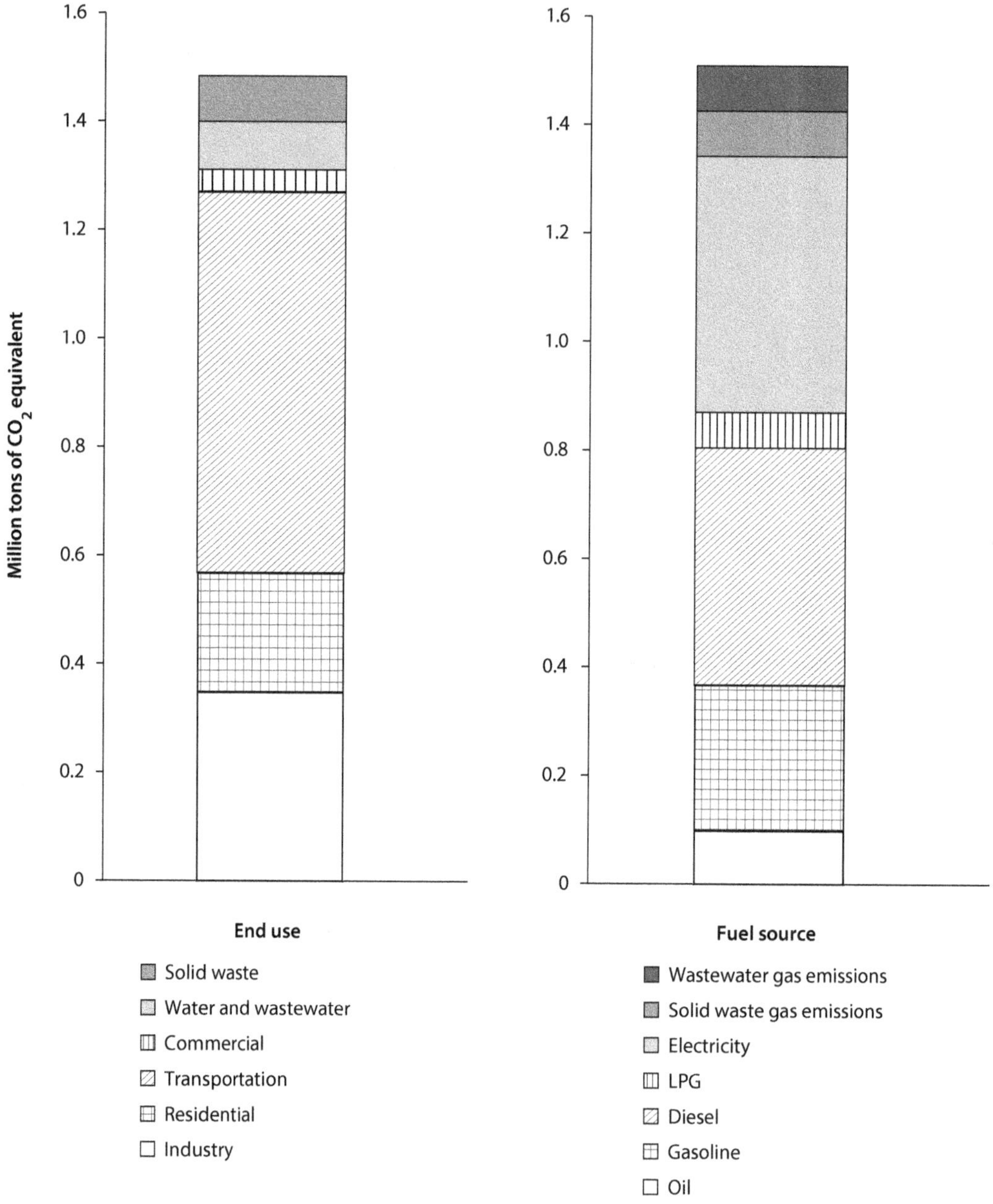

**End use**

- Solid waste
- Water and wastewater
- Commercial
- Transportation
- Residential
- Industry

**Fuel source**

- Wastewater gas emissions
- Solid waste gas emissions
- Electricity
- LPG
- Diesel
- Gasoline
- Oil

*Source:* Phase I pilot study.
*Note:* $CO_2$ = carbon dioxide; GHG = greenhouse gas; LPG = liquefied petroleum gas.

- Low per capita water consumption and relatively high water losses from the distribution system
- Midrange energy density of potable water production
- Low electricity consumption per light pole (although there is room for improvement in public lighting)
- Low but rising energy consumption in city buildings
- Very low level of recycling due to the absence of a formal recycling solid waste program
- Low levels of transmission and distribution losses

The TRACE analysis identifies priority areas in which significant energy savings are possible. Table 8.1 indicates the amount of energy spending in each of these sectors, the relative energy intensity (the percentage of energy that can be saved in each sector, based on the TRACE benchmarking), and the level of influence the city government has over these sectors. The savings potential is calculated by multiplying the three factors. The outcome of the TRACE analysis is a playbook of 58 energy efficiency recommendations applicable across all the analyzed sectors.[1] The recommendations are not meant to be exhaustive or normative. They simply outline a number of policies and investments that could help local authorities in Da Nang achieve higher energy efficiency standards.

Table 8.1 shows priorities with respect both to sectors over which the city authority has maximum influence and to citywide issues over which the authority has limited influence. The ranking suggests that Da Nang city government should prioritize street lighting, followed by city buildings, solid waste, and the water treatment system. In relation to the buildings sector, the results demonstrate a wide range of opportunity for all residential and commercial buildings in Da Nang although the TRACE analysis in the buildings sector focused on city buildings.

**Table 8.1  Da Nang Sector Prioritization Results**

| Priority ranking | Sector | 2010 energy spending (US$) | Relative energy intensity (%) | Level of city authority control[a] | Savings potential (US$)[b] |
|---|---|---|---|---|---|
| *City authority sector ranking* | | | | | |
| 1 | Street lighting | 1,200,000 | 78.2 | 1.00 | 939,141 |
| 2 | City buildings | 2,069,047 | 15.1 | 1.00 | 312,426 |
| 3 | Solid waste | 452,380 | 48.8 | 0.97 | 214,277 |
| 4 | Potable water | 564,349 | 26.4 | 0.96 | 143,163 |
| 5 | Wastewater | 95,000 | 11.1 | 0.96 | 10,133 |
| *Citywide sector ranking* | | | | | |
| 1 | Power | 54,285,714 | 33.8 | 0.38 | 6,973,725 |
| 2 | Private vehicles | 44,665,149 | 10.0 | 0.14 | 669,977 |
| 3 | Public transportation | 361,773 | 65.8 | 0.90 | 214,416 |

*Source:* Phase I pilot study.
a. 0 = no influence; 1 = maximum influence.
b. Based on TRACE (Tool for Rapid Assessment of City Energy) benchmarking; these figures are indicative of the quantum of savings that may be possible, but not necessarily practicable.

On a citywide basis, the power supply system is a clear area of focus, followed by both private and public transportation. Incorporating this information into a sector-by-sector analysis filters and narrows the recommendations to ensure that they are both viable and practical for Da Nang.

## Recommendations

The recommendations presented in this chapter focus on areas over which the city has direct influence and are, for the most part, derived from the TRACE. The recommendations were shared with the Da Nang City People's Committee and relevant agencies. It is expected that the recommendations will be further refined by the city and study team as further analysis and discussions are completed. Although the energy balance and GHG data discussed earlier are good indicators of the energy and emissions profile of Da Nang, additional analysis beyond the public sector is required to develop a sustainable urban energy and emissions plan that includes sectors outside the direct control of the city.

### Transportation

Although TRACE identified private vehicles and public transportation as the second and third priorities, respectively, on a citywide scale for immediate energy savings potential, the transportation sector has the highest potential for guiding future city growth in a sustainable manner. Public infrastructure investments are one of the main tools local authorities can use to encourage compact development (for example, development around transit hubs), encourage alternative modes of transportation (walking, biking, public transportation), and decrease local energy inputs (transportation is one of the most energy-intensive sectors). Da Nang has to be proactive in its thinking and must strategically use infrastructure to guide city growth from the current 0.9 million people to the estimated 1.65 million in 2020. In its planning, it should move away from a static model and focus on a dynamic model to accommodate the quickly changing characteristics of the city. To this end, integrated planning is especially important for transportation.

#### Public Transportation Development

Public transportation results in lower operating energy intensity and lower emissions per capita than private cars and has the potential to provide a more efficient transportation network. A reduction in the number of private vehicles in circulation can lower emissions and improve air quality. Bus rapid transit (BRT) has attracted a great deal of attention in cities around the world in recent years because of its ability to move large volumes of riders at a cost much lower than light and heavy rail systems. Da Nang would like to see a BRT system deployed in one or more parts of the city by 2016.

One of the Sustainable Urban Energy and Emissions Planning (SUEEP) team's most notable observations in the city is the need for an integrated land-use and transportation planning approach to ensure that plans for BRT, land use, street signals, parking policies, vehicle registration pricing, and sidewalk policies are coordinated and that there is an effective means for turning plan into practice. For example, Da Nang's central business district does not have a pedestrian-friendly environment—most sidewalks serve as motorbike parking lots or spots for food vendors to set up their operations. Similarly, encouraging the use of regular and electric bicycles alongside the development of a robust public transportation system will help to ease traffic flow in Da Nang, given increasing use of private cars. For such benefits to materialize, the Da Nang City Department of Transportation and City Vehicle Registration and Control Office would need to work together through city planning to encourage the use of public transportation and nonmotorized modes of transportation.

## Bicycle Use

Although it bucks the current local trend to abandon bicycles and take up motorized transportation, the lowest energy and carbon mobility path the city can pursue is to promote high rates of walking and cycling around the city. Such a strategy requires careful attention to the way in which new real estate development or neighborhood redevelopment takes place around the city to ensure a mixture of land uses that promotes relatively short travel distances between where people work, live, shop, attend school, relax, and so on. Da Nang's current land-use master plan emphasizes a mixture of land uses for development and redevelopment, thus facilitating low-energy, low-carbon mobility in those areas. Close coordination will be needed between the departments of Natural Resources and Environment, Construction, and Transportation to ensure the development of high-density, mixed-use areas, with space appropriately allocated for pedestrians and bicycles on local roadways and sidewalks. One alternative is dedicated pedestrian malls or other areas in which motorized vehicles are banned or restricted. Sufficient space for bicycle parking must also be provided, in locations that do not impede pedestrian traffic. Consideration must also be given to effectively linking bike use with public transportation, by creating large bike parking areas near bus stops.

## Electric Bicycles

Motorbikes already have achieved strong market penetration in Da Nang (and Vietnam as a whole), so it may be difficult to shift large numbers of users "backward" to nonmotorized bicycles. This is particularly true for motorbike users traveling long distances between home, work, school or university, and shopping. Therefore, the city may wish to consider policies and programs aimed at alternative-powered motorbikes such as electrified bicycles. This technology, which is extremely popular in China, could result in a shift away from gasoline to different forms of electric power generation. Large solar charging stations could be set up in different locations around the city to provide low- or no-cost battery charging. Battery-swap businesses could also be established around the city to allow riders to replace the batteries on their bikes quickly and continue on with their trips, rather than waiting for the battery to fully recharge. Electric bike owners could also recharge their batteries at home.

Energizing Green Cities in Southeast Asia • http://dx.doi.org/10.1596/978-0-8213-9837-1

### Land-Use Planning

People will generally walk a limited distance to a bus stop; anything farther is considered inconvenient, which will lead people to turn to alternative forms of transportation such as cars or motorbikes. To ensure high levels of bus ridership, it is important to increase the density of land development within this acceptable walk zone, thus increasing the likelihood of large numbers of bus users. To the extent possible, routes should be designated and land use in appropriate areas along the routes "upzoned," as soon as possible to allow for denser development. Taking this step well in advance of the deployment of the BRT system would allow a landowner to assess whether to increase the size of the building or sell it to other developers interested in some type of speculative development. Advance planning and zoning can help to ensure that by the time a BRT system is launched, residential and business density along the route will have increased to levels likely to support high ridership.

### Parking Policies

BRT is predicated on riders walking to and from the system to another destination. To support a BRT system, Da Nang should begin to reclaim the sidewalks around the city so they become more usable by pedestrians. Establishing designated motorbike parking areas and strictly enforcing their use will help accomplish this objective. Da Nang could also take a cue from other cities, where food vendors must apply for permission to occupy public sidewalks to ensure they are not located in areas that would obstruct pedestrians.

### Car-Minimization Strategies

The city should continue its program of car-minimization strategies. Limits on street parking, high vehicle registration fees, promotion of car-sharing programs, congestion pricing, and other strategies that discourage private vehicle use in certain areas or at certain times of the day or week can help make private vehicle use a less attractive option in Da Nang, potentially leading to higher rates of public transportation and BRT use.

### Bus System Design

Many elements of bus system design must be considered during the implementation planning process to encourage ridership. These include issues associated with ticket pricing and the availability of free transfers from one bus route to another. The city must also work with new BRT system operators to ensure high levels of rider comfort, longer operating hours, and an increase in the frequency of bus service along designated routes. All of these issues have been cited by the public as reasons for their limited use or overall dissatisfaction with the city's current bus system.

### Solid Waste

Residential, commercial, and industrial solid waste collection and management services in Da Nang are operated by Hanoi Urban Environment Company (URENCO), a state-owned enterprise. Waste collection is a massive endeavor for a city of this scale. The landfill in Da Nang is not currently set up to capture methane gas (thus missing an opportunity to make use of this resource), and is expected to reach capacity by 2025–30. The capacity problem is further exacerbated by low fees for residential waste collection, which provide no incentive for waste reduction. The following areas are identified for action by the city government or URENCO officials.

## Waste Combustion for Power Generation

Approximately 64 percent of Da Nang's waste is kitchen waste, which means it has a high moisture content that is unsuitable for most mass-burn type facilities, even those with energy recovery in mind. Some 28 percent of the waste stream is composed of highly combustible materials, while the remaining 8 percent is composed of inert, noncombustible materials. Both of the latter categories include materials that should be prioritized for recycling. High-moisture-content material can result in incomplete combustion, reducing the burn temperature and resulting power output. To the extent that nonrecyclable waste materials can be presorted into wet and dry streams, a more suitable mix of dry combustible materials may result.

## Landfill Gas Capture

Da Nang has proposed a Clean Development Mechanism (CDM) project aimed at generating electricity from methane captured at the landfill at Khanh Son. A successful landfill gas capture project would remove 140,000 tons of $CO_2e$ per year from Da Nang's GHG emissions profile and, if used to generate electricity, contribute to satisfying the city's electricity requirements. The initial analysis suggests the potential for 1 megawatt of power generation at the landfill given current methane gas levels. The Da Nang People's Committee could request information on anticipated gas availability levels over time from this facility. Most landfills experience peak gas availability 5–10 years after the facility is closed and capped; gas levels then decline until the quantity of gas that can be recovered is too low to support power generation or the quality of gas deteriorates and begins to degrade the power generation equipment.

## Truck Procurement Guidelines

URENCO hopes to upgrade its waste collection fleet in the next several years because most of its vehicles are 5–10 years old. No information was available on the current fuel efficiency of the fleet, but before any new vehicle purchases are made, Da Nang should work with URENCO to analyze waste collection vehicles available in the marketplace and then establish minimum fuel economy requirements for any vehicles purchased. Da Nang may wish to require that any purchases be subject to a life-cycle analysis that compares the upfront purchase cost, fuel purchases over the life of the vehicle, and maintenance expenditures to determine which vehicles are most appropriate or meet any desired cost-effectiveness threshold conditions.

## Progressive Tariff Structure

Da Nang's current waste collection tariff structure is volume based for business customers, but residential customers pay a flat fee. In other sectors, such as water, fees are structured to discourage excess consumption; in the case of waste, residential fees should be similarly structured to discourage excess waste generation. The Da Nang People's Committee should thus explore alternative rate structures when rates are next up for review in 2013. Should the city want to promote source separation of different materials to facilitate recycling, it could set rates to provide incentives for this practice. For example, the rate for clean organic waste could be very low (or even free), while rates for nonorganic, nonrecyclable materials could be much higher, encouraging households to reduce waste generation. Of course, in such systems, the city must take steps to ensure that households do not illegally dump waste materials to avoid payment.

## Water

In Da Nang, water losses from the system were approximately 25 percent in 2011, a decline from 2007 when losses were 40 percent. Da Nang Water Supply Company (DAWACo) indicated that old pipes are a primary cause of system losses and noted they are working to replace these pipes over time. DAWACo has also installed (and would like to install more) variable speed drive pumps, which adjust the pressure in relation to demand on the system. Pumps that consistently maintain high pressure when demand is low can result in leakage across the system.

### Active Leak Detection and Pressure Management System

Despite continuous improvements in the system, DAWACo has reported 25 percent physical water loss in the system, and the water that does eventually get to consumers requires energy-intensive pumping. An active leak detection and pressure management program could address both of these issues at once. Because the Da Nang water network falls under the authority of the Metro Da Nang Water District, technical interventions can only be leveraged by the city government's use of the planning system, for example, making land available for water reservoirs and distribution infrastructure to improve the network's pressure and energy performance.

DAWACo also mentioned that it would like to install Supervisory Control and Data Acquisition technology throughout its system to improve its real-time monitoring capability over the entire distribution network, but this is costly and would likely not be pursued without outside assistance from an international development aid organization.

Da Nang was the first city in Vietnam to have a wastewater management strategy. Currently, fewer than 20 percent of all residences are connected to the DAWACo system. The Da Nang Department of Construction established the policy that allows some buildings to have their own septic systems and requires others to connect to the citywide wastewater treatment system. Connected households pay for service according to a progressive rate schedule, whereby rates increase as usage of wastewater treatment services increases. This system provides an incentive to households to reduce wastewater levels, but may discourage connection to the system. Therefore, a progressive water tariff may be more effective at reducing wastewater although the rates charged are still quite modest.

From a more technical perspective, the current design of the wastewater treatment network does not allow for the capture and combustion of methane gas generated during the anaerobic phase of processing. If new facilities are constructed or any existing facilities are expanded, Da Nang may wish to encourage the inclusion of some type of electric power production technology that combusts the methane gas generated onsite.

In addition to developing a sludge beneficial reuse program, several actions could positively influence energy consumption levels across the DAWACo system.

### Increasing Connections

Increasing the number of buildings connected to the DAWACo wastewater treatment system would be important. Amending local policies and building codes to require any new development projects to connect to the system would reduce the amount of energy used per unit of wastewater treated. Increasing the volume of material in the system would provide important cobenefits by improving the biological oxygen demand concentrations at the treatment facilities, helping them to operate more efficiently and subsequently improving the quality of water released to local waterways.

### Demand-Side Measures

It is further recommended that Da Nang require water-harvesting or low-flow devices to be used in new construction projects. Cities are increasingly incorporating demand-side measures into their building codes, aiming to reduce the amount of material entering the wastewater treatment system from new construction projects. It does not appear as if Da Nang has any requirements for the use of low-flow toilets and showerheads or other fixtures that reduce water flow, and thus, wastewater discharge levels. The cost impact of such measures is quite small, but they provide long-term cost savings for households on both their water and wastewater bills. Systemwide benefits—particularly reducing water demand in the supply network—would also occur.

### Power

Da Nang Power has been charged with the implementation of the central government's Directive 171 requiring a 10 percent savings in overall electricity use, with a 1 percent savings from major industrial users. Da Nang Power has reportedly implemented several measures to meet the government targets, including (a) demand-reduction programs for factories, (b) end-use monitoring programs for the largest end users, (c) the use of compact fluorescent lighting, and (d) imposition of restrictions on when air conditioning units and building lighting systems can be used. In addition, a tariff structure that encourages energy use during off-peak hours has been established to reduce peak demand, and efforts have been put into more effective metering of the manufacturing sector. Although Da Nang Power already enjoys very low levels (4.2 percent) of transmission and distribution losses, the company plans to ground distribution lines, which will reduce nontechnical losses to an absolute minimum, in addition to improving safety and reliability.

Distribution and supply-side management are outside the remit of this study, so the SUEEP team's recommendations focus on the diversification of the city's power supply, particularly through local deployment of renewables. Demand management is primarily tackled through the buildings sector, and therefore is not explored in detail in this section. Da Nang Power and city leadership both highlighted the need for a master plan to enhance the reliability of electricity supply as well as a renewables master plan; this acknowledgment of the need for a renewables master plan is a very important development. In many cases, local officials do not have the clout to enact prorenewables policies similar to those put into place at the national level, such as feed-in tariffs, but Da Nang government officials have a number of opportunities for leveraging their authority, including (a) propagating buildings codes that include structural requirements to enable renewables retrofits, such as requirements that buildings be constructed to physically support the weight of renewables installations; (b) imposing green building codes or standards that encourage or require on-site renewables deployment; (c) introducing a green building rating system that rewards developers for incorporating renewables into their developments; and (d) embarking on pilot projects to demonstrate that renewables can be successful in practice.

The new Da Nang Wholesale Fish Market is an example of how the imposition of building codes can encourage the use of renewables. The fish market is a potential host of an on-site photovoltaic system to supply energy for ice making, water treatment, and lighting. The building is attractive for this purpose because of its large size (approximately 7,000 square meters [$m^2$]) and unobstructed roof design. If the highest efficiency monocrystalline or polycrystalline photovoltaic cells are installed on the fish market's roof, the SUEEP team's preliminary calculations indicate that all of the facility's energy needs could be met by the power generated over the greatest part of the year, with the exception of the monsoon months of September to November. It is unclear, however, whether the roof could support the weight of such a large installation, and thus it is more likely that lighter (but less efficient) thin film photovoltaics would be used. In this case, a significant portion, but not all, of the market's power requirements could be satisfied.

### Public Lighting

Da Nang has relatively low electricity consumption per light pole in comparison with other cities in the TRACE benchmark database, probably because the city uses low-energy fixtures and various dimming regimes have been implemented throughout the city. There is still room for improvement in the public lighting sector, including the mass roll out of light-emitting diodes (LEDs) and development of procurement codes with more stringent energy efficiency requirements. It is also important to consider the speed at which Da Nang is growing and its citizens' quality of life is improving. Lighting preferences could change in the future, putting more demand on the system to provide higher levels of lighting in more areas. Da Nang can prepare for this shift by continuing its excellent efficiency programs and pushing them even further.

Da Nang Bridge and Road Management Company (under the Department of Transportation) has already begun extensive measures to reduce energy consumption in public lighting, including installing LED traffic signals, replacing many of the existing mercury street lights with high pressure sodium luminaires, and replacing decorative halides with compact fluorescents. The agency is also piloting two lighting regimes to optimize street lighting and save energy. Forty of Da Nang's main streets operate under Regime 1, in which every third light is kept off from 6:30 pm to 11:00 pm, and every third light is kept on (with the other two off) from 11:00 pm to 5:00 am. Regime 2 is used for all other streets and all lights are on during night hours. These measures have reduced the electricity power requirement for public lighting by 28 percent across the city, according to city officials.

The Da Nang Public Lighting Operations and Management Company's goal is to reduce electricity power consumption by 40–50 percent for main streets. The city must continue its existing audit and retrofit program for public lighting. The city might like to reconsider the use of metal halides because these bulbs have high maintenance requirements, shorter life spans, and higher electricity demand than high pressure sodium bulbs or LEDs. It is noted that Da Nang is conducting a pilot test for the installation of LEDs on Tran Hung Dao Street and in Son Tra district, which will enable city officials to test their impact, technology, and aesthetics before the final decision is made.

## City Buildings

Energy consumption in Da Nang's city buildings is relatively low, most likely due to the limited funds available for energy expenditures. However, electricity use in city buildings is on the rise. A number of efforts in the city buildings sector have been undertaken to improve energy performance, such as lighting replacement programs and the implementation of air conditioning schedules. However, replacing old air conditioning units and other inefficient appliances, and improving the design and construction of building envelopes, provide additional opportunities for energy performance improvement. It is recommended that existing efforts be continued and a retrofit program be implemented through a city buildings energy efficiency task force.

### City Building Audit and Retrofit Program

To set a good example for other buildings in the city, local authorities should develop an audit and retrofit program for all the buildings the city owns. Such programs can help reduce energy bills and the carbon footprint of the city, and they offer a good knowledge basis for upgrading and updating city building codes.

The Da Nang city government could also benefit from the work of the Vietnam Green Building Council, which was launched in 2008 with the goal of promoting green building around the country. Work has begun on LOTUS, a Vietnam-specific green building rating system that drew inspiration from other building certification programs, including Leadership in Energy and Environmental Design in the United States, Building Research Establishment Environmental Assessment Method in the United Kingdom, and Green Star in Australia. Developed with voluntary contributions from experts in and outside Vietnam, a first set of guidelines for nonresidential facilities was released in late 2010. Six buildings have applied for certification thus far. There are nine categories in which points can be awarded for a project: energy, water, materials use, ecology, waste and pollution, health and well-being, adaptation and mitigation, community, and management.

Given the population and income growth anticipated in Da Nang over the next several decades, it is important that the Da Nang People's Committee begin work to address building-related energy consumption, both in city government buildings and in other buildings around the city. This is one of the few areas for which the city does not yet appear to have a comprehensive strategy, or to have conducted significant research on different policy or technology options. Taking action now can help the city lock in a lower energy use trajectory than will otherwise occur. There are several opportunities the city may wish to consider, discussed below.

### Lead by Example

The Da Nang People's Committee is reportedly planning to build a tall office tower that would bring together into a single building the local government departments currently dispersed across the city. By constructing a model green building that achieves LOTUS or other preeminent building performance standards, the People's Committee would send a powerful message to others of the importance of this type of design and its viability in Da Nang's economic climate and climatic zone. Work on such a high-profile project could provide training opportunities for businesses and individuals in the city, helping to jumpstart the creation of a local green building marketplace. When the Vietnam Green Building Council completes the LOTUS guidelines, the People's Committee should consider pursuing LOTUS Existing Buildings Operations and Maintenance certification for any government buildings that remain after completion of the new office tower.

### Voluntary or Mandatory Green Building Guidelines

The Department of Construction encourages investors to install energy efficient equipment and appliances when it issues construction licenses. A more aggressive approach could be used in advancing energy efficiency in the buildings sector. The fact that the LOTUS system now exists means the Department of Construction (DOC), the local agency with primary responsibility for enforcing compliance with all building codes, has a new tool in its green building education arsenal. The guidelines should prove very helpful in educating project developers and building operators about the full range of steps they can take to create a greener building. However, DOC's capacity to proffer its recommendations to enhance energy efficiency of building projects is limited, which has resulted in an uneven distribution of green features across Da Nang. The city may need to adopt a more aggressive tactic, requiring compliance with LOTUS (or other guidelines) as a condition of construction permit approval. The city could even consider imposing local building codes that are more stringent than national codes.

Da Nang need not impose these conditions on all buildings. In many cities, similar requirements are applied only to buildings exceeding a certain size. The bigger buildings play an iconic role and also tend to use the greatest amounts of energy. Targeting them can have an important effect in reducing local energy demand, improving local environmental quality, and creating new markets for green building products and services.

The city might also consider imposing such requirements sectorally, such as on the hospitality sector. Because Da Nang hopes to expand the local tourism industry dramatically and because hotels and resorts tend to use relatively more energy to provide numerous amenities and to ensure the comfort of their international clientele, focusing on the greening of that sector could have sizable and long-lasting benefits for the city. By partnering with the hospitality sector on these issues, the city could even turn this initiative into a marketing virtue, eventually allowing the city to promote itself as having the greenest hotels and resorts in Southeast Asia.

Sectoral strategies may require some tailoring of the LOTUS system or other best practice guidance to address the city's unique climate and waterfront location. This tailoring could be done directly by local government experts in Da Nang, in collaboration with the Vietnam Green Building Council. Alternatively, it could become a civil society initiative involving local university and business experts from around Da Nang, along with other technology, design, and engineering experts from Vietnam or the Southeast Asia region.

### Operational Efficiencies
The city building energy efficiency task force should be responsible for implementing and overseeing the appropriate energy efficiency initiatives in city buildings. The Da Nang Community Relations Department and DOC are well placed to take on responsibility for energy efficiency in building design and refurbishment. The buildings sector recommendations sheets in the TRACE provide further detail on implementation of individual measures.

### Computer Power Save Program
Although several good measures have been initiated in local government buildings, old computer workstations abound. These old computers use high levels of energy compared with newer models. Therefore, a computer power save program is deemed to be an effective recommendation for reducing energy consumption in city buildings.

Da Nang city government has already undertaken an impressive amount of work on energy matters, serving as a foundation for future efforts. This work has likely slowed the rate of energy demand growth; however, the anticipated population changes in the next 10–20 years require that much more action be taken. Utility stakeholders in the water, waste, and power sectors have done an excellent job at identifying opportunities for energy efficiency improvements, in addition to exploring ways to potentially capture energy from different renewable sources. These efforts clearly show the considerable talent the city can bring to bear on future energy policy and planning initiatives.

## Conclusion

In the future, energy governance should be prioritized because it will help strengthen the city's internal energy management practices as well as engage other key stakeholders who have not played a significant role in the city's energy planning efforts to date. Better governance practices include not just enhanced oversight and data tracking, but also improved procurement practices and a willingness to "lead by example" by showcasing best practice strategies for the benefit of local businesses and households. Therefore, it is important that Da Nang establish a citywide energy task force to improve coordination and establish a streamlined approach to energy. In keeping with Da Nang's current policy structure, this task force should operate under the auspices of the Department of Industry and Trade. The citywide task force should not limit its focus to government operations, nor should its membership be restricted to government officials. Business and real estate professionals, along with other members of industry, can provide valuable input and offer a fresh perspective on the energy challenges and opportunities facing Da Nang.

## Note

1. For further details on TRACE, see chapter 3.

# Sustainable Urban Energy and Emissions Planning Guidebook: A Guide for Cities in East Asia and Pacific

# CHAPTER 9

# Introduction to the Guidebook

## Introduction

Cities currently account for about two-thirds of the world's annual energy consumption and about 70 percent of the world's greenhouse gas (GHG) emissions. In the coming decades, urbanization and income growth in developing countries are expected to push cities' shares even higher. Urban growth will be particularly notable in Asia, where the urban population is expected to increase by 50 percent between 2000 and 2030, and the urban share of East Asia's total population is expected to rise from 46 percent in 2011 to 60 percent by 2030.

To this end, the Australian Agency for International Development (AusAID)–supported Sustainable Urban Energy and Emissions Planning (SUEEP) program in the East Asia and Pacific (EAP) region seeks to help city governments in the region to formulate long-term sustainable urban energy and low-carbon development strategies that can be integrated into existing development plans.

## Why Engage in the SUEEP Process?

Today's rapid population growth coupled with the increase in per capita energy consumption and urbanization taking place in the EAP region (map 9.1) mean that now is the time for government leaders to take action and future proof their cities against an unsustainable energy future. Current baseline projections for the region show a startling spike in energy demand on the horizon. However, the EAP region's rapid pace of construction, quickly shifting transportation sector, and growing industries give cities the unique opportunity to rein in energy-intensive development and nurture progressive energy and emissions policies that can help prevent the overly energy-dependent development suffered by many cities around the world.

**Map 9.1  East Asia and Pacific**

*Source:* World Bank.

## Overarching City Aspirations

Energy use and GHG emissions are inextricably linked with how well a city works overall. The SUEEP process is designed so that energy and emissions planning are aligned with overarching city goals, including the following:

- Improved quality of life
- Economic growth
- Environmental protection

A strong SUEEP process links these aspirations with actionable initiatives to improve energy and emissions performance. It also enables benefits such as local air quality improvements, financial savings, new jobs, local economic development, and new partnerships across city agencies and the private sector.

## Green Growth

Similarly, the SUEEP process is closely related to the fundamentals of green growth, that is, economic growth centered on sustainable use of natural resources.

Green growth is about making growth processes resource efficient, cleaner, and more resilient without necessarily slowing them.

Many cities around the world are now referencing the green growth model as they develop their economic plans. Fully addressing green growth strategies is beyond the scope of this Guidebook, but many green growth principles are demonstrated in the SUEEP process and can be considered comprehensively by city authorities.

## About This Guidebook

### What Is the SUEEP Process?

The purpose of the SUEEP process is to provide a comprehensive approach to planning to maximize energy efficiency across city sectors. The intent is to help cities to develop their own initiatives using different mechanisms and to help them to define a governance system for implementation, monitoring, and reporting. These are important outcomes because they improve energy governance in the city and create a common platform for collaboration between the city and donors, civil society, and the private sector. The SUEEP process also provides a framework that helps city governments prepare a series of investments in energy efficient infrastructure as well as mobilize green financing support. The strategic framework for sustainable urban energy development and the pipeline of bankable projects are to be identified in the city's energy and emissions plan.

Core components of an SUEEP process include the compilation of data on the city's energy and emissions baseline, involvement of stakeholders throughout the process, implementation of prioritized projects, and monitoring and reporting of outcomes. The data are used to set energy and emissions targets in line with the city's overall vision and goals.

### What Is the Purpose of This Guidebook?

This Guidebook is meant to provide a broad framework and an indicative step-by-step guide to help a city to develop its own energy and emissions plan. These guidelines are based on experiences in three pilot cities (Da Nang, Vietnam; Surabaya, Indonesia; and Cebu City, the Philippines) combined with best practices in sustainability planning in other cities. This document will undoubtedly be revised in the future after subsequent phases of the SUEEP process are implemented and the lessons learned are reviewed.

It is unlikely that a city will be able to use just these guidelines to undertake the complex, long-term process of energy planning. Many references are provided and components of the process that may require special support from local or international organizations are identified.

### Why Is This Guidebook Focused on the EAP Region?

Growing energy demand in EAP is expected to double the region's total carbon dioxide ($CO_2$) emissions by 2030. EAP is also home to the world's most rapidly expanding urban population. The region's growing middle class is accelerating the pace of construction and urban expansion. Energy use efficiency is influenced directly and permanently by urban form and density and by investment choices made today about urban infrastructure (transportation, water, energy) and capital, all of which will have a major impact on both energy demand and associated GHG emissions. For example, the recent explosive growth in personal motor vehicles and low-density housing, if untempered, foreshadows future development of high-energy-intensity urban scenarios as the energy demands of the middle- and upper-income residents increasingly mimic those of their counterparts in the developed world.

This Guidebook recognizes the unique conditions of EAP cities and provides information tailored to the region to enable cities to establish the programs and policies that can secure their energy and emissions futures. This document covers the entire planning process, starting with detailed guidance on articulating a vision, establishing energy governance, and engaging stakeholders—three components of sustainable urban energy typically not firmly established in EAP cities. In addition, special emphasis is given to the Clean Development Mechanism and alternative financing methodologies relevant to Asia because these incentives for green growth are strong opportunities for the region. Reference guides from Europe and the United States also provide technical guidance for specific challenges of the SUEEP process, and case studies from the EAP region and from around the world provide examples of how other cities have tackled the challenges of sustainable urban energy planning.

### Audience for This Guidebook

This document is targeted at mayors and city planning agencies in the EAP region, but it is also relevant for people working in government who are involved with the following:

- Utility services delivery
- Transportation
- Economic development
- Housing
- Poverty alleviation
- Environmental management
- Government facilities management
- Government procurement
- Financial planning
- Risk assessment
- Public health

### What Is in This Guidebook?

The Guidebook outlines the SUEEP process and describes a methodology cities can use to develop their own plans. The methodology is divided into six main stages, which are further divided into steps.

---

**Box 9.1  Supplementary Information Used in This Guidebook**

**Case Study:** The case studies presented throughout this Guidebook illustrate how cities and organizations around the world have implemented either part or all of the SUEEP process.

**Example:** Illustrative examples of key concepts or activities are provided throughout the Guidebook. Examples cover a wide range of information, but generally include information from cities that have already begun developing a component of an energy and emissions plan.

**Resources:** The Guidebook recognizes that some aspects of planning require more detailed implementation guidance. Resources point to various books, articles, references, and websites that provide further detail on a specific aspect of the SUEEP process.

**Technical Assistance Opportunity:** If technical assistance is available for complicated or specialized aspects of the SUEEP process, the Guidebook points the reader to resources for engaging assistance from external organizations.

**Tip:** The tips are short suggestions to help cities successfully implement the SUEEP process. Tips include lists of considerations, common pitfalls, key success factors, and the like.

**Toolkit Reference:** A Microsoft Excel–based toolkit supplements Stage II: Urban Energy and Emissions Diagnostics and Stage IV: Planning. The reader will be pointed to the specific tool referenced in the Guidebook. The applicable spreadsheets and templates are contained in the SUEEP Toolkit available at http://www.worldbank.org/eap/energizinggreencities.

---

## How to Use This Guidebook

Each stage and step describes the SUEEP process for the development of an energy and emissions plan. A number of examples, tips, case studies, and resources help focus the reader on specific aspects of the process. Explanations of each of these types of supplemental information are provided in box 9.1.

The processes and steps in this Guidebook are interrelated. Parts of the process sometimes rely closely on future or previous steps. See box 9.1 on supplementary information that points the reader to the related steps to help track associated information in the Guidebook.

## Role of the City and SUEEP Scope

The city plays a wide range of roles with respect to energy planning, including the following:

- Energy consumer
- Energy producer and supplier
- Regulator
- Motivator

The city's role as regulator gives it the most clout to manage energy and GHG on a citywide basis, but all four of the city's roles are important in the SUEEP process. For example, the city as energy consumer leads to an approach that is focused on internal operations and may include measures such as establishing an energy and GHG "champion," devising a sustainable procurement policy, retro-fitting buildings with energy-saving equipment, and educating employees about energy-saving practices. But for the city's role as energy producer and supplier, SUEEP activities might focus on developing efficient generation and distribution technologies, rolling out progressive tariff structures, and establishing effective city utility governance structures for energy efficiency and GHG management.

In addition to the city government's multiple internal roles, city leaders also need to work closely with the national government—many energy policies, plans, and programs implemented at the national level could affect local policies and investments.

## SUEEP Process Overview

The SUEEP process is divided into six main stages (see figure 9.1):

- Commitment
- Urban Energy and Emissions Diagnostics
- Goal Setting
- Planning
- Implementation
- Monitoring and Reporting

**Figure 9.1  The Stages and Steps of the SUEEP Process**

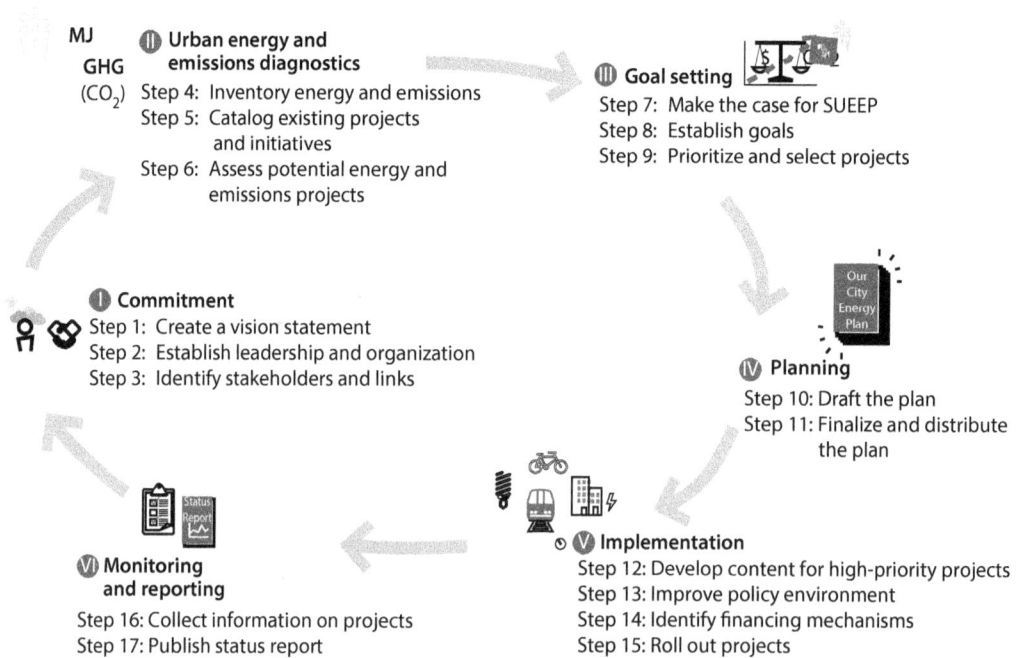

MJ
GHG
($CO_2$)

**II Urban energy and emissions diagnostics**
Step 4: Inventory energy and emissions
Step 5: Catalog existing projects and initiatives
Step 6: Assess potential energy and emissions projects

**III Goal setting**
Step 7: Make the case for SUEEP
Step 8: Establish goals
Step 9: Prioritize and select projects

**I Commitment**
Step 1: Create a vision statement
Step 2: Establish leadership and organization
Step 3: Identify stakeholders and links

**IV Planning**
Step 10: Draft the plan
Step 11: Finalize and distribute the plan

**VI Monitoring and reporting**
Step 16: Collect information on projects
Step 17: Publish status report

**V Implementation**
Step 12: Develop content for high-priority projects
Step 13: Improve policy environment
Step 14: Identify financing mechanisms
Step 15: Roll out projects

*Source:* Phase I pilot study.
*Note:* $CO_2$ = carbon dioxide; GHG = greenhouse gas; MJ = megajoule; SUEEP = Sustainable Urban Energy and Emissions Planning.

The steps within each stage, and the stages themselves, overlap on occasion—planning is not a neatly linear process. The SUEEP process is meant to give cities a framework for their energy- and emissions-planning activities, and is not designed to be overly rigid or formulaic. A city should work through the process in any way that suits its unique conditions.

The process begins with Commitment, the first stage in securing political and stakeholder support for the energy and emissions plan. A strong commitment from local leaders is essential to the plan's long-term success and lays the groundwork for future action.

During the second stage, Urban Energy and Emissions Diagnostics, a city collects the basic data needed to understand its energy and emissions baseline and identify areas of activity that are important and have the potential for improvement.

The third stage, Goal Setting, involves combining the city's overarching political priorities with the findings of the second stage to develop energy and emissions goals relevant to the city. Establishing a convincing story about the importance of energy and emissions planning and the way the city will benefit is crucial to the success of the process.

The fourth stage, Planning, brings together all the knowledge and thinking developed thus far into a documented plan that clearly expresses the city's strategic focus on energy, the initiatives that will help the city achieve its goals, and how progress will be monitored.

Implementation is the longest-running stage of the process, and will overlap with future iterations because some initiatives from this stage will span lengthy periods. The success of this stage directly depends on robust planning in the previous stage so that inadequate governance structures, a lack of financing, or a hostile policy environment do not derail implementation.

The Monitoring and Reporting stage takes stock of SUEEP progress and identifies components of the plan that need readjusting. This stage is crucial for establishing accountability and for refining the plan to continually improve its approach to energy efficiency and emissions reduction.

The Monitoring and Reporting stage provides crucial inputs into each successive SUEEP process—which ideally would occur every two to three years. After the first iteration of the process, the city will be able to reaffirm or revise its commitment, recast its vision and goals, if necessary, and begin the diagnostics to lay the foundation for its next energy and emissions plan.

## Multiple Levels of Engagement

The SUEEP process provides a comprehensive approach to integrating energy efficiency measures into a city's character, but cities have different capacities, resources, and priorities. To accommodate the individuality of cities, three levels of engagement are possible:

- A high-level and quick assessment of a city's energy efficiency measures (The Tool for Rapid Assessment of City Energy [TRACE], developed by the World Bank's Energy Sector Management Assistance Program, offers city governments quick diagnoses of energy efficiency performance across their systems and sectors. TRACE prioritizes sectors and presents a range of potential solutions. The tool includes embedded implementation guidance and illustrative case studies [for more details on TRACE, see the technical assistance opportunities in boxes 11.11 and 12.13]).

- Deeper sectoral engagement in selected areas to help finance energy projects and to bring technical expertise to projects (for example, public-private partnerships)
- Implementation of all stages of the SUEEP process, the success of which will depend on a city government's interest, commitment, and ownership.

**CHAPTER 10**

# Stage I: Commitment

*Commitment sets the stage for successful energy and emissions management by establishing high-level political buy-in to propel the rest of the process. Establishing political and stakeholder commitment to Sustainable Urban Energy and Emissions Planning (SUEEP) by developing the city's vision for energy and emissions is important, as is setting up governance structures for development and implementation of the plan. The key to success during the Commitment stage is communication.*

## Step 1: Create a Vision Statement

Articulating a clear and convincing vision statement for sustainable energy and emissions planning sets the political stage for securing buy-in from the wide array of stakeholders who play a part in the success of the SUEEP process.

### Developing Your Vision
### *What Is a Vision Statement?*

A vision is a concise statement that provides a picture of an ideal future condition the city may one day realize. A city's vision will generally include a vision statement (see examples in box 10.1) and a longer, more detailed explanation of why the vision is important and how it relates to the city. New York's *PlaNYC 2030* contains a good example of an SUEEP vision statement in its introduction (http://www.nyc.gov/html/planyc2030/html/home/home.shtml).

Developing a vision is a political process and should therefore link energy and emissions goals to the city's overall political priorities.

### *Why Is an Energy and Emissions Vision Important?*

The vision provides the principles underpinning development of the energy and emissions plan. An inspiring vision statement has the power to engage stakeholders by clearly communicating the purpose of the SUEEP process. A vision statement also provides a clear, ultimate goal that aligns the people, departments, organizations, academic institutions, and utilities that are

---

**Box 10.1  Example: Examples of Vision Statements**

"A climate resilient global city that is well positioned for green growth."

- *Singapore's National Climate Change Strategy 2012, as quoted in* Climate Change and Singapore: Challenges. Opportunities. Partnerships *(Singapore 2012)*

"A greener, greater New York."

- *New York Mayor, Michael R. Bloomberg, in* PlaNYC 2030 *(New York, New York 2007)*

"A clean, green, compact and connected city with an innovative smart economy and with sustainable neighbourhoods and communities."

- *Dublin City Development Plan, 2011–2017, as quoted in the "Dublin City Sustainable Energy Action Plan 2010–2020" (Dublin City Council and Codema 2010)*

---

---

**Box 10.2  Case Study: Da Nang, Vietnam—The Environmental City**

As part of its recent development efforts, the city of Da Nang established a vision for future development that focuses on remediating and celebrating the natural environment. The plan sets forth specific goals that link to the global sustainability agenda and the city's economic, social, and environmental aspirations.

The three overarching goals leading the plan include

- Establishing Da Nang as an environmental city, emphasizing land, water, and air quality while providing a safe and healthy environment for people, investors, and domestic and foreign tourists;
- Preventing environmental pollution and degradation while encouraging environmental rehabilitation; and
- Facilitating awareness of environmental issues among Da Nang's residents, international and local organizations, and individuals working in Da Nang.

The city developed its vision by collaborating with the German Organization for Technical Cooperation (GTZ) as part of a €1.5 million project implemented between mid-2010 and the end of 2012.

For more information on the city and administration of Da Nang, see http://www.danang .gov.vn. The environmental planning document is not available online, but is published in hard copy by the Da Nang Office of Natural Resources and Environment.

---

contributing to a city's energy and emissions plan. See case studies in boxes 10.2 and 10.3. Sustainable energy and emissions programs yield benefits beyond energy management and emissions mitigation, and these ancillary benefits can enhance the SUEEP vision. Espousing these benefits will help to gain stakeholder support for the program.

---

**Box 10.3 Case Study: North Vancouver, Canada—100 Year Sustainability Vision**

The City of North Vancouver and the University of British Columbia Design Centre for Sustainability worked together to prepare the city's 100 Year Sustainability Vision: "To be a vibrant, diverse, and highly livable community that provides for the social and economic needs of our community within a carbon neutral environment by the City's 200th birthday in 2107" (Smith 2009).

Operating under the themes of livability, sustainability, and resilience, this 100-year plan looks at likely scenarios, challenges, and opportunities in the coming decades, allowing the city to develop more forward-thinking policy planning and to be a better, stronger advocate for regional, provincial, and federal sustainability legislation. This long-range vision aims to guide the city toward carbon-neutral status by 2107, the city's 200th anniversary.

*Source:* http://www.cnv.org/Your-Government/Sustainability-in-the-City/City-Initiatives/100-Year-Sustainability-Vision.

---

*Economic Development.* A stable and reliable energy supply is essential for attracting businesses and growing city economies. Energy and emissions initiatives have the potential to create new, green jobs and to promote the development of new businesses such as energy services companies.

*Environmental Protection.* Improved energy and emissions management encourages the use of more efficient technologies and best practices that will strengthen environmental protection and contribute to the improvement of citizens' health.

*Social Equity.* Energy and emissions programs improve city energy operations and infrastructure, which are often linked to accessibility and reliability of supply for citizens.

*City Branding.* Energy and emissions programs can be used to attract investments through positive city branding.

*Improving Risk Management.* The data collection exercise required to develop the energy and emissions plan enables the city to better understand its current and projected energy use and take action to mitigate specific risks.

*Demonstrating Value to Society.* Energy and emissions programs enable cities to communicate to citizens the added value of energy-related government activities and clarify the rationale behind energy governance and rulemaking.

### Creating an Inspiring Vision

The SUEEP vision should relate to the city's character as well as its wider environmental, economic, social, and energy goals. The vision is critical to

---

**Box 10.4  Resources: Resources for Creating a Vision**

A number of useful resources illustrate the process of developing a vision:

- BELIEF (Building in Europe Local Intelligent Energy Forums) is a European project cofinanced by the European Commission under the Intelligent Energy–Europe program. BELIEF has published a document that gives cities guidance on establishing and running an Energy Forum. For further information, see http://www.managenergy.net/resources/916.
- Dialogue with the City, Perth, Australia. For more information, see "A Case Study in Deliberative Democracy: Dialogue with the City," Janette Hartz-Karp, 2005, *Journal of Public Deliberation,* Vol. 1, No. 1, Article 6.
- "Singapore Green Plan 2012" is a good example of a comprehensive vision document. See Singapore, Ministry of Environment and Resources, 2002, http://app.mewr.gov.sg/web/Contents/Contents.aspx?ContId=1342.

---

garnering support and maintaining stakeholder motivation to follow through with the energy and emissions plan in the long term. The elements of an inspiring vision are unique to each city and no prescriptive process can be used. See resources in box 10.4 for guidance.

### Engaging Stakeholders to Create a Vision

When developing the vision, a city should identify the energy and emissions plan's potential stakeholders and engage them early on to create buy-in for the ideas and support for the implementation of sustainable policies. The city's energy vision can be developed with stakeholders in a number of ways:

- Consultations and meetings
- Requests for feedback through traditional media, for example, newspapers
- Town hall meetings
- Energy forums
- Internet, mobile-based platforms, and texting

### Step 2: Establish Leadership and Organization

A city government's ability to formulate and implement sustainable energy policies will depend on its institutional structure, governance, and oversight function. Influence on stakeholders and specific sectors (for example, waste) provides cities with the capacity to implement climate change policies and action plans that can reduce energy use and greenhouse gas emissions. With strong leadership and good governance, city governments can encourage an inclusive approach in the SUEEP process and garner wide support for its policies to tackle climate change.

## Mapping Institutional Structures

Establishing good energy governance requires first understanding how energy issues are dealt with internally and which lines of communication are most important. This understanding can be achieved through an institutional map that highlights agencies and individuals integral to the energy planning process.

## The City Government's Roles and Responsibilities

Energy efficiency cuts across sectors and extends through most areas of public service provision and private enterprise. Cities often have the most direct line of public communication to residents, businesses, and industries, which means that education and incentives are most efficiently delivered through city governments. These factors, combined with the city government's overall picture of city development and its possession of the tools to influence or regulate sectors, means that it should be responsible for developing a comprehensive, sustainable plan for the city. The city government should take the lead in energy and emissions planning and in advocating for and implementing changes that advance the city's goals. One of its responsibilities would thus be to set up an organizational structure for energy planning that includes national, regional, and sublocal governments as well as the key stakeholders in the SUEEP process.

In development planning, the government must recognize the influence it has in implementing policies at the city level. The city government has to be fully responsible for sectors it has significant influence over (for example, street lighting, water supply, and wastewater treatment). For sectors in which national policies affect the city, the city government should work closely with the national government to seek support or financing for measures that are aligned with national goals or to ensure that city policies are not negatively affected by national ones.

## The National Government's Roles and Responsibilities

Many aspects of energy consumption, such as power sources, gasoline subsidies, household appliance energy efficiency standards, and vehicle fuel efficiency standards, are influenced by national energy policies. The expected increase in energy consumption caused by rapid urbanization in the East Asia and Pacific region means that national governments will have to take the lead in implementing policies (through regulations or incentives) to promote the efficient use of energy. However, the efforts of city governments will be essential to achieving national policy goals and targets, so national governments should work closely with city governments. For example, the Ministry of Energy, Ministry of Transportation, and Ministry of Environment should be included in city-level SUEEP discussions.

The national government should provide clear guidance to cities about the direction it will take with regard to sustainable development to allow cities to plan and, where possible, cooperate in spheres in which national and city goals are aligned. In these areas, policies implemented by the city and national governments can serve to reinforce each other, making efforts to develop the city sustainable and more effective.

### Establishing Energy Governance
#### Adapting City Structures

Establishing city structures that support energy governance helps cities manage their energy planning process and increases the likelihood of a successful outcome. See the case study in box 10.5. Good energy governance includes the following:

- An energy and emissions champion
- Formal city groups dedicated to the SUEEP process, for example, a steering committee and working groups (WGs)
- A communications strategy to inform government employees of the SUEEP process
- Training opportunities with links to local educational institutions where required

City departments such as planning, land-use planning, and transportation planning will want to integrate energy and emissions planning into their functions—after all, they will be affected by it.

#### Energy and Emissions Champion

The energy and emissions champion will be responsible for overseeing the SUEEP process and making the high-level decisions associated with it. This person should be assigned a team of direct reports or should be given the authority to coordinate various groups undertaking the SUEEP work. The energy and emissions champion must advocate for the benefits of the SUEEP process and understand the various roles that stakeholders can play in providing inputs to the planning process.

---

**Box 10.5  Case Study: The Barcelona Local Energy Agency**

The Barcelona Local Energy Agency came into being on May 14, 2002, in response to the European Union's Green Paper (EC 2000) and White Paper (EC 1997) on Energy. These documents recognized the role of local authorities in energy administration and the positive value of exchanging experiences.

The agency comprises city agencies involved in energy and environmental management along with local educational institutions. The Barcelona Local Energy Agency aims to promote Barcelona as an exemplar city with respect to energy and environmental protection.

The Energy Agency is the focal point of Barcelona's Energy Improvement Plan, which sets forth a series of ambitious targets for energy management in the city.

For further information, see http://www.barcelonaenergia.cat.

---

## Steering Committee and Working Groups

The steering committee is made up of city administrators and stakeholders. These people provide strategic direction and technical support to the SUEEP process. See case studies in boxes 10.6 and 10.7. Other stakeholders (for example, nongovernmental organizations [NGOs] and representatives of the private sector) should also be given the opportunity to be part of the steering committee because this will enable them to buy into and take ownership of the recommendations put forward by these committees.

---

### Box 10.6  Case Study: Singapore Inter-Ministerial Committee on Climate Change

Singapore's Inter-Ministerial Committee on Climate Change (IMCCC) was established in 2007 to oversee interagency coordination on climate change. As of January 2012, the IMCCC has been chaired by Teo Chee Hean, Deputy Prime Minister, Coordinating Minister for National Security, and Minister for Home Affairs. The other members of the IMCCC are ministers from the Finance, Trade and Industry, National Development, Environment and Water Resources, Foreign Affairs, and Transport ministries. The IMCCC is supported by an executive committee (Exco) comprising the permanent secretaries of these ministries. The secretariat is from the National Climate Change Secretariat (NCCS), which was set up as a dedicated agency under the Prime Minister's Office in July 2010 to coordinate Singapore's domestic and international policies, plans, and actions on climate change.

The Exco oversees the work of three WGs (figure B10.6.1):

• The International Negotiations Working Group defines Singapore's strategy in the international climate change negotiations under the United Nations (UN) Framework Convention on Climate Change.

**Figure B10.6.1  Basic Structure of Singapore's Climate Change Framework**

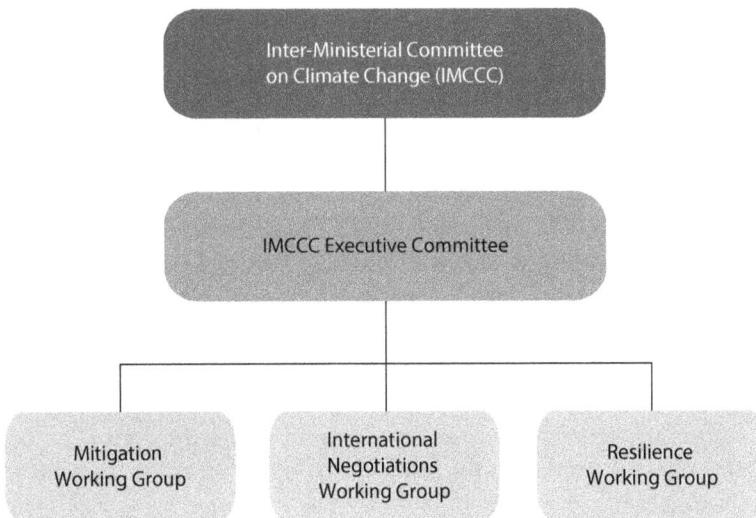

*box continues next page*

**Box 10.6  Case Study: Singapore Inter-Ministerial Committee on Climate Change** *(continued)*

- The Mitigation Working Group establishes the suite of domestic measures to mitigate carbon emissions. Members include the permanent secretaries of ministries such as Trade and Industry, Transport, National Development, Environment and Water Resources, as well as management from statutory boards such as the Energy Market Authority, the National Environment Agency, and the Economic Development Board.

- The Resilience Working Group studies Singapore's vulnerability to the adverse effects of climate change and develops long-term plans to ensure that Singapore is able to cope with climate change. Members include the deputy secretaries of ministries such as National Development, Environment and Water Resources, Finance, and Health as well as management from the Building and Construction Authority, Maritime and Port Authority of Singapore, Energy Market Authority, and the Public Utilities Board, among others.

In addition, NCCS uses two platforms, the Climate Change Network and the Climate Change Forum, to nurture dialogue on climate change–related issues. The Climate Change Network, which comprises distinguished members from the media, business, and academic communities, serves as a platform for representatives from the public and private sectors to meet, network, and exchange information on climate change issues. The Climate Change Forum allows government agencies to exchange information, share best practices, and update each other regularly on climate change–related events or forums they host or participate in.

For more information, see http://app.nccs.gov.sg/page.aspx?pageid=47.

---

**Box 10.7  Case Study: Bangkok's Steering Committee for Global Warming**

Bangkok's Steering Committee for Global Warming and its five WGs have been leading the way to developing a sustainable approach to emissions planning for the city.

As part of the city's commitment to improved mitigation of climate change, Bangkok established a global warming steering committee supported by five WGs: the WG for Improvement of Transportation System, the WG for Promotion of Renewable Energy, the WG for Energy Conservation and Building Retrofit, the WG for Solid Waste and Wastewater Management, and the WG for Expansion of Green Areas (figure B10.7.1). These five groups provide the steering committee with technical and policy advice to support its decisions about sustainability in the city.

Part of the steering committee's core responsibility is the city's 5-Year Action Plan for Global Warming Alleviation. The plan includes concrete goals and actions to ensure the goals are met.

*box continues next page*

**Box 10.7  Case Study: Bangkok's Steering Committee for Global Warming** *(continued)*

**Figure B10.7.1  Structure of Bangkok's Global Warming Steering Committee**

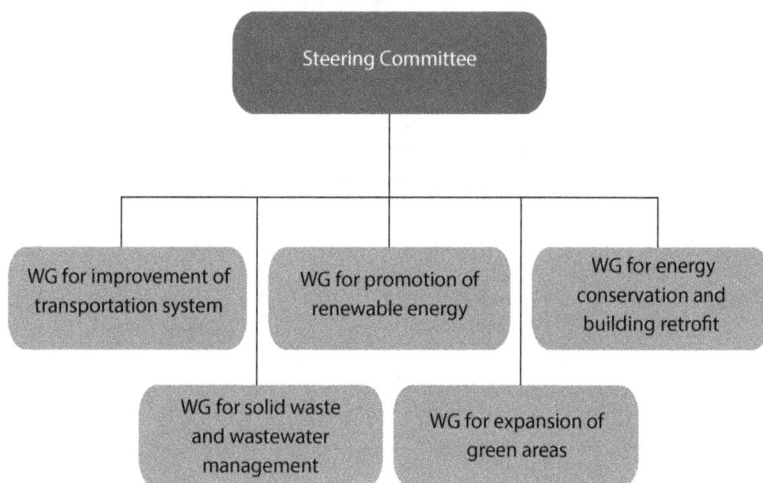

*Source:* Suwanna Jungrungrueng. 2011. "Low Carbon Target and Activity in Bangkok." Department of Environment, Bangkok Metropolitan Administration, Bangkok, Thailand. http://lcc.ait.asia/upload/activities/BMA_20May2011.pdf.
*Note:* WG = working group.

WGs are made up of skilled people from the various city departments, public agencies, and potentially, utilities. Other stakeholders can be included, especially those who would be most affected by policies arising from the SUEEP process, as well as those whose support would be required in implementing policies. WGs are responsible for the technical and production work required for the SUEEP process.

### Internal Communication

Energy and emissions planning is a multidisciplinary exercise and requires coordination and communication among numerous internal government agencies so that

- Politicians and administrators who are not directly part of the SUEEP process can be educated about what it involves so they have the opportunity to point out synergies or conflicts that may arise within their areas of work; and
- Politicians and administrators who are directly part of the SUEEP process have a broad awareness of the issues involved in energy planning and understand their colleagues' roles, including points of contact if questions or issues arise.

Internal communication can be difficult in large organizations such as governments—one of the energy champion's most important responsibilities is to take a strategic approach by establishing a communications plan early on. There is no specific formula for a communications plan, but it can include actions such as periodic team meetings, information sharing, using email lists, and so on.

**Figure 10.1  Potential Structure for an Energy and Emissions Task Force**

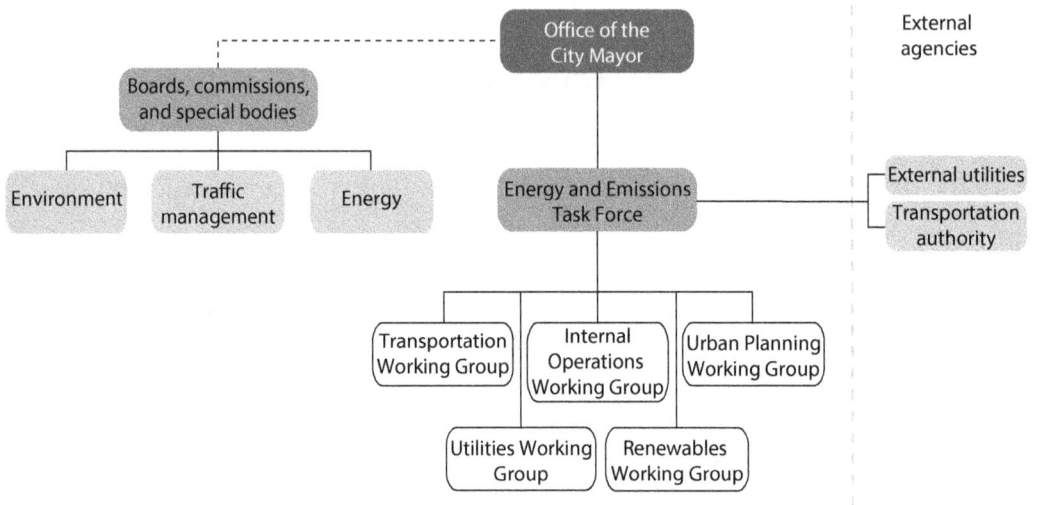

## Training

A common challenge in the energy and emissions planning process is that skilled personnel and other resources are not available within the city government. A city can find creative ways to augment its capacity, such as internal resourcing adjustments, training programs, reaching out to other cities, partnering with academic institutions, or forming internal groups for workshop activities.

## Energy and Emissions Task Force

An energy and emissions task force is a small team of dedicated city government staff members whose predominant job is to develop the energy and emissions plan and manage its implementation. (See figure 10.1 for a possible organizational structure for such a task force.) Whether the task force leads the SUEEP process depends on the city's organizational structure and specific requirements. The task force's scope of responsibility can be determined based on the city's needs, but generally will include overseeing and executing data collection and analysis, ensuring the appropriate stakeholders are brought into the SUEEP process, taking on city-led project development and project implementation, and monitoring and reporting. This list of responsibilities is not exhaustive and will be determined based on the needs of the city.

## Step 3: Identify Stakeholders and Links

Engagement of stakeholders underpins the long-term success of the SUEEP process in several ways. First, it improves the quality, effectiveness, and legitimacy of the plan by allowing a broad consensus to be reached. Second,

**Box 10.8 Tip: Characteristics of Inclusive Governance**

According to the UN Economic and Social Commission for Asia and the Pacific, good governance is characterized by eight traits (http://www.unescap.org/huset/gg/governance .htm).

City leadership should bear these characteristics in mind when designing the SUEEP governance structure (figure B10.8.1).

**Figure B10.8.1  Inclusive Governance**

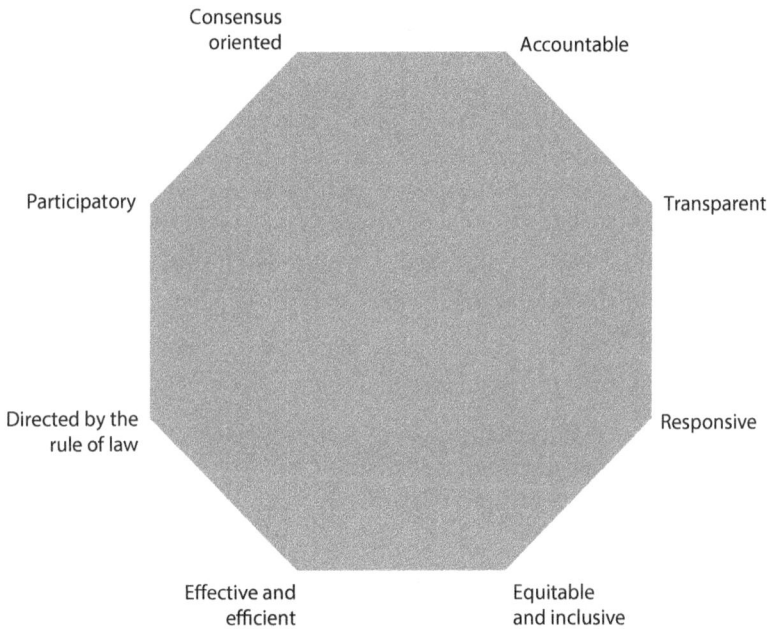

Source: UNESCAP undated.

it encourages transparency and innovation by incorporating inputs from stakeholders with different perspectives. Finally, it ensures the long-term acceptance, viability, and support of strategies and measures recommended. See the tip in box 10.8.

Given the importance of stakeholder engagement to the SUEEP process, a city will find it essential to create a methodical and comprehensive strategy to engage stakeholders and provide sufficient budgets to carry out that strategy. Stakeholders should be brought on board early in the SUEEP process because gaining understanding and buy-in up front will help to break down the barriers that can lead to failed projects.

### Mapping Stakeholders

During Step 1, the city identified and made a list of appropriate stakeholders to provide inputs to the city's vision. In Step 2, the city considered which stakeholders (both government and external) should be part of the steering committees, WGs, and the energy and emissions task force. This mapping of stakeholders will continue throughout the development of the energy and emissions plan as city governments identify parties who could provide the necessary data and information.

Stakeholders can be involved in the SUEEP process to varying degrees (see the tip in box 10.9). When developing an energy and emissions plan, an engagement strategy should be devised that determines the level of input needed from each

### Box 10.9  Tip: Understanding Stakeholders

Understanding your stakeholders will allow you to develop the best consultation strategy for their varying degrees of influence and interest. A stakeholder map (figure B10.9.1) is a useful way to define your city's different types of stakeholder as the basis for your consultation strategy. Stakeholders with low levels of influence and low interest should be kept informed of the SUEEP process but do not need to be involved further. However, stakeholders with high levels of influence and strong interest are key players and need to be persistently and actively engaged. Stakeholders that fall in between can be engaged with varying levels of participation.

#### Figure B10.9.1  Stakeholder Map

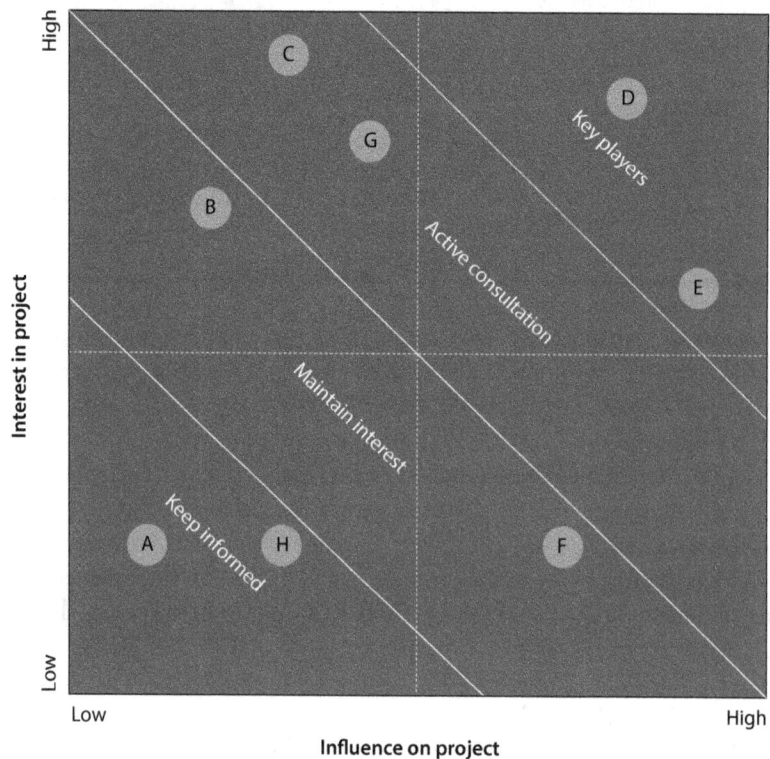

stakeholder. This strategy should identify the key audience, specify the message to be transmitted and the desired outcome, and establish a set of indicators to evaluate the impact of the communication (headcount, surveys, website hits, feedback, and the like). In addition, the city's unique circumstances will have to be considered and the strategy tailored to meet the needs and desires of the audience. Details on key elements of a communication strategy, including everything from background notes to case studies, can be found at the Covenant of Mayors website (http://www.eumayors.eu).

## Engaging Internal Support

Building strong internal political support for the SUEEP process is just as important as establishing external support. See the case study in box 10.10. The mayor and SUEEP team must drive the SUEEP process and develop an environment that enables city government agencies to understand their roles in attaining the city's goals and vision.

Engaging internal support involves establishing a suitable institutional structure and building internal knowledge through communication and personal relationships.

Strategies to attain support for the energy and emissions plan include the use of informational presentations, email, postings, and gatherings; establishing an internal forum or task team; and getting government funding for training activities associated with the energy and emissions plan.

See the example in box 10.11 for potential internal and external stakeholders.

---

**Box 10.10  Case Study: Tshwane, South Africa—Joint Stakeholders Partnership**

Since August 2003, Tshwane has participated in the Sustainable Energy for Environment and Development program, an initiative that focuses on building capacity in cities to address energy issues. This program aims to promote the integration of sustainable energy and environmental approaches and practices into all operations of the city.

As part of the program, Tshwane established an interdepartmental steering committee to run activities of the Sustainable Energy for Tshwane (SET) program. SET received political support from high-level officials and technical assistance from Sustainable Energy Africa. In addition, the city brought in a political champion, a lead city government agency, the SET committee, and a nonconventional energy forum to help create awareness of energy approaches and practices in city operations.

The SET program initially faced challenges from a lack of commitment by some departments that did not have decision-making powers to advance the agenda. These challenges were overcome through interdepartmental communication, and the steering committee's success has been maintained through ongoing strategic workshops.

For more information, see the SET website: http://www.tshwane.gov.za/Services/EnvironmentalManagement/Pages/default.aspx.

---

**Box 10.11  Example: Potential Stakeholders**

**Internal**

- City mayor
- City-controlled utilities
- City planners and zoning committee
- Department of transportation
- City procurement office
- Department of construction
- Chamber of commerce
- City budget office
- Department of economic development
- Civil servants from relevant city administrations
- Representatives from city neighborhoods or divisions

**External**

*Government institutions*

- National and regional politicians
- Planning managers
- Energy utilities in the city
- Regional and local energy agencies

*Industry*

- Industrial energy experts
- Consultants
- Business community
- Trade unions
- Developers

*External organizations*

- Representatives from cooperatives and foundations
- Representatives from relevant interest organizations
- Academic institutions
- Schools
- Nongovernmental organizations

*The Public*

- Citizens

## Engaging External Support

Establishing external support for the SUEEP process needs to be undertaken on a larger scale, given the wide spectrum of stakeholders that the city government will have to engage. Significant time and resources will need to be allocated to gaining external support.

Energizing Green Cities in Southeast Asia  •  http://dx.doi.org/10.1596/978-0-8213-9837-1

Stakeholders have different aspirations, expectations, and needs, so a considered approach needs to be taken on the strategies to be used in engaging them. The strategy will ultimately depend on the city's context, but some of the most commonly used strategies include the following:

- Task and partnership teams
- Meetings with key leaders
- Focus group workshops and public meetings
- Energy forums
- Media

### Partnerships

Bringing in the right external partners helps fill gaps in city capacity and adds valuable knowledge and experience to the SUEEP process. See the case studies in boxes 10.12 and 10.13.

The right partnerships can be established if you carefully consider which external parties can be most helpful and make sure there is no conflict of interest (for example, a supplier or manufacturer of energy efficiency equipment could influence the SUEEP process toward their specific product or service). A city needs to consider partnerships with people and organizations that may lie outside its usual set of contacts.

Technical assistance through NGOs or development aid projects, capacity building in conjunction with other cities that have already undertaken the energy planning process, and participation in national and international energy and climate change programs are useful strategies. In addition to partnering with official organizations, public-private partnerships have been gaining popularity and innovative financing mechanisms are available to harness their potential. See box 10.14 for resources on stakeholder management and external assistance.

---

**Box 10.12  Case Study: London Energy Partnership**

The London Energy Partnership is a crucial element in London's response to the challenges of climate change, reliability of energy supply, and fuel poverty. It aims to transform London into a world-class city for sustainable energy by bringing together a range of sectors and organizations to deliver energy more effectively. The partnership is made up of a consortium of businesses, government, and public bodies. Acting as an independent organization, it uses the power of partnership to promote sustainable energy solutions in London.

For example, the London Energy Partnership worked with the London Borough of Dagenham to develop "A Guide to the Barking Town Centre Energy Action Area," an implementation plan that sets out a strategy for reducing carbon emissions from new developments. This is one of many partnerships that the London Energy Partnership has engaged in. For more information, see its website: http://www.lep.org.uk.

---

Energizing Green Cities in Southeast Asia · http://dx.doi.org/10.1596/978-0-8213-9837-1

## Box 10.13  Case Study: Surabaya, Indonesia—Bappeko

Bappeko is the long-term planning agency for the city government of Surabaya. The head of the agency reports directly to the mayor, and the purpose of the agency is to create long-term plans with the participation of multiple city agencies through wide stakeholder engagement. This agency creates the integrated land-use plan, which requires input from the Department of Transportation (traffic, public transportation), the City Water Company, Cleansing and Park Department (street lighting, waste, wastewater), and a number of other city agencies. Bappeko also works externally with local universities, neighborhood leaders, and private sector businesses and developers.

*Source:* http://bappeko.surabaya.go.id.

## Box 10.14  Resources: Stakeholder Engagement and External Assistance

### Stakeholder Engagement

- Bristol Environment Agency guidance on public participation techniques: Judith Petts and Barbara Leach. 2000. *Evaluating Methods for Public Participation: Literature Review*, Bristol Environment Agency.
- Partners Foundation for Local Development. 2010. *Handbook 3: The Councilor as Decision Maker*, http://www.fpdl.ro/publications.php?do=training_manuals&id=1.
- For a comprehensive guidebook produced by the Energy Model Project, see http://www.energymodel.eu/spip.php?rubrique100.

### External Assistance

- World Bank Group
- Asian Development Bank
- UN-Habitat (United Nations Human Settlements Program)—Sustainable Urban Development Network
- ICLEI—Local Governments for Sustainability
- United States Energy Association Energy Partnership Program
- South Asia Regional Initiative for Energy
- Private energy services companies

*Source:* For stakeholder engagement, Covenant of Mayors (http://www.eumayors.eu).

# References

Barcelona, Spain. undated. "Plan for Energy Improvement in Barcelona." Summary (in English): http://www.barcelonaenergia.cat/document/PMEB_resum_eng.pdf. Complete document (in Catalan): http://www.barcelonaenergia.cat/document/PMEB_integre_cat.pdf.

City of New York. 2007. *PlaNYC 2030: A Greener, Greater New York*. New York: The Mayor's Office. http://nytelecom.vo.llnwd.net/o15/agencies/planyc2030/pdf/full_report_2007.pdf.

Covenant of Mayors. 2010. "How to Develop a Sustainable Energy Action Plan (SEAP)—Guidebook." Publications Office of the European Union, Luxembourg.

Dublin City Council and Codema. 2010. "Dublin City Sustainable Energy Action Plan 2010–2020." Dublin. http://www.dublincity.ie/WaterWasteEnvironment/Sustainability/Documents/SEAP-FINAL%20version%20for%20website.pdf.

EC (European Commission). 1997. "Energy for the Future: Renewable Sources of Energy. White Paper for a Community Strategy and Action Plan." COM(97)599. http://europa.eu/documents/comm/white_papers/pdf/com97_599_en.pdf.

———. 2000. "Green Paper, Towards a European Strategy for the Security of Energy Supply." COM(2000)769. http://europa.eu/legislation_summaries/energy/external_dimension_enlargement/l27037_en.htm.

Hartz-Karp, Janette. 2005. "A Case Study in Deliberative Democracy: Dialogue with the City," *Journal of Public Deliberation* 1 (1).

Partners Foundation for Local Development. 2010. *Handbook 3: The Councilor as Decision Maker.* http://www.fpdl.ro/publications.php?do=training_manuals&id=1.

Petts, Judith, and Barbara Leach. 2000. *Evaluating Methods for Public Participation: Literature Review.* Handbook 4. Bristol, U.K.: Environment Agency Partners Foundation for Local Development. http://www.fpdl.ro/publications.php?do=training_manuals&id=1.

Singapore. 2012. *Climate Change and Singapore: Challenges. Opportunities. Partnerships.* National Climate Change Secretariat, Singapore.

Singapore, Ministry of the Environment and Resources. 2002. "Singapore Green Plan 2012." Singapore. http://app.mewr.gov.sg/web/Contents/Contents.aspx?ContId=1342.

Smith, Suzanne. 2009. "100 Year Sustainability Vision and Concept Plan." *Planning West* 51 (1): 4–7.

UNESCAP (United Nations Economic and Social Commission for Asia and the Pacific). undated. "What Is Good Governance?" Bangkok, Thailand. http://www.unescap.org/huset/gg/governance.htm.

**CHAPTER 11**

# Stage II: Urban Energy and Emissions Diagnostics

*This stage summarizes the creation of baseline diagnostics that provide the foundation for deciding which projects to implement. Creating the energy balance and greenhouse gas (GHG) emissions inventory as well as cataloging past and ongoing energy efficiency and energy planning initiatives help a city to identify major trends and opportunities. This information allows a city to assess the potential for energy and emissions reduction projects so its sustainability goals can be achieved. This stage also refers to the Sustainable Urban Energy and Emissions Planning (SUEEP) Toolkit spreadsheets and templates that explain how to assess the potential of projects to reduce energy and emissions. Suggestions on how to collect the data, do the calculations, and bring the data together as a whole—as well as common pitfalls—are also discussed in this stage.*

## Step 4: Inventory Energy and Emissions

The energy and emissions baseline inventory provides the foundation of data on which the energy and emissions plan is based. Major trends and opportunities are quantified during the baseline inventory and subsequent monitoring inventories.

The energy and emissions inventory gathers the technical information and data needed to develop two key components of the energy planning process: the energy balance and the GHG inventory.

### Methodology for the Baseline Inventory

The methodology used in this Guidebook is based on the Local Governments for Sustainability's (ICLEI) "International Local Government Greenhouse Gas Emissions Analysis Protocol" (2009), which follows principles of the Intergovernmental Panel on Climate Change's (IPCC) "2006 Guidelines for Greenhouse Gas Inventories" (2006) and has been revised by the authors for application by cities. Further elaboration on the methodology can be found in the

---

**Box 11.1  Example: Typical Fuel Categories for Energy and Emissions Diagnostics**

Transportation

- Diesel and gasoline
- Compressed natural gas
- Liquefied petroleum gas (LPG)
- Electricity (for electric bus or rail)

Industrial

- Electricity
- Diesel and gasoline
- Natural gas

Commercial

- Electricity
- Natural gas

Residential

- Electricity
- Natural gas and LPG
- Other cooking fuels

---

Covenant of Mayors (2010) "How to Develop a Sustainable Energy Action Plan (SEAP)—Guidebook" (http://www.eumayors.eu/IMG/pdf/seap_guidelines_en.pdf). The World Resources Institute and the World Business Council for Sustainable Development are also working with the C40 Cities Climate Leadership Group (C40; a network of 40 of the world's large cities, plus affiliate cities, committed to implementing meaningful and sustainable local climate-related actions that will help address climate change globally) to develop a consistent protocol for determining the urban GHG inventory and energy balance.

Typical categories for the inventories are outlined in the example in box 11.1, though some cities may consume other types of fuel in addition to those in the example.

The broad methods for calculating the energy balance and GHG inventory follow:

- For the energy balance:
  - Define the city and data boundary.
  - Define the baseline year.
  - Define the sectors of study.
  - Define fuels for each sector.
  - Collect fuel sales and consumption data.

- For the GHG inventory:
  - Gather the information from the energy balance.
  - Define emissions factors for each fuel.
  - Define sectors whose emissions are not related to the use of fuel.
  - Define emissions factors for nonfuel sectors.
  - Collect data as defined above for nonfuel sectors.

The inventory should be as complete and accurate an assessment of the city's energy and emissions as possible so that short-term and long-term energy policies can be developed to support the city's economic development and enhance the quality of life of its citizens. Because the focus of this Guidebook is on supporting cities that have not begun the inventory process, more detailed and complex aspects of GHG inventory methods (such as life-cycle assessments and embodied energy of materials) are not included.

### Data Collection

Data collection is an iterative process requiring multiple requests, clarifications, and approvals before an energy and emissions inventory is completed. This section provides direction on what information to request and where data can usually be found. Further details on calculation methods, specific data elements, and conversions can be found in the "Energy Balance and GHG Inventory Spreadsheet" in the SUEEP Toolkit available at http://www.worldbank.org/eap/energizinggreencities.

The most common source of base energy data is information for electricity, gas, and fuel sales by utilities and national fuel companies. A national regulatory body that oversees multiple private energy providers can also be a good source of consumption data. Beyond fundamental energy and fuel consumption data, a variety of contextual data, such as power plant fuels and regional electrical grid distribution, is required to provide accurate emissions factors.

Step 6, during which energy and emissions reduction projects are assessed, requires a much broader mix of data, such as power plant combustion technology, motor vehicle fleet data, street lighting lamp type inventory, and many other behavioral and technological data. So collecting these types of data while you are collecting the base energy and fuel sales data can save you some time.

### Data Reliability

Data collected for all sectors should use the same physical boundaries, calendar year, and collection methodology to provide the level of accuracy and reliability required to meet global standards for energy planning and GHG inventories. This allows comparability between cities and better sharing of data and projects, in addition to consistency between your own inventory years.

### Data Collection Process

To collect energy data, you will need diligence and perseverance because there are multiple sources of data and they are often not in the format, units, year, or boundary definitions needed for an inventory. You can ease the data collection process by developing relationships with key individuals in each organization (see the example in box 11.2 on typical organizations) to minimize the number of contacts (or even narrow it down to a single point of contact within each organization) required to collect the variety of information needed. Organizations often ask for an initial meeting at which you will make the data request before they will begin to collate the base consumption and sales data and the contextual data. A follow-up meeting is often necessary to confirm the context, boundary, detail, and subtleties of the data provided. It is important that you pin down any areas of uncertainty, for example, physical boundaries, calendar year, and collection methodology, to make sure that the data are reliable. A letter from the mayor's office is often helpful for explaining the purpose of the data requests and to reassure the data provider that the information will be kept confidential and will not be used for any purpose other than for energy planning and policies.

### Boundary Issues

Clear and consistent boundary definitions for the collected data will provide more accurate consumption and emissions totals. Cities often choose the mayoral

---

**Box 11.2  Example: Typical Sources for Data**

Transportation

- National petroleum company
- Department of transportation

Industrial

- Electrical utility, industrial customers
- Natural gas utility, industrial customers
- National petroleum company

Commercial

- Electrical utility, commercial customers
- Natural gas utility, commercial customers
- (Natural gas, liquefied petroleum gas [LPG], liquefied natural gas [LNG])

Residential

- Electrical utility, residential customers
- Natural gas utility, residential customers
- (Natural gas, LPG, LNG)

City

- Electrical utility, city as customer
- Natural gas utility, city as customer
- (Natural gas, LPG, LNG)

---

geopolitical jurisdiction as the boundary for the energy and emissions plan. However, boundaries should be extended given that local governments are responsible for policies of subnational regions that influence the flow of energy and materials and because a city's energy consumption is affected by national and regional decisions and policies. Although this approach is preferred, collecting data on cross-boundary flows of vehicles, fuels, and energy can be difficult for cities because vehicle fuel purchases occur both inside and outside city boundaries. It is recommended that each city that embarks upon a GHG inventory process identify the most reliable data source for each cause of cross-boundary emissions and use these sources consistently for every GHG inventory update.

### Stakeholder Engagement

Most data come from stakeholders outside the city government's authority, such as the electrical utility and the natural gas utility. Bringing these stakeholders into a technical working group or even the energy task force to gain their buy-in and understanding of the context of the data requests can be helpful. (See the technical assistance opportunity in box 11.3.) Representatives from these organizations are often more aware of how to gain approvals from within the organization and know the right people to approach to collect and summarize the requested data. On top of this, utilities and departments that provide energy and emissions data may also be the entities that will eventually roll out energy projects. In light of these factors, engaging stakeholders is critical.

### The Energy Balance

The energy balance illustrates the flow of energy into and out of a city. It is presented in a constant unit, typically joules (megajoules, MJ; gigajoules, GJ; terajoules, TJ; or petajoules, PJ), even though the data are generally collected in the same units in which the energy is sold (kilowatt-hours [kWh] of electricity, liters of gasoline, cubic meters of liquefied petroleum gas [LPG], and so on). So you will need to convert energy sales data into a consistent unit of energy to be able to compare the scale of energy provided by each fuel type.

An energy balance shows energy by primary fuel, purchased energy, and useful energy, as well as wasted energy from conversion processes, and is often summarized in a Sankey diagram as shown for Surabaya, Indonesia, in figure 11.1.

---

**Box 11.3  Technical Assistance Opportunity: Energy and Emissions Data Collection**

Technical consultants who specialize in the energy field can be hired to collect and analyze the data for the energy balance and GHG inventory. Although specialists can calculate the energy balance in their respective sectors, it is important for one individual to be in charge of collecting and synthesizing all data. The consultant who conducts the first inventory may train a member of the energy task force to do future inventories. Some cities eventually employ full-time staff to do the annual energy balance and GHG inventory.

---

Energizing Green Cities in Southeast Asia • http://dx.doi.org/10.1596/978-0-8213-9837-1

**210**

**Figure 11.1  Energy Balance Sankey Diagram**

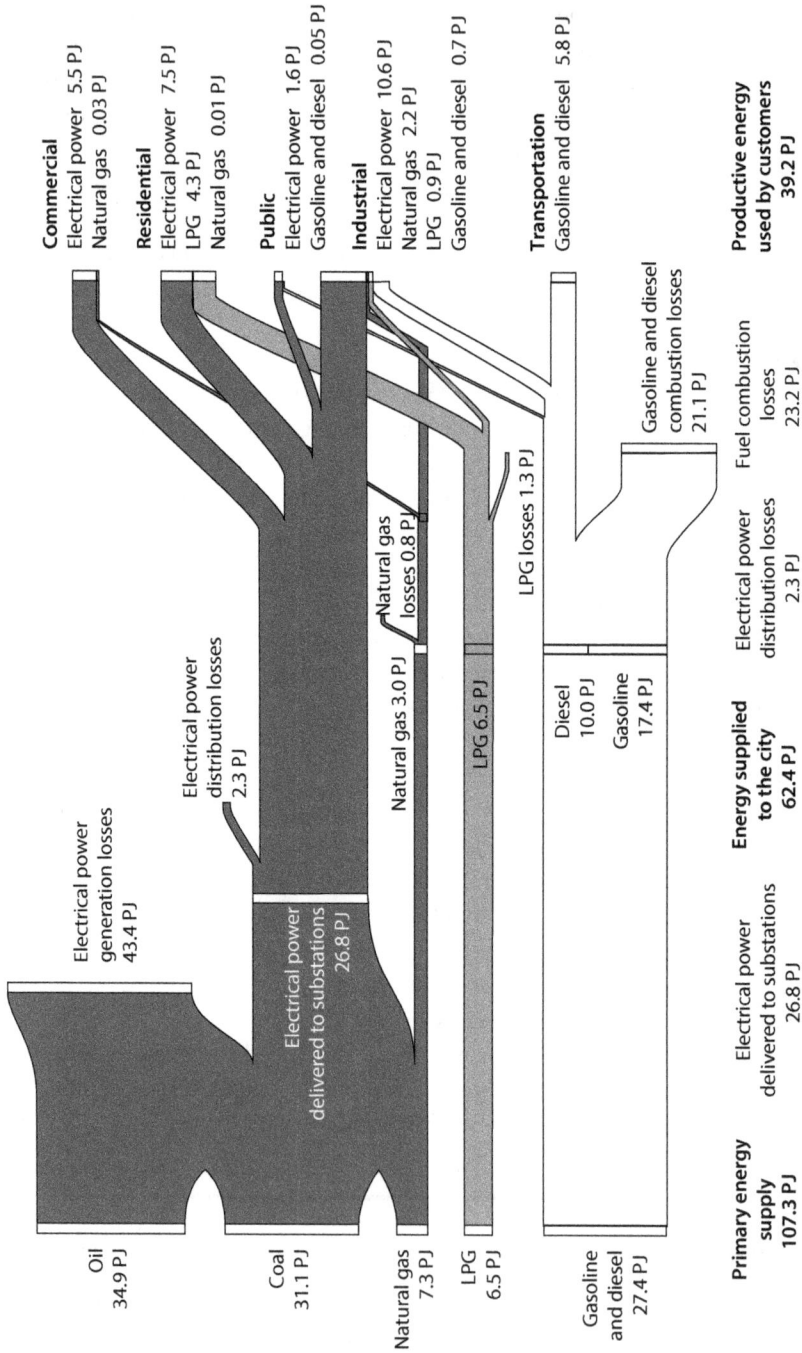

**Commercial**
Electrical power  5.5 PJ
Natural gas  0.03 PJ

**Residential**
Electrical power  7.5 PJ
LPG  4.3 PJ
Natural gas  0.01 PJ

**Public**
Electrical power  1.6 PJ
Gasoline and diesel  0.05 PJ

**Industrial**
Electrical power  10.6 PJ
Natural gas  2.2 PJ
LPG  0.9 PJ
Gasoline and diesel  0.7 PJ

**Transportation**
Gasoline and diesel  5.8 PJ

**Productive energy used by customers  39.2 PJ**

Electrical power generation losses  43.4 PJ

Electrical power distribution losses  2.3 PJ

Electrical power delivered to substations  26.8 PJ

Natural gas 3.0 PJ

Natural gas losses 0.8 PJ

LPG 6.5 PJ

LPG losses 1.3 PJ

Gasoline and diesel combustion losses  21.1 PJ

Fuel combustion losses  23.2 PJ

Electrical power distribution losses  2.3 PJ

Diesel  10.0 PJ

Gasoline  17.4 PJ

**Energy supplied to the city  62.4 PJ**

**Electrical power delivered to substations  26.8 PJ**

Oil  34.9 PJ

Coal  31.1 PJ

Natural gas  7.3 PJ

LPG  6.5 PJ

Gasoline and diesel  27.4 PJ

**Primary energy supply  107.3 PJ**

*Source:* Phase I pilot study.

*Note:* LPG = liquefied petroleum gas; PJ = petajoule. "Public" includes the end-use energy of city buildings, street lighting, city vehicles, water, wastewater, and solid waste management.

The energy balance helps to identify the largest energy users, which will help in the goal setting and the prioritization of projects in Step 9. The energy balance also shows the primary fuel types and will inform the GHG inventory calculations.

## Calculating the Energy Balance

The energy balance calculation is summarized below. This process requires familiarity with technical energy unit conversions and an understanding of site energy versus source energy (see tip in box 11.4).

1. Determine all significant fuel types for each end-use sector (box 11.1).
2. Collect fuel sales data for each fuel type for each end use (box 11.1).
3. Convert fuel sales data into a common energy unit (MJ, TJ) (box 11.4).
4. Calculate total site energy consumption by fuel type.
5. Calculate (or estimate) primary source energy fuel consumption for electricity consumption (see box 11.4).
6. Calculate total primary fuel type energy consumption.

This process is elaborated in the "Energy Balance and GHG Inventory Spreadsheet" in the SUEEP Toolkit. See the toolkit reference in box 11.5.

See the case study in box 11.6 for one city's experience determining its energy balance.

---

### Box 11.4 Tip: Energy Balance Calculations

**Energy Conversions**

One of the most common mistakes in energy planning occurs when the original data for energy sold, which come in a wide variety of units, are converted into a common unit such as MJ. Fuel sales data may be in units of volume (liters of gasoline, cubic meters of natural gas, and so on) or units of weight (tons of LPG, or kilograms [kg] of coal). Average energy densities must be used to convert volumetric or weight-based units (MJ per liter of gasoline, MJ per kg of coal) into units of energy (MJ, GJ, TJ). Converting from kWh or thousands of British thermal units (BTU) to MJ is also necessary. These conversions require care and attention to detail—particularly if data are provided in monthly or daily rates—to ensure that the energy balance is in MJ for the entire year.

**Site Energy versus Source Energy**

Another common mistake in energy planning is disregarding the differences between source energy and site energy. Site energy is electricity or fuel consumed within a property boundary or vehicle. Source energy is the initial fuel consumed to produce either electricity or transportation fuel. Typically, three units of source energy are consumed to produce one unit of site electricity for electricity generated by fossil fuels. So 1 MJ of site electricity is not equivalent to 1 MJ of transportation fuel because it takes 3 MJ of source fossil fuel energy to generate 1 MJ of site electricity energy.

---

**Box 11.5  Toolkit Reference: Energy Balance and GHG Inventory Spreadsheet**

The energy balance spreadsheet lays out the step-by-step process for calculating the energy balance and includes sample calculations for converting input data into energy balance flows in a single common unit (MJ). Typical conversion factors for major fuel types are provided. A template for the calculation method used to determine the values for the Sankey diagram for Surabaya (figure 11.1) is also included in the spreadsheet. The spreadsheet requires some limited knowledge of energy units, conversions, and how to use a spreadsheet program, but otherwise, it provides an easy-to-use way to do the complex calculations for the Sankey diagram. The SUEEP Toolkit is available at http://www.worldbank.org/eap/energizinggreencities.

**Box 11.6  Case Study: Surabaya, Indonesia—Energy Balance 2010**

The city of Surabaya developed an energy balance and GHG inventory in 2010 with the support of the World Bank under the first phase of the SUEEP process. Bappeko, an agency that develops long-term plans for the city, was appointed to be the lead agency. A local external consultant was hired to support the data collection across city agencies and utilities.

These agencies contributed data to the Surabaya energy balance and GHG inventory:

- Surabaya Department of Transportation
- PERTAMINA—state oil company
- DKP—Cleansing and Park Department
- PLN—state-owned electricity utility
- Department of Finance
- PDAM—regional drinking water utility
- PGN—state-owned natural gas company

In addition to input from the city agencies, the Japan International Cooperation Agency performed a transportation study in 2009.

Figure 11.1 summarizes Surabaya's energy balance. It shows that electricity is generated mainly from oil and coal, as well as a small amount of natural gas. It also shows that the predominant transportation fuels are gasoline and diesel.

*Source:* http://www.iea.org/media/workshops/2011/ipeecweact/Ostojic.pdf.

A Sankey diagram illustrates a city's energy balance and gives a snapshot of the largest energy consumers and the city's primary fuels. The diagram reads from left to right, with the width of each bar showing the amount of energy for that end use or fuel type. Energy coming into the city or into power plants as primary source energy is shown on the left and is typically categorized by fuel type. The bars flow to the right and show how fuel is used. For example, some natural gas flows directly to industrial customers for useful heating energy, but some natural gas flows and merges into the primary input energy for electrical power generation, along with coal and oil.

Energy conversion in the thermodynamic cycle of electrical power generation, resulting in the loss of 60–70 percent of the energy value of primary fuel, is also illustrated on the top left of the diagram.

### The GHG Inventory

A GHG inventory is a snapshot of all of the GHGs emitted by a city in a year. It includes emissions from fossil fuel combustion in electrical power plants, cars, and trucks as well as other emitted GHGs such as methane and hydrofluorocarbons.

The governing equation for calculating a GHG inventory is below. Each relevant activity is calculated separately.

- *Activity Data × Emissions Factor = Emissions.*
- *Activity Data* = fuel or material consumption. For example, electricity consumption in kWh, or coal consumption in tons (t).
- *Emissions Factor* = Factor based on the carbon content of the fuel or material. For example, carbon content of electricity in kg of carbon dioxide ($CO_2$) per kWh, or carbon content of coal in kg of $CO_2$ per ton of coal.

Activities include the combustion of fossil fuels, emission of methane from solid waste landfills or the sanitary wastewater treatment processes, emission of hydrofluorocarbons from industrial activities, and others. Emissions factors can be found in the United Nations (UN) Framework Convention on Climate Change (http://unfccc.int/) or the 2006 IPCC Guidelines for National Greenhouse Gas Inventories (IPCC 2006). If sufficient data are available, the actual emissions factor for the mix of electricity generation for a city can be calculated. The Energy Balance and GHG Inventory Spreadsheet in the SUEEP Toolkit provides a framework for the required data inputs. See the toolkit reference in box 11.7.

An important aspect of the GHG inventory is identifying Scope 1, Scope 2, and Scope 3 emissions, which are outlined below:

- **Scope 1.** GHG emissions that occur within the physical boundary established for the inventory.

---

**Box 11.7 Toolkit Reference: Energy Balance and GHG Inventory Calculator**

The GHG inventory calculator provides a step-by-step process for determining the city's annual carbon emissions following ICLEI's industry standards for an urban GHG inventory. This method relies on the data gathering and calculations in the energy balance calculator, so these two operations must be done consecutively. The output from this spreadsheet provides the city's total $CO_2$ equivalent emissions by end use and by fuel type. The SUEEP Toolkit is available at http://www.worldbank.org/eap/energizinggreencities.

---

- **Scope 2.** Indirect emissions that occur outside the city boundary as a result of activities that occur within the city, limited to electricity consumption, district steam, and district cooling.
- **Scope 3.** Other indirect emissions and embodied emissions that occur outside the city boundary as a result of activities conducted by the city, including electrical transmission and distribution losses; solid waste disposal; waste incineration; wastewater handling; aviation and marine activities; and embodied emissions in fuels, construction materials, water, imported food, and upstream of power plants (such as emissions from fossil fuel extraction).

### *Output from a GHG Inventory*

The result of a GHG inventory is a city's total annual GHG emissions in tons of $CO_2$ equivalent (see the tip in box 11.8 for an explanation of $CO_2$ equivalent). The inventory results provide a breakdown of the total emissions into end-use sectors, fuel types, and month-by-month or even daily data (box 11.9). See the case study in box 11.10 for how Seoul used the data from its GHG inventory.

---

**Box 11.8  Tip: $CO_2$ Equivalent**

Another commonly overlooked technical aspect of a GHG inventory is the common unit of "$CO_2$ equivalent." Many gases can cause a heat-trapping phenomenon similar to that caused by $CO_2$, but they have varying degrees of what is called "global warming potential" (GWP). $CO_2$ is the most common GHG, so all other GHGs are converted to the common unit of $CO_2$ equivalent.

The GHG inventory should include measurements for the gases shown in table B11.8.1 according to the UN Framework Convention on Climate Change:

**Table B11.8.1  Global Warming Potential of Various Greenhouse Gases**

| GHG | Formula | GWP |
|---|---|---|
| Carbon dioxide | $CO_2$ | 1 |
| Methane | $CH_4$ | 56 |
| Nitrous oxide | $N_2O$ | 280 |
| Sulphur hexafluoride | $SF_6$ | 16,300 |
| Hydrofluorocarbons (HFCs) | | |
| HFC-23 | $CHF_3$ | 9,100 |
| HFC-32 | $CH_2F_2$ | 2,100 |
| Perfluorocarbons (PFCs) | | |
| Perfluoromethane | $CF_4$ | 4,400 |
| Perfluoroethane | $C_2F_6$ | 6,200 |
| Perfluoropropane | $C_3F_8$ | 4,800 |
| Perfluorobutane | $C_4F_{10}$ | 4,800 |
| Perfluorocyclobutane | $C_4F_8$ | 6,000 |
| Perfluoropentane | $C_5F_{12}$ | 5,100 |
| Perfluorohexane | $C_6F_{14}$ | 5,000 |

*Source:* UN Framework Convention on Climate Change.
*Note:* GHG = greenhouse gases; GWP = global warming potential (20 years).

---

## Box 11.9  Example: Typical Results from a GHG Inventory

The total emissions for a city are often broken down by end-use sectors and fuels as shown in figure B11.9.1.

**Figure B11.9.1  Illustrative Results of a GHG Inventory**

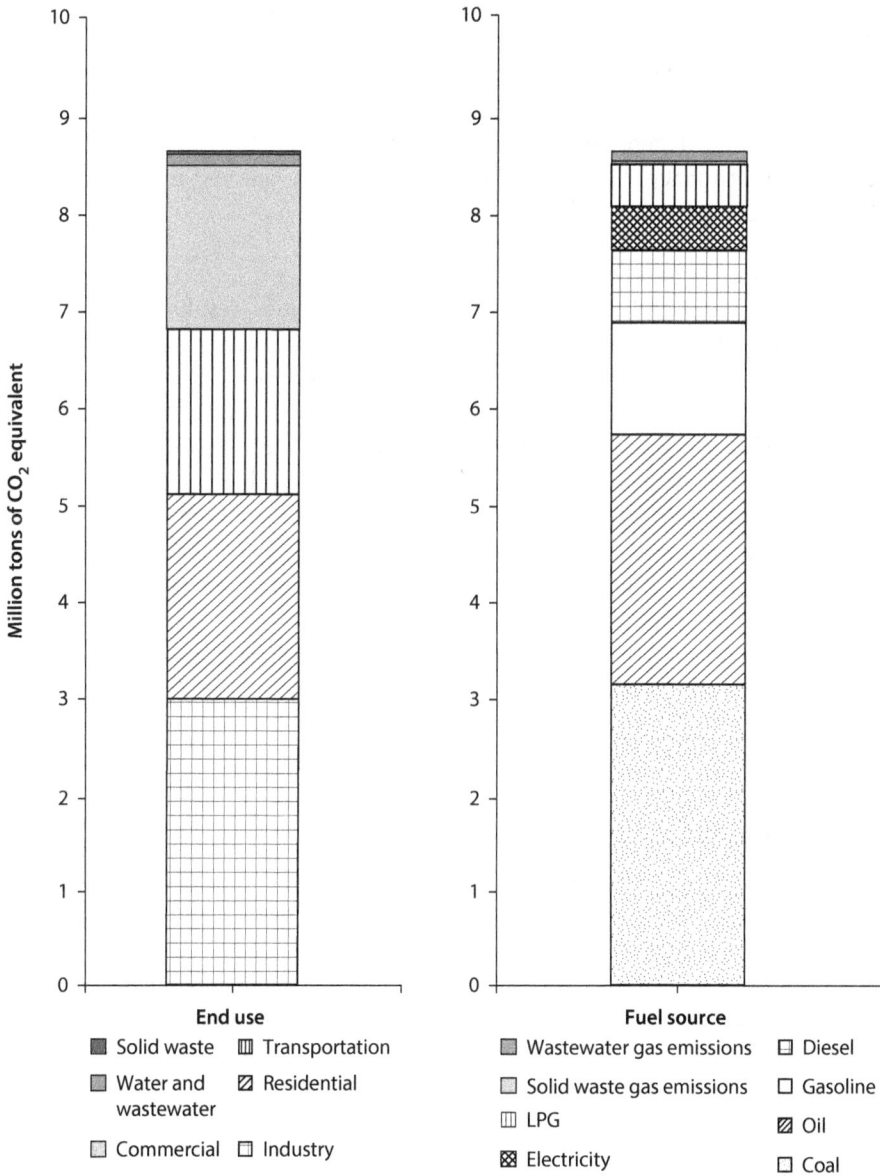

**End use**
- ■ Solid waste    ▥ Transportation
- ■ Water and      ▨ Residential
  wastewater
- ▢ Commercial     ▢ Industry

**Fuel source**
- ▨ Wastewater gas emissions    ▢ Diesel
- ▢ Solid waste gas emissions   ▢ Gasoline
- ▥ LPG                         ▨ Oil
- ▨ Electricity                 ▢ Coal

*Note:* $CO_2$ = carbon dioxide; GHG = greenhouse gas; LPG = liquefied petroleum gas.

---

**Box 11.10  Case Study: Seoul, the Republic of Korea—GHG Inventory 2008**

In 2008, Seoul undertook a citywide GHG inventory exercise for "planning to realize all-around green innovation ranging from building, urban planning and transportation to daily life by 2030 to become a city with world-leading green competitiveness." The energy and GHG information has been used to develop better building codes and a more energy efficient transportation network to justify the increase in the share of renewable energy sources and to streamline the waste collection operations of the city.

*Source:* Carbon Disclosure Project 2011.

---

The actual single data point of emissions in one year does not tell a story or show levels of performance. This datum becomes more relevant and meaningful only if it can be compared with data for previous years.

### Benchmarking Results

Comparing energy and emissions results with other cities' results or global averages can provide significant insights into how a city uses energy and how it stacks up against its peers or against other cities that governments aspire to emulate. Benchmarking can be based on citywide metrics such as tons of $CO_2$ per capita or tons of $CO_2$ per unit of gross domestic product (GDP) to show macro trends and comparisons. A sectoral comparison of energy use and emissions across cities is a useful indicator of the efficiency of a city's sectors in energy use and emissions. In Step 16: Collect Information on Projects, results from individual energy and emissions reduction projects can be compared with the results of similar projects in other cities.

Benchmarking can be done against peer cities, or against cities that have been working to reduce carbon emissions for many years or have higher levels of economic development (see the technical assistance opportunity in box 11.11). The resources in box 11.12 provide additional detail on energy diagnostics.

## Step 5: Catalog Existing Projects and Initiatives

Once the inventories of energy consumption and GHG emissions are complete, you will want to catalog current and past energy initiatives to learn about the factors that enabled past projects to succeed and the hurdles they faced. This step is an important learning exercise for SUEEP leadership and for the energy task force. They will be able to take actions to build on current and past successes, address roadblocks in future projects and avoid the mistakes made in the past. But care should be taken not to discount a project because it was not implemented or not successful; circumstances may have changed.

### Survey Stakeholders

The best way to catalog past energy projects is through face-to-face interviews. The energy task force should interview relevant stakeholder

---

**Box 11.11  Technical Assistance Opportunity: Tool for Rapid Assessment of City Energy (TRACE) Benchmarking Tool**

The TRACE online tool developed by the World Bank Energy Sector Management Assistance Program allows a city to enter its energy consumption data and see how it ranks against a database of more than 64 cities and 28 Key Performance Indicators. This database is based on publicly reported energy consumption figures and GHG inventory data along with background data such as population, GDP, and so forth. A wide variety of cities from around the world are included, which allows comparisons to be made with cities of similar size, region, level of development, and institutions. Information on larger, more developed cities that a city may aspire to emulate is also included in TRACE. Benchmarking provides an important snapshot of sectors that are performing well and of those that have the opportunity to improve and thereby change the course of a city's energy consumption.

*Source:* TRACE, http://www.esmap.org/esmap/node/.

---

**Box 11.12  Resources: Energy and Emissions Diagnostics**

These resources can be consulted for further information on diagnostics:
- ICLEI, 2009, "International Local Government GHG Emissions Analysis Protocol (IEAP)" (http://www.iclei.org/index.php?id=ghgprotocol).
- IPCC, 2006, "2006 Guidelines for National Greenhouse Gas Inventories" (http://www.ipcc-nggip.iges.or.jp/public/2006gl/index.html).
- Covenant of Mayors, 2010, "How to Develop a Sustainable Urban Energy Plan (SEAP)—Guidebook, Part II, Baseline emissions inventory" (http://www.eumayors.eu/IMG/pdf/004_Part_II.pdf).

---

agencies—including those employees who have an interest in energy issues as well as those who have been at the agencies for many years and have built up stores of institutional knowledge about past projects. Interviewing multiple representatives within an agency is useful because departments often work independently and may not be aware of projects undertaken by other departments in the same agency.

Stakeholders should be asked the following questions to get a sense of the context and circumstances they faced when previous projects were implemented. These questions will help you find out the critical success factors or failure points:

- Which agency led the project?
- Who were the stakeholders?
- Which stakeholders should have been involved but were not?
- How much time was spent implementing the project?

- How much money was spent on the project?
- Was funding sufficient?
- What data on the impact of the project were collected?
- Did social or cultural norms prevent uptake of the project? If so, what were they?
- Which stakeholders benefited the most and who lost out from the project?

In addition to interviews, gather project implementation plans, status reports, and any other project documentation you can find. Collate and organize it so the information will be useful for future projects.

### Extract the Lessons Learned

After collecting the data and identifying factors that contributed to the success or failure of past projects, the energy task force should summarize its findings and highlight both positive and negative themes. Examples of lessons learned include the benefits of a thorough cost-benefit analysis; the importance of following city procurement guidelines; the delays caused by insufficient staff time allocated to implementing the project; and the disincentives arising from the creation of too much documentation to prove that energy efficiency equipment was actually installed.

### Document in Project Assessment Sheet

The results of the interviews should be documented in the Project Assessment Sheet (discussed in detail in Step 6, and available in the SUEEP Toolkit at http://www.worldbank.org/eap/energizinggreencities), which will contribute to the prioritization and development of all projects in Step 9.

The Project Assessment and Prioritization Toolkit (Tab 3) outlines assessments of 78 typical energy and emissions reduction projects undertaken by cities. These assessments are based on an understanding of urban energy issues common in the East Asia and Pacific region and on experiences with similar projects implemented around the world. If a project that your city has undertaken is not among the 78 projects, it can be assessed and considered for inclusion as part of your energy and emissions plan.

### Local Energy Initiatives

Virtually all cities have undertaken energy efficiency projects of some sort (for example, street lighting lamp replacement and vehicle emissions standards and testing). To learn the most about these past and current projects, identify individuals linked to every sector that potentially has undertaken energy projects, such as transportation, buildings, industry, and city operations. If your city already has dedicated energy teams or energy efficiency champions in each department, identifying the people who will be able to provide a complete list of projects undertaken by their departments will make the process of collating the information on projects much more efficient.

## National Energy Initiatives

National initiatives are generally broader and affect cities at various levels. Examples of national initiatives include subsidies on gasoline, diesel, or electricity, and nationally funded energy research projects on renewable energy, transportation, or buildings energy efficiency (see the case study in box 11.13). Although a city may not be able to replicate or directly improve a national program, it may be able to participate in the program, expand implementation in its jurisdiction, or even obtain funding. Thus, city governments need to be aware of national policies on urban energy. City programs can build upon national ones and can even set a higher regulatory standard.

Some countries are taking on carbon emissions reduction targets following the Kyoto Protocol and the outcomes of recent Conference of the Parties summits, such as Durban 2011. Such actions can help support Step 7: Make the Case for SUEEP.

## Non-Energy Initiatives

Many of the drivers of energy consumption growth may not be directly related to energy, for example, land-use planning, transportation planning, and cultural habits and norms. Many projects and policies pursued by cities and countries have significant impacts on the future energy path of a city. It is important for a city to assess its land-use planning and transportation policies and the implications of those policies for the use of high-capacity public transportation versus personal cars. Initiatives to increase the safety and comfort of walking and biking should be inventoried. This exercise should be undertaken for all sectors, but the

---

**Box 11.13  Case Study: Quezon City, the Philippines—National Standards That Influence the City**

During a 2009 World Bank mission to test the TRACE audit tool, a GHG inventory and energy initiative stock-taking were performed in Quezon City. A wide variety of energy initiatives developed by numerous stakeholders were discovered during the process:

- A national vehicle emissions standard was imposed by the Department of Transportation.
- A national building code existed and was to become the baseline for a local green building guideline.
- A national refrigerator energy code required energy consumption standards for all imported refrigerators.
- National funding was used for the expansion of a bus rapid transit line through Quezon City.
- A national gasoline subsidy still existed, but was gradually being reduced.
- National goals and funding for street lighting expansion for new and existing streets applied.

*Source:* ESMAP 2010.

---

focus should be on the transportation sector because it often has the most energy implications from non-energy-related policies.

## Step 6: Assess Potential Energy and Emissions Projects

This step creates an inventory of the qualities of potential energy projects and ranks the projects based on their relevance to city energy goals and likely performance levels. This is a diagnostic exercise, not a decision step or a prioritization step, which are discussed in Step 9. In this step, projects need to be carefully examined to ensure the assessment is accurate for your city's energy and emissions plan.

### The Purpose of the Assessment

The purpose of the assessment is to help you to start building your list of high-priority projects from a long and varied list of possible projects. The assessment provides the input data required for Step 9: Prioritize and Select Projects. Only after the projects' characteristics (energy savings potential, implementation cost, level of city control, and so forth) have been determined and goals have been identified in Step 8: Establish Goals, can the projects be prioritized in Step 9. The assessment does not require detailed technical development, budgeting, and research. Each project just needs to be put into general categories.

The projects and assessments provided in this Guidebook are meant to be a starting point, and not a definitive analysis of projects for all cities. Details about each pre-developed project can be changed to suit your city, and the pre-set assessment fields that define project characteristics can be revised to be more specific and relevant for your city's energy and emissions plan. See the toolkit reference in box 11.14.

Brief descriptions of each assessment category follow. (See the technical assistance opportunity in box 11.15 and the example in box 11.16.) These fields provide criteria to be evaluated according to the goals that will be established in Step 8.

### Project Description

Describe the project briefly to give a general understanding of what changes the project will make to reduce or change the path of energy consumption in the city.

---

#### Box 11.14  Toolkit Reference: Project Assessment and Prioritization Toolkit

The Project Assessment and Prioritization Toolkit in the SUEEP Toolkit contains a list of 78 common energy projects and an initial ranking of the assessment categories. It also includes calculation methods to help estimate energy, cost, and GHG savings potential. All projects in the toolkit can be seen at once, and each project has examples and case studies for further reference. This list is meant to be a starting point—be prepared to add projects, adjust them, and make different calculations. The SUEEP Toolkit may be accessed at http://www.worldbank.org/eap/energizinggreencities.

---

---

**Box 11.15 Technical Assistance Opportunity: Project Assessment Consultant**

Highly specialized knowledge is required to assess the wide variety of energy efficiency and energy reduction projects referenced in this Guidebook. Few cities have in-house staff with knowledge of the cost, difficulty, and potential impact to be able to complete the project assessments necessary for the projects to be prioritized. Universities, energy consultants, or energy-focused nongovernmental organizations can often be tapped to help make the preliminary assessments suggested in this step of the SUEEP process.

---

*Sector*
Assign the project to one of the five major sectoral categories. Some projects may seem to fall into more than one category (for example, an industrial truck emissions reduction project). In these cases, it is helpful to choose the sector in which implementation is likely to happen. For industrial truck emissions, the "industry" sector would be used instead of the "transportation" sector because the project would be rolled out by working directly with industrial businesses.

*Project Type*
Assign the project to one of the four types of project categories. This will give the city a sense of the expertise required to develop the project.

*Energy Savings Potential*
Estimate the project's potential annual energy savings by referencing case studies similar to the particular project, or by using a calculation method in which end-use energy consumption (residential lighting energy in the city, for instance) is multiplied by the capture rate of the project (households per year) and efficiency savings per household (30 percent lighting savings). An estimate of potential annual energy savings does not need to be precise. It only needs to show low, medium, or high energy savings. The values in the Project Assessment Sheet are recommended, but can be revised based on local experience or issues local to the city. Choose a time horizon for the savings that includes full implementation (for example, two to five years), but take care to use the same time horizon for the cost of the project. Not all projects will use the same time horizon. Some projects can be completed in one year while others will need 10 years of planning and implementation to achieve their full energy savings potential.

*Fuel Type Savings*
Identify the type of fuel that is reduced or affected by the project. For electricity savings, mark grid electricity, not the underlying primary fuel for power plants in the region.

*GHG Savings Potential*
Using emissions factors from the GHG inventory spreadsheet in the SUEEP Toolkit, multiply the energy or fuel savings estimated in the previous categories

**Box 11.16  Example: Project Assessment Sheet**

**Project Name: Street Lighting Audit and Retrofit Project**

**Project description:** Existing public lighting is often highly inefficient, using high energy consumption technologies and lacking strategic coordination of placement and operation. An audit of the existing stock and an assessment of operations and maintenance will help identify appropriate measures to significantly increase energy efficiency. Interventions that include new technologies and retrofitting will also increase the design life of luminaires, which reduce both the requirements and costs of maintenance. The aim of this recommendation is to enable a comprehensive assessment of the lighting system to identify areas for improvement across the network.

| **Sector:** | **Implemention cost:** |
|---|---|
| □ Transportation | □ High (>$5,000,000/year) |
| □ Industry | □ Medium ($100,000–$5,000,000/year) |
| □ Commercial | □ Low (<$100,000/year) |
| □ Residential | |
| □ City | |

| **Project type:** | **Estimated cost savings:** |
|---|---|
| □ Incentive project | □ High (>$100,000/year) |
| □ Major project | □ Medium ($10,000–$100,000/year) |
| □ Organizational development project | □ Low (<$10,000/year) |
| □ Policy project | |

| **Energy savings potential:** | **Recipient of savings:** |
|---|---|
| □ High (>10,000,000 megawatt hour [MWh]/year) | □ City government |
| | □ City residents |
| □ Medium (10,000–10,000,000 MWh/year) | □ Energy services company |
| □ Low (<10,000 MWh/year) | □ Utility or private entity |

| **Fuel type savings:** | **Likelihood of funding:** |
|---|---|
| □ Grid electricity | □ High |
| □ Motor gasoline/diesel | □ Medium |
| □ Natural gas | □ Low |
| □ Liquefied petroleum gas | |

| **GHG savings potential:** |
|---|
| □ High (>10,000 tons/year) |
| □ Medium (1,000–10,000 tons/year) |
| □ Low (<1,000 tons/year) |

*box continues next page*

**Box 11.16  Example: Project Assessment Sheet** *(continued)*

---

**Ease of implementation:**
☐ Easy
☐ Medium
☐ Hard

**Timing of project implementation:**
☐ < 1 year
☐ 1–10 years
☐ > 10 years

**Level of city control:**
☐ High: Budget, regulatory
☐ Medium: Regional stakeholder
☐ Low: National stakeholder

**Stakeholders:**
• Cleansing and Park Department
• Department of Transportation
• PLN (national electricity company)
• Leaders of barangay communities without street lighting
• Mayor's office

**Potential funding:**
• Annual budget for Cleansing and Park Department
• Annual budget for Department of Transportation
• Mayor's office discretionary budget
• Donor-funded loan program for energy efficiency
• Clean Development Mechanism

**Previous attempts at similar projects:**
• None

---

by the fuel emissions factor. The estimate for this category does not need to be precise because this step is only meant to categorize the savings potential as low, medium, or high so as to rank the potential projects.

*Implementation Cost*
Mark the cost to implement the project. The estimate should include costs incurred by both the public and private sectors where applicable. An estimate of potential annual costs incurred does not need to be precise. It only needs to show low, medium, or high costs.

*Estimated Cost Savings*
Based on the energy savings potential, energy saved for that sector should be multiplied by the cost per unit of energy. Keep in mind that the price of electricity for city governments is often lower than commercial rates, and diesel fuel purchased by large industries can sometimes be purchased at bulk rates that are lower than rates at public gas stations.

*Recipient of Savings*
Mark the recipient of savings from the project. This is a critical aspect of energy planning and is often not recognized. Many projects are funded by one source but the benefits accrue to another. This is acceptable as long as all parties acknowledge this situation.

### Likelihood of Funding

Taking stock of all the previous assessment categories and an initial review of costs, benefits, control, timing, and stakeholders, assess the likelihood of obtaining funding for the project. Step 14 provides an extensive discussion on funding and financing.

### Ease of Implementation

Based on case studies and direct knowledge of implementing projects within a particular sector or city agency, identify whether the project is likely to be relatively easy or difficult to implement. This evaluation can include a variety of factors such as (a) the requirement for legislation (for example, building codes); (b) the requirement for national government investment (for example, a new bus rapid transit line or power plant); and (c) the city's ability to fully develop, finance, and implement the project.

### Timing of Project Implementation

Estimate the length of time required to design, implement, and collect data on the effectiveness of each project. This is not meant to be a specific time schedule, but a way to identify each project as either short or long term. In the prioritization process, you will want to have a mix of quick-win and long-term projects.

### Level of City Control

Identify whether the project lies within the jurisdiction of the city government, or if it will require approvals, financing, or involvement from the private sector, national government, international donor agencies, or other stakeholders over which the mayor has limited direct authority or control.

### Stakeholders

Identify the project's major stakeholders. The initial list of stakeholders will consist of departments and organizations; specific stakeholders should be added later.

### Potential Funding

Identify the potential sources of funding. The initial list could comprise local departments, the national government, or nongovernmental organizations.

### Previous Attempts at Similar Projects

Describe any previous energy efficiency projects that the city, electrical utility, or national government has studied or implemented that are similar to the project. Note whether the project was successful. If it was not successful, describe the factors that made the project difficult.

See the resources in box 11.12 for additional information.

## References

Carbon Disclosure Project. 2011. *CDP Cities 2011: Global Report on C40 Cities*. London: Carbon Disclosure Project. http://c40citieslive.squarespace.com/storage/CDP%20 Cities%202011%20Global%20Report.pdf.

Covenant of Mayors. 2010. "How to Develop a Sustainable Energy Action Plan (SEAP)—Guidebook." Publications Office of the European Union, Luxembourg.

ESMAP (Energy Sector Management Assistance Program). 2010. "Field Testing of the Tool for Rapid Assessment of City Energy (TRACE) in Quezon City, Philippines and Recommendations." ESMAP, World Bank, Washington, DC. http://www.esmap.org/ node/1310.

ICLEI (Local Governments for Sustainability). 2009. "International Local Government GHG Emissions Analysis Protocol (IEAP)." Bonn, Germany. http://carbonn.org /fileadmin/user_upload/carbonn/Standards/IEAP_October2010_color.pdf.

IPCC (Intergovernmental Panel on Climate Change). 2006. "2006 IPCC Guidelines for National Greenhouse Gas Inventories." Prepared by the National Greenhouse Gas Inventories Programme, Hayama, Japan. http://www.ipcc-nggip.iges.or.jp/public/ 2006gl/index.html.

# CHAPTER 12

# Stage III: Goal Setting

*Armed with an inventory and an understanding of potential projects, city leaders can establish a direction for the city. Projections of energy use and emissions will need to be made based on the city's growth and urbanization forecasts, and various energy scenarios should be compared with status quo policies and organization. This will help city leaders make the case for committing time and resources to an energy and emissions program. Clear goals will need to be set that outline energy and carbon emissions savings, along with mechanisms for verifying and reporting success. Once projects are assessed, the next task is to prioritize them for implementation and begin to line up the resources needed to tackle the initiatives that align most closely with the city's aspirations, targets, and goals.*

## Step 7: Make the Case for Sustainable Urban Energy and Emissions Planning (SUEEP)

Once the energy and emissions diagnostic information is complete, the next step in the SUEEP process is to develop the arguments about why an alternative energy pathway is important to the city's wider vision and goals. This step lays out common arguments that can be adapted to each city's particular needs.

### Areas of Concern in the East Asia and Pacific (EAP) Region

Carbon dioxide ($CO_2$) levels in our planet's atmosphere have reached the highest levels in recent history. The burning of fossil fuels such as oil, coal, and natural gas is the main reason behind the increase in $CO_2$. Increases in global temperature associated with these human-caused releases of $CO_2$ and greenhouse gases (GHGs) are now better understood, and the questions that global entities such as the World Bank are asking are no longer "what is causing it," but "what can we do to curb it." More evidence of changing climatic conditions, such as increased flooding, drought, and hurricane activity, is visible, further elevating climate change concerns to the top of geopolitical agendas. These factors will affect energy, urban growth, and economic policies given their impact on the growth of GHG emissions.

The rapid urbanization rate in the EAP region makes the increase in energy consumption—and corresponding GHG emissions—particularly pressing. In May 2010, the World Bank and the Australian Agency for International Development (AusAID) released *Winds of Change: East Asia's Sustainable Energy Future*, which outlined issues faced by the EAP region, underscoring the need for energy planning in the region. See the resources in box 12.1.

For the last three decades, the region has experienced the strongest economic growth in the world, with a tenfold increase in GDP. This growth is expanding urban centers, creating new growth in suburban areas, and significantly increasing the demand for energy. Although GHG emissions per capita are still low compared with developed countries, the region is expected to catch up, resulting in significant impacts on the local and global environment.

The *Winds of Change* report sends a clear message that the region has options for mitigating emissions growth without sacrificing economic competitiveness. If this path can be followed, the impact of fast-growing economies and urbanization can be balanced against the increasing need for energy reliability and environmental sustainability.

In the EAP region, long-term energy planning must be integrated with the wider planning processes for the land use, transportation, and buildings sectors. These are the major sources of energy and emissions growth in the region, and

---

**Box 12.1  Resources: *Winds of Change: East Asia's Sustainable Energy Future* (World Bank 2010)**

This report lays out alternative future scenarios for energy consumption. Based on these scenarios, the report posits a number of conclusions relevant to this Guidebook:

1. It is within the reach of East Asia's governments to maintain economic growth, mitigate climate change, and improve energy reliability.
2. To achieve these goals, governments must take immediate action to transform their energy sectors toward much higher energy efficiency and more widespread use of low-carbon technologies.
3. This shift to clean energy requires major domestic policy and institutional reforms.
4. Developed countries need to transfer low-carbon technologies and provide substantial financing for these technologies.

The World Bank is committed to scale up policy advice, knowledge sharing, and financing in sustainable energy to help the region's governments make such a shift. The combination of the *Winds of Change,* this Guidebook and its accompanying SUEEP Toolkit and direct technical and economic assistance are key interventions to help EAP cities build a sustainable energy future.

---

cities hold the key for regulating and changing the direction of growth toward a lower carbon pathway.

### Triple Bottom Line Thinking

Making the case for an energy and emissions plan is not only about justifying reduced energy costs and lower emissions. To capture citizens' hearts and minds, the SUEEP process will have to establish a connection to the city's larger plan, tying into its aspirations across a variety of sectors.

This concept is called "triple bottom line thinking." The idea is to establish ties not only to climate change aspirations (reduced carbon emissions) but also to align the initiatives with other environmental factors, economic development, and healthy social networks (figure 12.1). See the case study in box 12.2. For example, a project established under triple bottom line thinking might include the following components:

- **Economic:** Distributed generation techniques reduce energy costs for developers, creating more attractive long-term leases.
- **Social:** Reduced numbers of four-wheel vehicle result in cleaner air and reduces asthma cases in children.
- **Environmental:** Reduced cooling demand in buildings will reduce water consumption and align with future water-reduction goals.

**Figure 12.1  Triple Bottom Line Returns**

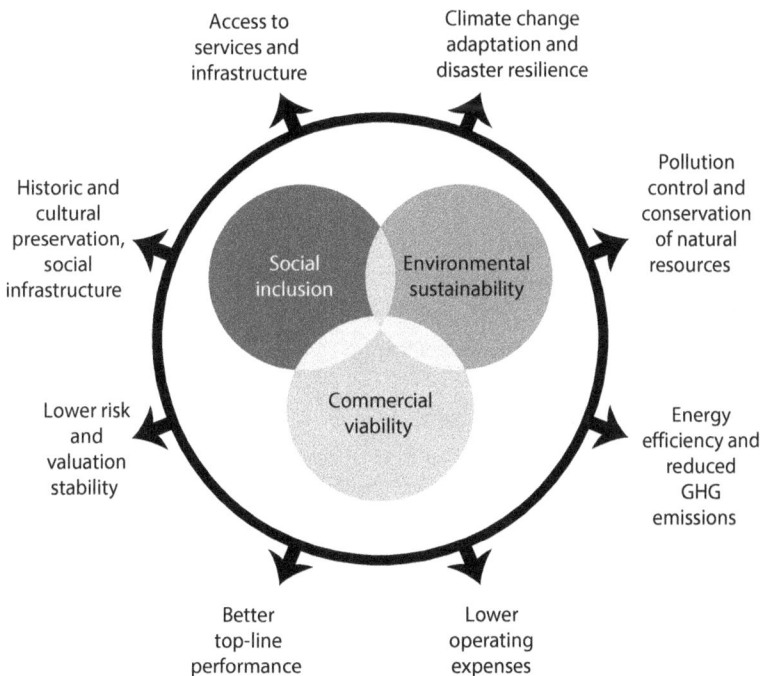

*Source:* Lenora Suki, © Smart Cities Advisors. Reproduced with permission; further permission required for reuse.
*Note:* GHG = greenhouse gas.

**Box 12.2  Case Study: Philadelphia, Pennsylvania, USA—Mayor Nutter Makes the Case**

In Mayor Michael Nutter's January 2008 inaugural address, he pledged to make Philadelphia the number one green city in America. His energy goal was "to reduce Philadelphia's vulnerability to rising energy prices." To make good on his pledge, he created the Mayor's Office of Sustainability.

The Mayor's Office of Sustainability spent a year researching city sustainability, talking with residents, and drafting Greenworks Philadelphia. The ambitious plan sets 15 sustainability targets in the areas of energy, environment, equity, economy, and engagement to make Philadelphia the greenest city in America by 2015.

Greenworks Philadelphia was released in spring of 2009, and in the first year of implementation the Mayor's Office of Sustainability and its partners made great strides toward making Philadelphia more sustainable.

*Source:* http://www.phila.gov/green/greenworks/index.html.

### Establish Links to Citywide Goals

Gaining support for the SUEEP process can be accomplished by developing synergies between energy savings and the qualities of a successful city. Find those interconnections, and give voice to the idea that this energy plan is not just about a greener planet, but that it is about the city's aspirations and its unique qualities. See the example in box 12.3.

These synergies and links can alter the perceived definition of an "energy" project (a project in which the benefits are mainly reduced energy costs) to one that enables the city to attain its vision. These links may be internal to the city, but might also reach beyond its boundaries. Stories about the extent of the project's reach may be most beneficial politically.

### Policies and Aspirations

City agencies have wide-ranging agendas and often inconsistent messages between departments. The creation of an energy and emissions plan can align goals across departments, for example:

- **Economic:** A bus rapid transit project needs more momentum and organization than just the Department of Transportation to obtain the required funding, so linking to energy and carbon can provide the support needed for implementation.
- **Social:** The departments of Parks, Transit, Buildings, and Health all agree that air quality concerns are important. The energy and emissions plan can serve as a bridge between these departments to provide unity and consistency.
- **Environmental:** Poor water quality in visible public places is recognized as a detriment to businesses near the contamination. An energy and emissions plan can be a vehicle for driving wider environmental issues to the top of policy agendas.

---

**Box 12.3  Example: Sample Links to Wider City Goals**

The following are a few examples of connections between energy projects and other city initiatives or aspirations.

1. Bus Rapid Transit (BRT) Boosts Local Businesses

   Markets, shops, and restaurants on the new BRT lanes have prospered in cities because bus riders are more likely to patronize shops than are drivers of cars who are unable to find parking spots.

2. Street Lighting Saves Lives

   Crime rates have decreased dramatically in areas with new street and site lighting, making energy savings a secondary issue to the increased safety associated with this project.

3. Parking Restriction Cleans Up City

   New parking projects have allowed for more efficient trash collections, with greater access at lower costs. Besides reduced congestion and lower emissions, the streets have become cleaner and more walkable.

---

## Understanding Current Progress

Cities may not realize that existing policies already employ some form of triple bottom line thinking. For example, congestion taxes may have been put in place as a result of scarce public parking, but the taxes, in turn, increase demand for more public transportation, which reduces overall GHG emissions, improves air quality, and reduces congestion on the city's roads.

Establishing connections to what has already been done will not only help tell the story, but may also reveal data that can be used to assure stakeholders that the projections for various scenarios are accurate. City governments should engage with various department leaders and local organizations to understand the current progress and, where politically appropriate, make the case for the energy and emissions plan by aligning it with successful initiatives and declaring that the plan will create continued synergies with these projects.

## The Price of Doing Nothing: Scenario Analysis

Making the case for a sustainable energy future requires an understanding of the issues that will occur in the upcoming years and decades. The costs of the SUEEP process and its organization and implementation may at first appear out of reach; however, the costs of business as usual will likely be much higher. The impact of doing nothing is considerable. The first task in making the case for an SUEEP process is understanding existing practices and what those practices are likely to lead to if maintained. See the example in box 12.4.

### Understanding Existing Policies

Gather information from administrative and sectoral leaders about current policies as they relate to energy and carbon concerns. Has anything changed in recent years?

Energizing Green Cities in Southeast Asia • http://dx.doi.org/10.1596/978-0-8213-9837-1

**Box 12.4  Example: *Winds of Change* GHG Scenarios**

The study underpinning this Guidebook examined two energy scenarios up to 2030: (a) a reference scenario, which features a continuation of current government policies (REF scenario); and (b) an alternative scenario of sustainable energy development (SED scenario), which aims to put the energy sectors on a sustainable path. Two separate studies projected the growth of energy demand in the transportation and household sectors. The transportation study examined the potential to reduce transportation fuel consumption through fuel economy standards, public transportation, urban planning, and pricing policies. The household study explored the potential to reduce residential electricity consumption through appliance efficiency. The potential emissions reduction is shown in figure B12.4.1.

**Figure B12.4.1  Emissions Gap between REF and SED Is Large, but Can Be Bridged by Energy Conservation and Low-Carbon Technologies, 2009–30**

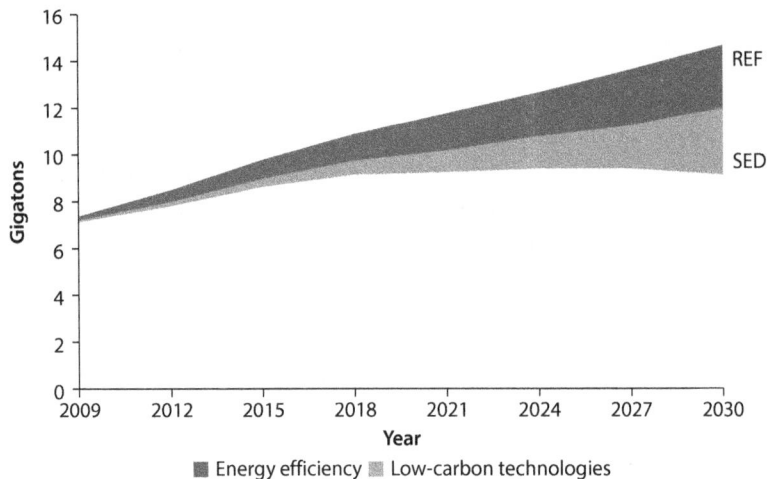

*Source:* World Bank 2010, 4.
*Note:* REF = reference scenario; SED = sustainable energy development scenario.

*Forecasting Growth*

Gather best estimates of the city's population growth. What are the demographic expectations for the new population?

*Recognizing Supply-Side Practices*

Review current supply-side energy approaches. How is the fuel mix likely to change? Will surrounding population growth and changes in city boundaries affect supply distribution in the region? What influence does the city government have with suppliers of energy and drivers of national and regional energy policies?

## Increasing Energy Intensity

Assess the impacts on the city of economic growth and urbanization. Will economic growth comprise more-energy-intensive businesses that replace less-energy-intensive business or industrial sectors? What will the urbanization trends mean for transportation end uses?

## Qualities of the City

Determine the social aspects of the city that are attractive and must remain, as well as the aspects that are unattractive or harmful and must be addressed. Will urbanization and increased energy consumption affect the livability of the city if policies and practices are unchanged?

## Energy Risks

Document the potential risks facing the city if current practices continue. What are the energy reliability concerns associated with regional growth and urbanization? Will increased carbon emissions diminish the attractiveness of the city to various business entities?

With a grasp of these issues, city governments can tell a "story" about where the city currently stands, the direction of energy use and emissions in the city if current practices remain unchanged, and where it could potentially go if actions were taken to tackle energy use. (See the technical assistance opportunity in box 12.5.) The adverse outcomes that accompany inaction will help to garner support for the SUEEP process.

### Find Compatible National Energy and Emissions Goals

Most countries in the EAP region have signed and ratified the Kyoto Protocol and adopted climate action plans. Most large-scale utility providers in the region recognize the risks associated with business as usual but do not have a clear understanding of the steps needed to change their trajectory.

Many nonprofit groups and climate change advocates are making the case for region-wide transformative practices and supply-side projects. Politically speaking, establishing connections with these groups can align a city with

---

### Box 12.5 Technical Assistance Opportunity: Energy Planner for Scenario Development

Developing models to project energy growth under various development scenarios may not be within the current skill sets and resources available to city administrations. Energy planning consultants from local technical universities or international energy consultants can undertake the technical calculations and work with city leadership to define various development scenarios, including the business-as-usual base case.

---

Energizing Green Cities in Southeast Asia • http://dx.doi.org/10.1596/978-0-8213-9837-1

these causes, giving the entire process momentum and high-level national support.

These ties will also help to create change when implementation funding, resources, or partnerships are needed. A mayor's sphere of influence extends far beyond the city walls.

## Step 8: Establish Goals

Now that the energy inventory is complete and projections for future growth of energy use have been made, the next step is to create goals. Although the primary objective of the SUEEP process is to improve energy efficiency across sectors, the SUEEP process provides flexibility should a city decide that its goal is to minimize GHG emissions. In such cases, other tools, such as the McKinsey cost curve (see the tip in box 12.6), would need to be used to perform a cost-benefit analysis.

The intent of establishing goals is to articulate targets based on key performance indicators in critical energy use sectors. Achieving the targets will define success. The goals should support the vision statement developed in Step 1 and should use simple overarching statements to frame a desired direction.

### Aggressive but Viable

Targets must be based on city-specific empirical data and analysis rather than simply the numbers used by neighboring or peer cities. Based on projections of various scenarios and assessments of either existing or potential projects, a city government would have a range of targets that it could strive to achieve.

The targets should be aggressive but viable and meaningful to stakeholders. See the case study in box 12.7. Although cities are expected to stick to their targets, changes to energy targets and goal statements are acceptable if new data come to light, or if external factors suggest that targets need to change.

---

### Box 12.6  Tip: McKinsey and Company GHG Abatement Cost Curve

McKinsey and Company's global GHG abatement cost curve (see figure B12.6.1) provides a quantitative basis for discussions about actions that would most effectively deliver emissions reductions and what they might cost. The cost curve in figure B12.6.1 shows the range of emission reduction actions possible with current technologies or with technologies likely to be available between now and 2030. The height of each bar represents the average cost of avoiding one ton of $CO_2$ emissions by 2030 through that technology. The width of each bar represents the potential of that technology to reduce GHG emissions in a specific year compared with business as usual.

*box continues next page*

**Box 12.6  Tip: McKinsey and Company GHG Abatement Cost Curve** (*continued*)

**Figure B12.6.1  Global GHG Abatement Cost Curve beyond Business as Usual—2030**

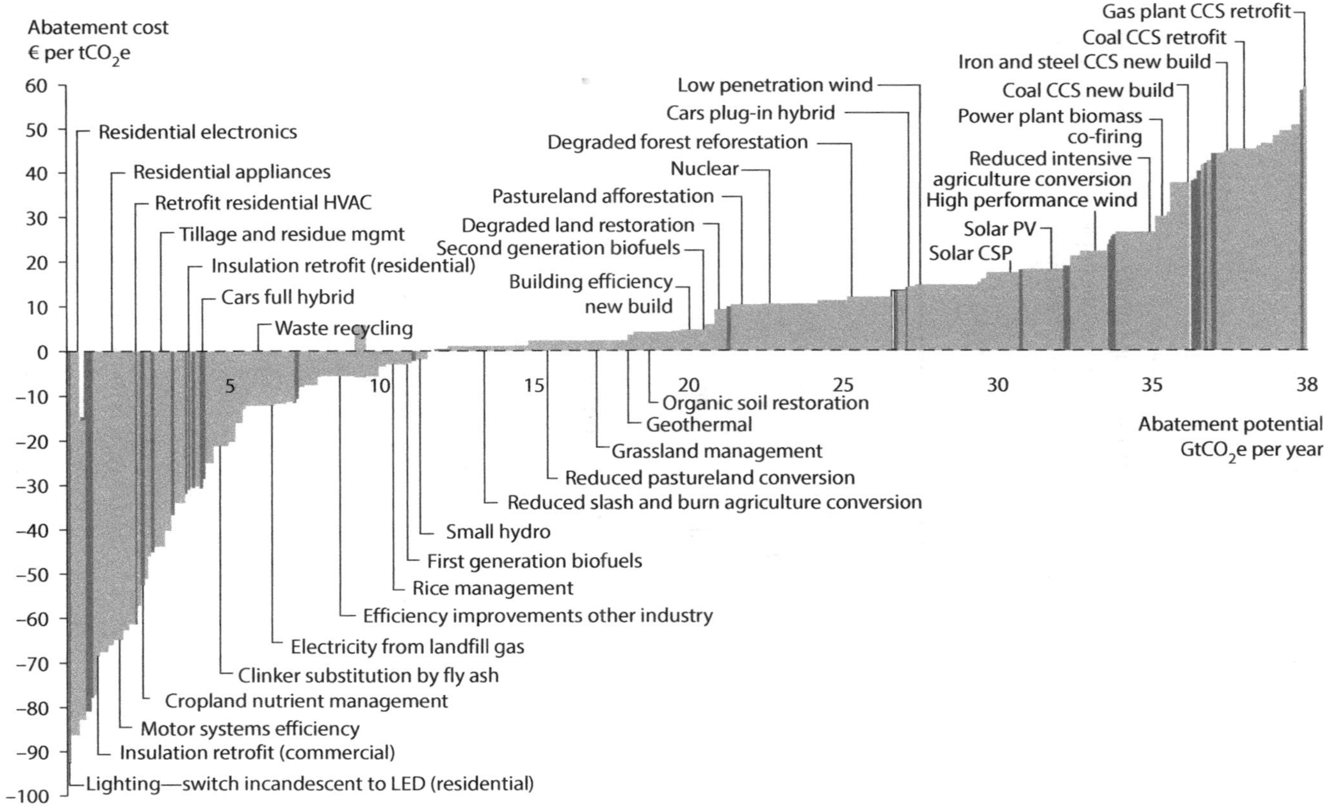

*Source:* © McKinsey and Company. Reproduced, with permission, from McKinsey and Company 2009; further permission required for reuse.

*Note:* The curve presents an estimate of the maximum potential of all technical GHG abatement measures below €60 per tCO$_2$e if each lever was pursued aggressively. It is not a forecast of what role different abatement measures and technologies will play.

CCS = carbon capture and storage; CSP = concentrating solar power; GtCO$_2$e = gigatons of carbon dioxide equivalent; HVAC = heating, ventilation, and air conditioning; LED = light-emitting diode; mgmt = management; PV = photovoltaic; tCO$_2$e = tons of carbon dioxide equivalent.

**Box 12.7  Case Study: Shanghai, China—Shanghai Sticks to Energy Goals**

Shanghai's target is to reduce energy usage by the equivalent of 800,000 tons of standard coal in 2010 with the eventual aim of cutting the standard coal equivalent by as much as 1 million tons.

The city government will strictly control the number of new projects with high energy consumption and curb excessive growth of industries with high emissions. Projects not in line with energy efficient standards will not be approved.

"It is austere to achieve 2010 targets to reduce emissions and save energy," Mayor Han said. "We have to rigorously stick to this goal and ensure the implementation of rules."

In 2010, Shanghai reduced energy use by 6.2 percent, beating a 3.6 percent target set at the start of the year.

*Source:* http://www.shanghai.gov.cn/shanghai/node27118/node27386/node27387/node27388/userobject22ai38370. html.

## Choosing Energy Units

Target statements used for quantifying a city's energy future can use several different units depending on whether energy costs, energy usage (such as million British thermal units or megajoules), or carbon emissions are the important factors for the city. Remember that GHGs are emitted from a variety of sources, not just the energy sector (for example, methane from agricultural practices is typically excluded from energy action plans), so noting the extent of the carbon commitment and the emissions sources used in the analysis are important when developing the SUEEP.

## Vision Statement Versus Target Statements

The city government should align its energy vision with its larger aspirations. Although the vision statement can be general, target statements should be quantifiable, trackable, and measurable. See the examples in boxes 12.8 and 12.9.

## Interim Targets

Goal statements look toward long-term targets, but short-term and mid-range targets ensure that long-term targets are met. These targets might be linear, that is, if the goal is a 20 percent reduction in 20 years, then 10 percent in 10 years might be appropriate. However, it is more likely that external factors and project momentum will build over time, and a project's effects may be compounded. The path from current energy use to future projections will probably not be linear, so understanding synergies between projects and the compounding of benefits are necessary for city governments to set realistic interim goals. See the case study in box 12.10 and the example in box 12.11.

## Feasibility of Targets

A host of factors affect the city's ability to meet its established targets, including changes in who holds the position of mayor from the time the SUEEP process commences to the future year in which targets should be attained. It is in the interest of the city government to establish targets that are within its means and that will be just as feasible for future leaders.

---

### Box 12.8  Example: Vision Statement versus Target Statement

**VISION**: A statement of the city's aspirations:
"A greener, greater city."

**GOAL**: A qualified statement beginning with an action:
"Maximize use of renewable energy."

**TARGET:** A quantifiable objective aligned with the goal:
"Reduce GHG emissions by 40 percent by 2030."

#### What makes a good target statement?

The following qualities have led to successful target statements:

- **Clear and concise:** Avoid verbose statements that are difficult to follow.
- **Not too technical:** Keep it simple. Use terms and units that the general public understands.
- **Add target date:** Establish a year that is practical for achieving the target.

---

### Box 12.9  Example: Disseminating Energy Goals throughout a City's Sectors

Figure B12.9.1 represents a sample city's concept of organization. The energy and emissions plan derived from the SUEEP process is a critical part of the city's overall vision, but it is not inclusive enough for the city as a whole.

This Guidebook outlines an energy planning framework that ties into other aspects of a city through linkages and synergies, but focuses on energy. The energy framework will have its own process and organization.

Energy goals are outlined for the city as a whole and should be connected to the city's vision. Energy goals should lead with an action word such as "Reduce," "Minimize," or "Align" and should support the city's vision.

Targets should be established that commit to achievement of quantifiable objectives by a certain date. Key performance indictors should be created to ensure that targets are monitored and that the city is on track to achieve them.

*box continues next page*

**Box 12.9  Example: Disseminating Energy Goals throughout a City's Sectors** *(continued)*

**Figure B12.9.1  Link between Energy Plan and City's Vision**

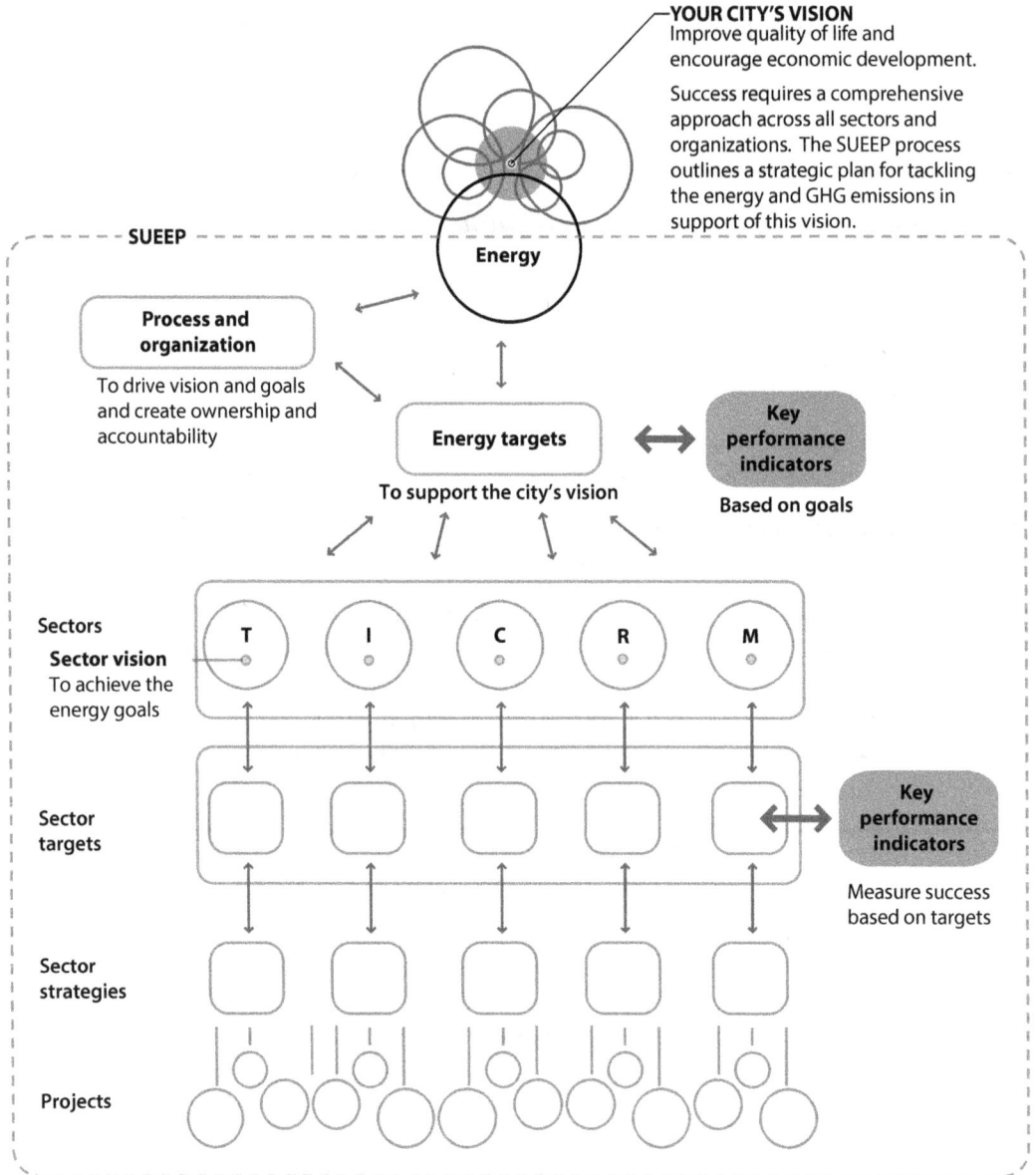

*Note:* GHG = greenhouse gas; SUEEP = Sustainable Urban Energy and Emissions Planning; T = Transportation; I = Industry; C = Commercial buildings; R = Residential buildings; M = City buildings, site lighting, water, waste.

**Box 12.10  Case Study: Barcelona, Spain—Targets for GHG Emissions Reductions**

The Barcelona Energy Improvement Plan (PMEB) forms the general framework for the work of the Barcelona City Council in matters of energy policy and its environmental impact on the city. Within this context, the PMEB includes an energy-related and environmental diagnosis of the present-day Barcelona and its future trends (to 2010), which allows the prediction of the increase of the city's energy consumption and its repercussions according to different scenarios. As a result of this analysis, the PMEB established a set of local action measures addressed to the achievement of a more sustainable city model, while reducing the environmental impact through energy savings, an increase in the use of renewable energies, and energy efficiency.

In the Target Scenario, GHG emissions were to be reduced by 30.3 percent compared with the Trend-Based Scenario for 2010, resulting in emission cuts of 2.76 tons of $CO_2$ equivalent per capita. To achieve this, the city rolled out the PMEB, which comprised 55 projects that were assessed from the energy-related, environmental, and economic standpoints.

*Source:* Barcelona, Spain undated.

## City Boundaries

Boundary issues were discussed in Step 4. Several factors will lie outside the control of a city's SUEEP process, given the flow of energy across boundaries and sectors:

- Energy grids will be affected by regional energy policies (for example, regional-level promotion of renewable energy and more-efficient and less-GHG-emitting technologies).
- Transportation networks are funded and maintained by national resources but affect the commuting patterns of a city because they cross city boundaries.
- Water and wastewater treatment facilities in outlying areas of a city may be outside a city government's physical and jurisdictional boundaries.

Understanding these external effects on energy use and emissions in the city will be useful. Similarly, adjacent cities that are altering their emissions and physical or political connections will affect the boundary. Projections will need to be based on the best assumptions that can be made about these external policies and regulations.

## How Granular Should Targets Be?

Although the level of detail contained in goals established under the SUEEP process will vary, all successful plans rely on measurable targets. For the first effort at an energy and emissions plan, targets should be overarching

## Box 12.11  Example: Sample Reduction Targets

Figure B12.11.1 shows examples of city government operations emissions reduction targets from CDP. The bars in the figure show the baseline year and target year; the number in the bar is the percentage of reduction.

The figure illustrates the wide range of reduction targets and time scales for the various cities.

Understanding the diversity of commitments is important.

**Figure B12.11.1  City Government Operations Emission Reduction Targets, by City**

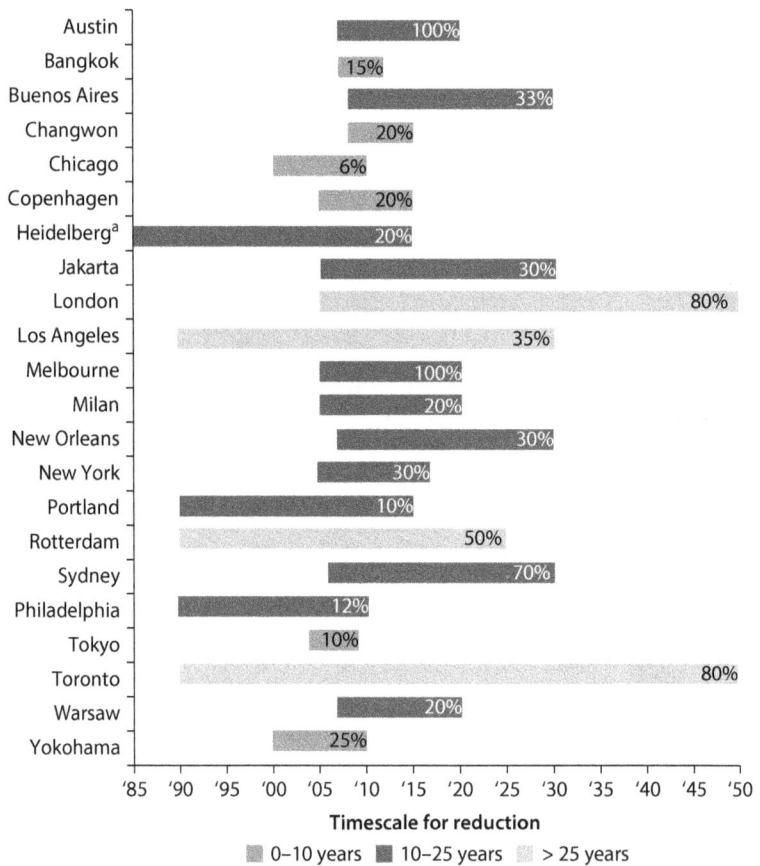

Timescale for reduction

▨ 0–10 years   ▇ 10–25 years   ▨ > 25 years

© 2011 Carbon Disclosure Project

*Source:* Carbon Disclosure Project 2011. Reproduced with permission.
a. City did not disclose a baseline.

(that is, applicable to sectors, not to specific projects) until more is understood about the potential opportunities and constraints in each sector.

Good examples of measurable targets follow:

- Reduce energy consumption by 20 percent by 2030.
- Reduce city transportation energy consumption by 20 percent by 2030.
- Reduce waste vehicle fleet petroleum consumption by 20 percent by 2030.

See the example in box 12.12 for an illustration of one city's sector-level targets that will help it to "bend the curve" of citywide energy consumption.

## Step 9: Prioritize and Select Projects

Inventories have been established, existing conditions and policies have been reviewed, potential projects have been assessed, and goals have been stated. The case has been made for change rather than a business-as-usual approach. Now initiatives and projects must be implemented to get the city from point A (today) to point B (a target tied to the goal statement).

Prioritizing projects takes a combination of common sense, quantifiable targets aligned with goals, and judgment by city leaders who understand the city's issues and politics.

This section outlines a three-stage approach to project prioritization. (Also see the technical assistance opportunity in box 12.13.) Any of these stages in

---

**Box 12.12  Example: Bending the Curve**

The city's overall SUEEP target is to reduce primary energy consumption by 18 percent from 2010 levels by 2020. Figure B12.12.1 illustrates how "bending the curve" is possible by reducing projected energy use across various sectors.

**How Is the Target Set?**
Energy growth by sector was projected based on forecasts of economic trends and population growth. Assuming that the city government continues with current policies, a business as usual scenario is established. Based on policies that the city government has the ability and influence to enact, a "moderate" projection was established that determined the potential energy savings that could be expected across sectors (that is, buildings, transportation, and so forth). This was then used to guide the target statement.

**Why 2020?**
Based on the task force's work with utility providers and key stakeholders, 2020 was set as the target because the parties could envision results from policies to stem energy growth coming to fruition by then. The targeted year was also far enough in the future that projects could be planned and implemented—but not so far that the year seemed out of reach and too long term.

*box continues next page*

**Box 12.12  Example: Bending the Curve** *(continued)*

**Figure B12.12.1  Illustration of Bending the Curve**

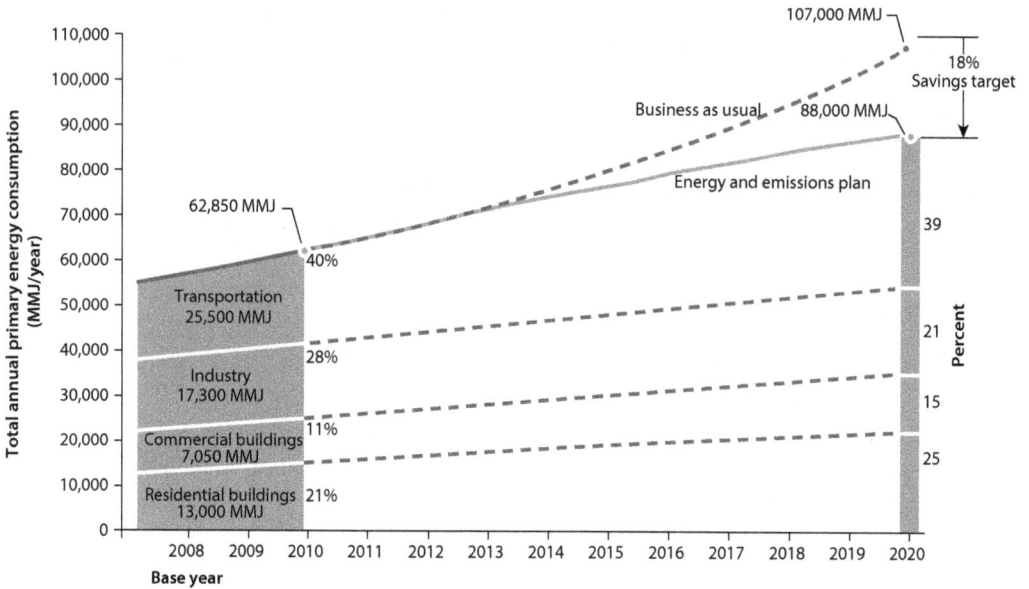

*Note:* MMJ = million megajoules.

isolation will likely lead to misuse of resources and long-term failure to achieve the energy goals. All three should be undertaken using a combination of analysis and judgment:

- Stage 1. Eliminate: Remove obvious projects from priority list.
- Stage 2. Prioritize: Create an analytical approach to understanding which projects have the highest potential to meet the city's goals and targets and rank them.
- Stage 3. Select: Qualitatively review the prioritized projects to select the ones to implement.

See figure 12.2 for an illustration of this process.

Most cities will categorize projects and targets into sectors, therefore, five sectors are used on the energy demand side (the letter in parentheses will be used throughout this section to denote a specific sector):

(T) = Transportation
(I) = Industry
(C) = Commercial buildings
(R) = Residential buildings
(M) = City buildings, site lighting, water, waste

## Box 12.13  Technical Assistance Opportunity: Multiple Levels of Engagement

The SUEEP process is based on a comprehensive approach, but the process outlined in Step 9 could be perceived by cities to be too detailed or more complex than necessary for a city's current stage of energy planning. Depending on a city's capacity, resources, and priorities, it could engage in energy planning at levels other than the SUEEP. A high-level, rapid assessment could be performed as an introduction (for example, the World Bank's Tool for Rapid Assessment of City Energy [TRACE]) or deeper sectoral engagements in a few selected areas (for example, public-private partnerships and sector-wide interventions). TRACE, developed by the World Bank's Energy Sector Management Assistance Program, offers cities a quick and easy way to assess their energy efficiency and identify underperforming sectors for possible improvement. This tool prioritizes sectors with significant energy savings potential, and identifies appropriate energy efficiency interventions across six city services—urban passenger transportation, buildings, water and wastewater, public lighting, solid waste, and power and heat.

The opportunities for energy efficiency in individual sectors are determined on the basis of the product of the following three factors built into the TRACE software:

- **Energy spending information.** This information is obtained either directly from the six sectors and city budget offices or through the conversion of energy use across the city into energy spending per sector.
- **Energy efficiency opportunity.** Opportunities to increase energy efficiency are determined using key performance indicators chosen from the TRACE benchmarking process that are most indicative of energy use across a particular sector or subsector. To define opportunity, the mean value of sectoral energy use of the better-performing cities in the peer group is calculated, and the difference between this value and the city's current performance provides an improvement target for the city; this is termed the "relative energy intensity" of the sector.
- **The control or influence of the city government.** This is determined by establishing the extent of influence that the city government has in each sector. This ranges from minimum (national government has greater or even full control) to maximum (city has full budgetary and regulatory control).

The sector prioritization process in TRACE is quick and automatic, but not as flexible as the one in the SUEEP Toolkit. TRACE identifies sectors that a city should prioritize, as compared with the SUEEP Toolkit, which identifies high-priority projects across sectors. It is not possible to change or add projects in the TRACE software as can be done in the Project Assessment and Prioritization Toolkit. So TRACE is not compatible with the comprehensive SUEEP process, but it may be a realistic alternative that can move the energy planning process forward. (See http://www.esmap.org/esmap/node/235 for details of the TRACE process.)

**Figure 12.2  Elimination, Prioritization, and Selection Process**

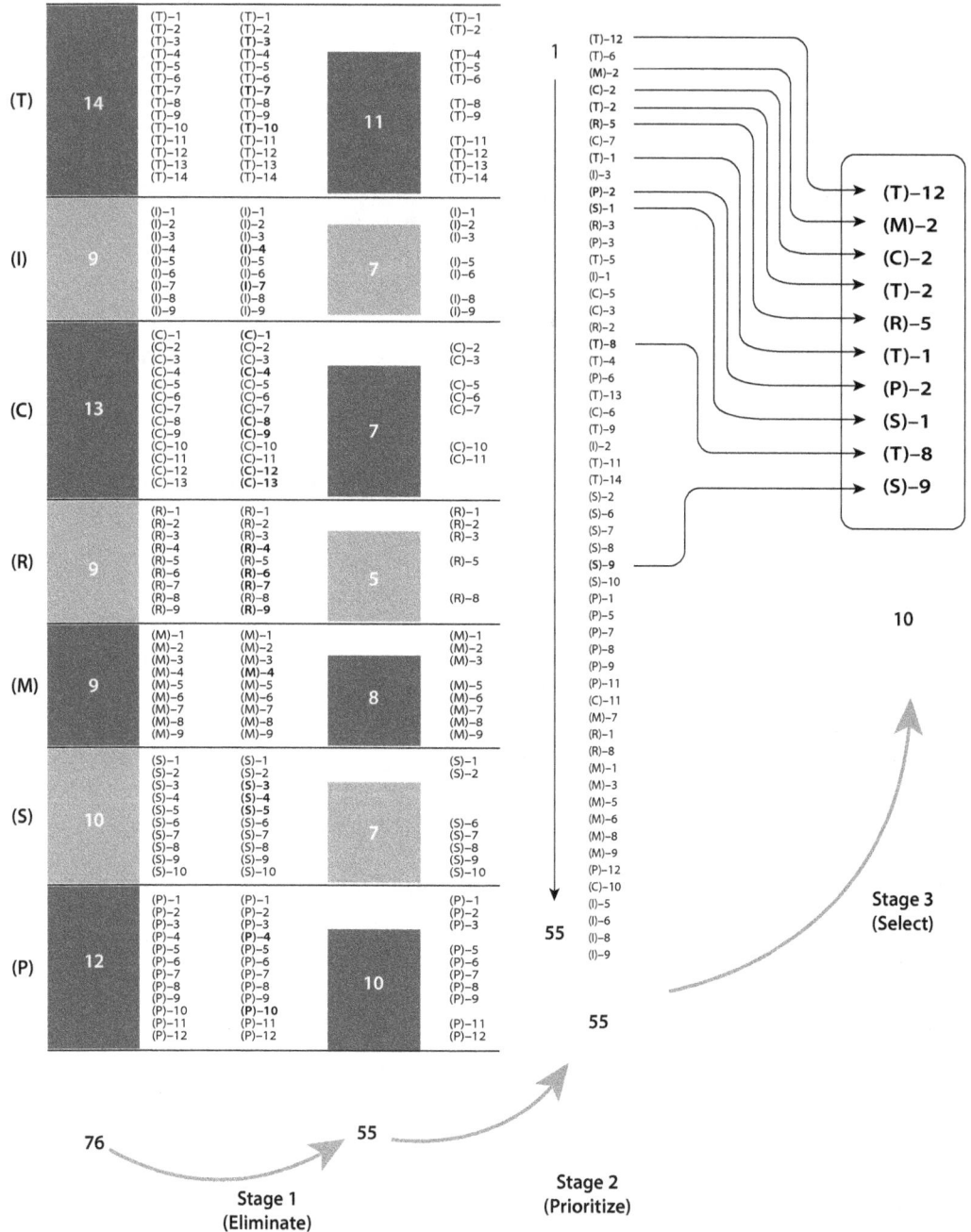

*Note:* T = Transportation; I = Industry; C = Commercial buildings; R = Residential buildings; M = City buildings, site lighting, water, waste; S = Supply; P = Process.

Two other sectors can be useful in organizing projects associated with the supply of energy and the process or organizational structure for developing the energy and emissions plan:

(S) = Supply
(P) = Process

### Toolkit Reference
#### Project Assessment and Prioritization Toolkit
The Project Assessment and Prioritization Toolkit, available at http://www .worldbank.org/eap/energizinggreencities, outlines 76 projects across the five demand-side sectors, as well as supply and process projects. Project characteristics are provided along with the formulas that allow the user to calculate a Project Score (PS) (explained in this Step).

### Stage 1: Eliminate
*Remove obvious projects from priority list.*
This first step is a nontechnical rejection of projects that are obviously inapplicable to the city or are clearly nonstarters (for example, because of a lack of resources or funding). This stage results in an immediate elimination of approximately 30 percent of the projects (76 down to 55).

Through all three stages, the city's goals and the energy goals for each sector should be clearly understood and kept in mind. For example, assume the goals in the transportation sector are as listed:

#### Transportation Sector Goals
1. Reduce the cost of energy used for transportation.
2. Minimize GHG emissions from transportation.
3. Maintain reliable energy supply to the transportation sector.

#### Strategies for Achieving the Sector Goals
The Project Assessment and Prioritization Toolkit outlines strategies for enhancing energy efficiency across different sectors. Strategies in the transportation sector comprise the following:

(T)A: Shift to nonmotorized transportation modes.
(T)B: Shift to public transportation.
(T)C: Improve efficiency of existing vehicles.
(T)D: Improve efficiency of new vehicles.

Table 12.1 shows the possible projects associated with the strategies for the transportation sector. The city may have lists of potential projects or it can use the list in the Project Assessment and Prioritization Toolkit available at http://www .worldbank.org/eap/energizinggreencities. The table shows which projects were eliminated in this example, and why. This stage reduces potential projects in the transportation sector to 11. These 11 projects now move to Stage 2.

**Table 12.1  Elimination of Projects in Stage 1**

|  |  | Transportation sector projects | Eliminate in Stage 1? |
|---|---|---|---|
| A. Shift to nonmotorized transportation modes | (T) A.1 | Bike lane expansion and improvement project | Keep |
|  | (T) A.2 | Mixed-use planning initiatives | Yes—This exercise was completed recently, and politically, it cannot be revisited. |
|  | (T) A.3 | Streetscape improvement project | Keep |
| B. Shift to public transportation | (T) B.1 | Bus rapid transit lines | Keep |
|  | (T) B.2 | Expand rail services | Keep |
|  | (T) B.3 | Carpool project | Keep |
|  | (T) B.4 | Parking restriction project | Keep |
|  | (T) B.5 | Park and ride project | Keep |
|  | (T) B.6 | Improve and expand ferry service | Yes—Landlocked city has no ocean or rivers. |
| C. Improve efficiency of existing vehicles | (T) C.1 | Vehicle emissions testing and compliance | Yes—Resources needed to administer and enforce this are not available. |
|  | (T) C.2 | Motorcycle (2–4 stroke) upgrade | Keep |
| D. Improve efficiency of new vehicles | (T) D.1 | Angkot replacement project | Keep |
|  | (T) D.2 | Four-wheel vehicle fuel efficiency tax | Keep |
|  | (T) D.3 | Taxi replacement and efficiency project | Keep |

*Note:* T = Transportation.

### Stage 2: Prioritize

*Use an analytical approach to understand which projects have the highest potential to meet the city's goals and targets.*

This stage is more detailed and can take several forms. This Guidebook contains a framework for quantifying the usefulness of the remaining projects for achieving the goals and aligning with the unique qualities of the city. This methodology ranks (from 1 to 55) projects across all sectors based on the set of characteristics reviewed in Step 6. More on this process and examples are included in the following pages. The characteristics discussed in Step 6 are listed below:

(C1) Energy savings potential
(C2) Fuel type savings
(C3) GHG savings potential
(C4) Implementation cost
(C5) Estimated cost savings
(C6) Recipient of savings
(C7) Likelihood of funding
(C8) Ease of implementation
(C9) Timing of project implementation
(C10) Level of city control

#### Review Project Score Criteria

Table 12.2 shows that a Project Score (PS) Criteria score has been assigned to each characteristic. The PS criteria will be used to rate projects on each characteristic, from 0 (low) to 10 (high). Taking the energy savings potential characteristic

**Table 12.2 Project Score Criteria**

| Characteristic (C) | | Project Score (PS) Criteria (0–10) | | | |
|---|---|---|---|---|---|
| | | The higher the PS, the better the project is aligned with the city's SUEEP goals | | | |
| (C1) Energy savings potential | Rank | Low | Medium | High | |
| | Definition (kWh/year) | <10,000 | 10,000–10,000,000 | >10,000,000 | |
| | PS | 0 | 1–9 | 10 | |
| (C2) Fuel type savings | Fuel | Motor gasoline/diesel | LPG | Grid electricity | Natural gas |
| | PS | 4 | 2 | 2 | 4 |
| (C3) GHG savings potential | Rank | Low | Medium | High | |
| | Definition (tons/year) | <1,000 | 1,000–10,000 | >10,000 | |
| | PS | 0 | 1–9 | 10 | |
| (C4) Implementation cost | Rank | High | Medium | Low | |
| | Definition ($) | >5,000,000 | 100,000–5,000,000 | <100,000 | |
| | PS | 0 | 1–9 | 10 | |
| (C5) Estimated cost savings | Rank | Low | Medium | High | |
| | Definition ($/year) | <10,000 | 10,000–100,000 | >100,000 | |
| | PS | 0 | 1–9 | 10 | |
| (C6) Recipient of savings | Saver | Energy services company | Utility or private entity | City residents | City government |
| | PS | 1 | 3 | 6 | 8 |
| (C7) Likelihood of funding | Rank | Low | Medium | High | |
| | PS | 0 | 1–9 | 10 | |
| (C8) Ease of implementation | Ease | Hard | Medium | Easy | |
| | PS | 0 | 1–9 | 10 | |
| (C9) Timing of project implementation | Timing | Slow | Medium | Fast | |
| | Definition (years) | >10 | 1–10 | <1 | |
| | PS | 0 | 1–9 | 10 | |
| (C10) Level of city control | Rank | Low | Medium | High | |
| | Definition | National stakeholder | Regional stakeholder | Budget and regulatory | |
| | PS | 0 | 5 | 10 | |

*Note:* GHG = greenhouse gas; kWH = kilowatt-hours; LPG = liquefied petroleum gas; SUEEP = Sustaining Urban Energy and Emissions Planning.

as an example, projects with the potential to provide savings of more than 10,000,000 kilowatt-hours per year (kWh/year) are scored "10" whereas those with the potential to provide savings of less than 10,000 kWh/year are scored "0." For the fuel type savings characteristic, a lower score is given if availability of primary fuel is not an issue in the city. For example, liquefied petroleum gas (LPG) is given a low score of 2 because there is no shortage of LPG in the city, which means that projects that could affect the use of LPG are not as important as those that could affect natural gas, the supply of which is more limited.

### Assign Project Scores

Table 12.3 shows the process for assigning scores to a specific project. As an example, the table uses Project (T) A.1 *Bike lane expansion and improvement project.*

Energizing Green Cities in Southeast Asia • http://dx.doi.org/10.1596/978-0-8213-9837-1

**Table 12.3  Assigning Scores to Project (T) A.1 Bike Lane Expansion and Improvement**

| Characteristics (C) | Project score and description |
|---|---|
| (C1) Energy savings potential | 4—Estimated to be 100,000 kWh/year |
| (C2) Fuel type savings | 4—Motor gasoline/diesel |
| (C3) GHG savings potential | 4—Roughly 4,000 tons/year |
| (C4) Implementation cost | 5—Estimated to be ~$1 million |
| (C5) Estimated cost savings | 4—Roughly $40,000/year |
| (C6) Recipient of savings | 6—City residents |
| (C7) Likelihood of funding | 8—Support groups have been raising funds |
| (C8) Ease of implementation | 4—Resources are in place, and this is moderately easy |
| (C9) Timing of project implementation | 5—Estimated total of 2–5 years for entire project |
| (C10) Level of city control | 8—Mostly within the city's control, with some minor private entity collaboration |

*Note:* GHG = greenhouse gas; kWH = kilowatt-hours.

The methodology should be applied to each project. Rating the projects according to a standard set of characteristics ensures that all projects are rated consistently.

### Assign a Weight to Each Characteristic
Table 12.4 is a characteristic weighting table, which provides a way to emphasize the importance of various characteristics for achieving the city's goals. Because the PS will be multiplied by the characteristic weight (CW), the higher the number, the more the city values the characteristic.

### Calculate a Total Weighted Project Score (TWPS)
Each PS is multiplied by the CW, then summed to provide a Total Weighted Project Score (TWPS). Table 12.5 shows the process and calculations, leading in this example to prioritization of Projects 2, 1, and 3 (from greatest to least TWPS). Table 12.5 shows only three characteristics (of 10) to clarify the process for calculating the TWPS. A more comprehensive example appears in box 12.14.

In addition to a sector-by-sector calculation, a PS should be determined for each project across all sectors in Stage 2, that is, run two exercises: First, prioritize projects by individual sector, which will allow you to select the top two or three projects in each sector. Second, combine all projects from all sectors into one larger grouping and prioritize across sectors. The projects with the highest scores from each of these prioritization exercises are the projects that should advance to Stage 3 for final selection.

It is not unusual for a single sector to be heavily weighted in the TWPS process. For example, if 11 Transportation projects and 7 Industry projects make it past Stage 1 (Eliminate), it may prove in Stage 2 that all 11 Transportation projects score higher than any of the Industry projects. Understanding the reasons for this as well as the forces that lead to this ranking in Stage 2 is important.

Energizing Green Cities in Southeast Asia  •  http://dx.doi.org/10.1596/978-0-8213-9837-1

**Table 12.4  Weighting Characteristics**

| Characteristic weight (CW) | Importance |
|---|---|
| 0 | No importance |
| 1–4 | Increasing importance |
| 5 | Critical to the city |

**Table 12.5  Calculating TWPS**

| | (C1) | (C2) | (C3) | (C4) … (C10) | |
|---|---|---|---|---|---|
| Characteristic weight (CW) (1–5) | 5 | 2 | 3 | | Total Weighted Project Score (TWPS) |
| | Project Score (PS) (1–10) | | | | |
| Project 1 (Ex. [T] A.1) | 4 | 4 | 4 | | (5 × 4) + (2 × 4) + (3 × 4) = 40 |
| Example—Project 2 | 9 | 3 | 3 | | (5 × 9) + (2 × 3) + (3 × 3) = 60 |
| Example—Project 3 | 3 | 5 | 2 | | (5 × 3) + (2 × 5) + (3 × 2) = 31 |

City governments should take a comprehensive view of the projects, considering the unique attributes of their cities (including constraints faced and the structure of their economies) to achieve a deeper understanding of the factors that could result in the outcomes of Stage 2. A lower-ranked project (an Industry project that does not score as high as a Transportation project) may ultimately be selected for implementation based on the judgment of city leaders. This process occurs in Stage 3.

Based on the analytical exercise in Stage 2, the transportation sector projects in the table in the example in box 12.14 were calculated to have the highest potential to meet the city's goals. The ultimate rankings are shown in table 12.6.

Now doing the analysis across all sectors will result in the ranking of projects across a wide variety of end uses, demand sectors, and organizations. See the technical assistance opportunity in box 12.15.

### Stage 3: Select
*Provide a Qualitative Review and Select High-Priority Projects*
After eliminating (Stage 1) and prioritizing using quantitative analysis (Stage 2), it is time to select high-priority projects (Stage 3). This stage requires the judgment of city leadership (mayor, task force, stakeholders, consultants) to choose projects based on the political and socioeconomic realities of the city.

Highly ranked projects emerging from Stage 2 may not necessarily be implemented because internal and external forces can compel the selection of lower-ranked projects. Such forces will be based on the challenges, constraints, and pressures that the city faces. Sometimes these forces are so powerful or important to the city that an additional "characteristic" should be added to Stage 2 to address these concerns.

For example, in the hypothetical city's selection process in table 12.7, (T) A.1 *Bike lane expansion and improvement project* received the second-highest TWPS of all the projects considered. Project (C) E.3 *Public office buildings audit*

# Box 12.14 Example: Prioritization of Transportation Sector Projects Using the TWPS Method (Table B12.14.1)

## Table B12.14.1 Project Prioritization

| | (C1) | (C2) | (C3) | (C4) | (C5) | (C6) | (C7) | (C8) | (C9) | (C10) | Total |
|---|---|---|---|---|---|---|---|---|---|---|---|
| | Energy savings potential | Fuel type savings | GHG savings potential | Implementation cost | Estimated cost savings | Recipient of savings | Likelihood of funding | Ease of implementation | Timing of project implementation | Level of city control | Weighted Project Score (TWPS) |
| Characteristic weight | 5 | 2 | 3 | 4 | 3 | 2 | 3 | 4 | 2 | 2 | |
| Transportation sector projects | | | | | Project Score | | | | | | |
| (T) A.1 Bike lane expansion and improvement project | 4 | 4 | 4 | 5 | 4 | 6 | 8 | 4 | 5 | 8 | 150 |
| (T) A.3 Streetscape improvement project | 2 | 4 | 2 | 3 | 2 | 3 | 9 | 2 | 3 | 7 | 103 |
| (T) B.1 Bus rapid transit lines | 6 | 4 | 6 | 2 | 6 | 3 | 6 | 2 | 3 | 7 | 134 |
| (T) B.2 Expand rail services | 8 | 4 | 8 | 1 | 8 | 3 | 2 | 1 | 1 | 8 | 134 |
| (T) B.3 Carpool project | 4 | 4 | 4 | 8 | 4 | 3 | 7 | 8 | 9 | 5 | 171 |
| (T) B.4 Parking restriction project | 3 | 4 | 3 | 8 | 3 | 3 | 4 | 7 | 7 | 5 | 143 |
| (T) B.5 Park and ride project | 3 | 4 | 3 | 8 | 3 | 3 | 3 | 8 | 8 | 7 | 150 |
| (T) C.2 Motorcycle (2–4 stroke) upgrade | 3 | 4 | 3 | 6 | 3 | 3 | 2 | 6 | 1 | 3 | 109 |
| (T) D.1 Angkot replacement project | 3 | 4 | 3 | 5 | 3 | 8 | 3 | 5 | 5 | 3 | 122 |
| (T) D.2 Four-wheel vehicle fuel efficiency tax | 2 | 4 | 2 | 7 | 2 | 3 | 3 | 8 | 7 | 3 | 125 |
| (T) D.3 Taxi replacement and efficiency project | 3 | 4 | 3 | 6 | 3 | 8 | 2 | 7 | 7 | 5 | 139 |

*Note:* C = characteristic; GHG = greenhouse gas; T = Transportation.

**Table 12.6  Ranking by TWPS**

| Project | | TWPS | Rank |
|---|---|---|---|
| (T) B.3 | Carpool project | 171 | 1 |
| (T) B.5 | Park and ride project | 150 | 2 |
| (T) A.1 | Bike lane expansion and improvement project | 150 | 2 |
| (T) B.4 | Parking restriction project | 143 | 4 |
| (T) D.3 | Taxi replacement and efficiency project | 139 | 5 |
| (T) B.1 | Bus rapid transit lines | 134 | 6 |
| (T) B.2 | Expand rail services | 134 | 6 |
| (T) D.2 | Four-wheel vehicle fuel efficiency tax | 125 | 8 |
| (T) D.1 | Angkot replacement project | 122 | 9 |
| (T) C.2 | Motorcycle (2–4 stroke) upgrade | 109 | 10 |
| (T) A.3 | Streetscape improvement project | 103 | 11 |

*Note:* T = Transportation; TWPS = Total Weighted Project Score.

---

**Box 12.15  Technical Assistance Opportunity: Prioritizing Projects**

City governments, task forces, and stakeholder groups are the best people to provide the details for developing projects because they understand the internal issues, challenges, and opportunities of each project. But technical advisers who have gone through the project prioritization and selection process before have the best skills and experience for estimating energy savings, costs, and the impacts of each project and thus for assisting with prioritizing projects. Support from industry experts and planners should also be sought.

---

*and retrofit* was ranked 21, but the mayor believed that this was a crucial first step, despite its score.

A number of these guiding principles apply to the EAP region and are outlined in the *Winds of Change* (World Bank 2010), referred to in Step 7. See the example in box 12.16 for some of the principles city leadership may use in project selection.

### Finding the Balance

The city's aspirations and the project opportunities should both be clear after completion of the previous steps. Complementarities between the energy sector and broader-reaching engagements will have come to light. Most cities will begin by implementing low-cost, easy-win projects that require relatively minimal resources and have traditionally led to high energy reductions in a short time. Examples include

- Congestion pricing
- Building codes
- Street lighting

**Table 12.7  Project Selection**

| Rank | | Project | TWPS | Selected at Stage 3 |
|---|---|---|---|---|
| 1 | (T) B.3 | Carpool project | 171 | Yes |
| 2 | (T) B.5 | Park and ride project | 150 | Yes |
| 3 | (T) A.1 | Bike lane expansion and improvement project | 150 | No |
| 4 | (C) C.1 | Commercial building operator awareness training | 148 | Yes |
| 5 | (T) B.4 | Parking restriction project | 143 | Yes |
| 6 | (C) D.1 | Tenant metering project | 140 | Yes |
| 7 | (T) D.3 | Taxi replacement and efficiency project | 139 | Yes |
| 8 | (R) B.3 | Code compliance improvement project | 137 | Yes |
| 9 | (T) B.1 | Bus rapid transit lines | 134 | No |
| 10 | (T) B.2 | Expand rail services | 134 | Yes |
| 11 | (R) C.1 | Residential unit metering project | 130 | No |
| 12 | (T) D.2 | Four-wheel vehicle fuel efficiency tax | 125 | No |
| 13 | (T) D.1 | Angkot replacement project | 122 | No |
| 14 | (C) B.3 | Code compliance improvement project | 118 | No |
| 15 | (C) E.1 | Schools audit and retrofit project | 111 | Yes |
| 16 | (T) C.2 | Motorcycle (2–4 stroke) upgrade | 109 | Yes |
| 17 | (M) A.1 | City building audit and retrofit project | 106 | No |
| 18 | (T) A.3 | Streetscape improvement project | 103 | No |
| 19 | (C) B.1 | Update commercial building energy code | 100 | No |
| 20 | (P) A.1 | City government energy task force | 98 | Yes |
| 21 | (C) E.3 | Public office buildings audit and retrofit project | 98 | Yes |
| 22 | (P) D.4 | Partner with grants-funding agencies | 98 | No |
| 23 | (C) D.2 | Tenant behavior and energy efficiency awareness | 98 | No |
| 24 | (R) C.2 | Residents behavior and energy efficiency awareness | 96 | No |

*Note:* TWPS = Total Weighted Project Score; T = Transportation; C = Commercial buildings; R = Residential buildings; M = City buildings, site lighting, water, waste; P = Process.

---

## Box 12.16  Example: Some Guiding Principles in the Selection Stage

### 1. Fuel type stress

External supply constraints on a particular fuel may lead to the prioritization of projects that address this constraint, despite potentially higher capital costs. The costs of not implementing these projects will be higher than taking on projects that may have scored high in Stage 2.

*Example:* Natural gas constraints drive a low-ranked energy efficiency residential air conditioning project.

### 2. Political expedience

Sometimes people speak out, protest, or petition, and selection of a lower-ranked project is the politically wise action. The selection and successful implementation of one project instead of another may allow for political leverage.

*Example:* Concerns about air quality around a school yard leads to support for regulations to ensure industrial process efficiency.

---

*box continues next page*

**Box 12.16 Example: Some Guiding Principles in the Selection Stage** *(continued)*

### 3. Strong donor opportunity
A third-party group or private investor may push for a public-private partnership in which both groups benefit economically and the city grows closer to reaching its goals.
*Example:* A developer wants to invest in a transit hub on its new property, supporting transit projects that could not previously be funded.

### 4. Synergy with other urban problems
The mayor or other senior leader recognizes connections with other city qualities that can benefit from the implementation of projects.
*Example:* Safety issues drive a streetscape plan that includes street lighting and bike lanes.

However, all cities' aspirations, resources, and commitments are different, so the choice of projects to meet energy goals will be unique to the city. If carbon reduction is a priority, resources should focus on shifting fuel types, providing incentives for distributed generation, and perhaps encouraging renewable energy solutions. If air pollution is a major issue, projects could be focused on congestion pricing, constructing new bike lanes, or upgrading building codes for exhaust locations.

### Stick to the Vision
Prioritizing means choosing what is right for the city within the resources available. When in doubt, the city government should return to its vision and goal statements. Projects should be chosen to achieve the goal and communicate the success using the goal statement; for example, "New Dedicated Bike Lane Project Positions City to Achieve Carbon Emissions Reduction Targets."

### Easy Wins versus Repositioning for Change
Short-term wins make for good press. They give credibility to the SUEEP process and to the city's goals. Quick achievements should continue to be exploited, discussed, and communicated as successes. But prioritizing projects should not just mean selecting projects that give the best short-term results—those that position the city for long-term success are also crucial.

### Know the Strengths of the City and the People
What can finally be implemented depends on what the administrative groups can achieve based on their resources, knowledge base, and influence. The success of the projects and the overall SUEEP process will be determined by the groups designated to lead and administer these projects.

Understanding the current strengths of the departments, the public sector, or the nonprofit presences should influence project prioritization. Relying on an unmotivated or unqualified individual to lead a priority project will not work. Empowering strong leaders will increase the odds of success.

## References

Carbon Disclosure Project. 2011. *CDP Cities 2011: Global Report on C40 Cities.* London: Carbon Disclosure Project. http://c40citieslive.squarespace.com/storage/CDP%20Cities%202011%20Global%20Report.pdf.

Barcelona, Spain. undated. "Plan for Energy Improvement in Barcelona." Summary (in English): http://www.barcelonaenergia.cat/document/PMEB_resum_eng.pdf. Complete document (in Catalan): http://www.barcelonaenergia.cat/document/PMEB_integre_cat.pdf.

McKinsey and Company. 2009. *Pathways to a Low-Carbon Economy, Version 2 of the Global Greenhouse Gas Abatement Cost Curve.* Washington, DC: McKinsey and Company. https://solutions.mckinsey.com/ClimateDesk/default.aspx.

World Bank. 2010. *Winds of Change: East Asia's Sustainable Energy Future.* Washington, DC: World Bank.

# CHAPTER 13

# Stage IV: Planning

*This stage outlines the process for compiling the energy and emissions plan. This document will represent the city's vision and will be what the public and the international community will see. Therefore, the document should be clearly written, summarizing the city's current situation and its aspirations for achieving a sustainable energy future. The energy and emissions plan should summarize why an alternative energy future is important, and it should make the case for the specific initiatives and projects that will be implemented in the coming years.*

*A good energy and emissions plan can garner internal support from throughout the city government and help to motivate city workers to support the mission and the purpose of new programs. A good energy and emissions plan can convey the city's sophistication and organization to international funding agencies and private sector energy businesses. This is a critical part of the Sustainable Urban Energy and Emissions Planning (SUEEP) process and should be developed with care and attention.*

## Step 10: Draft the Plan

This section describes the crafting of the SUEEP process into a public document that summarizes and synthesizes the planning efforts made thus far. The energy and emissions plan is the midpoint of the planning process—it marks the end of the analysis and inventory phase and the beginning of the implementation and action phase.

### Purpose of the Energy and Emissions Plan

The energy and emissions plan synthesizes the work undertaken on energy planning and outlines the steps a city will take to alter the course of its energy consumption.

The document should inform the public of new initiatives, explain convincingly why an alternative energy future is important, inspire city residents and businesses to take action and contribute to improving the city, and show the international donor community that the city is organized and serious about implementing energy projects. The energy and emissions plan is typically not a legal document, although the energy plans it espouses may be formally adopted by legislative bodies.

The energy and emissions plan is different from the SUEEP status report outlined in Step 17 (see chapter 15 of this volume). The energy and emissions plan is created before projects are implemented and sets out the long-term vision, goals, and strategies of the city's leadership. In contrast, the status report provides updates about progress toward the goals set out in the energy and emissions plan.

### Audience for the Energy and Emissions Plan

The energy and emissions plan has a wide audience including city residents, city workers, private sector businesses, international energy services companies, financial institutions, international donor agencies, and peer cities that may follow in your city's footsteps. Because this audience is wide and diverse, the document must be straightforward and intelligible, not heavily weighted with data, but also not simply a compendium of generalities and platitudes. It should be easy to read and contain specific goals and targets that are achievable and relevant to stakeholders.

### Who Should Write the Energy and Emissions Plan?

The lead agency or energy task force within the mayor's office will be the central point of contact and coordination for the entire SUEEP process. Thus, this lead agency should pull together all the data, projects, benchmarking, and background required for the energy plan. This agency should also have close ties to the mayor or the city council, who will ultimately approve the energy and emissions plan. The agency should also seek stakeholder input on the document before it is finalized. The actual writing of the energy and emissions plan may be done by staff of the agency, or a local or international consultant may write and compile the document depending on the capacity, budget, and time available to the agency. The first energy and emissions plan is more likely to be written by nonpermanent staff, whereas future energy and emissions plans may be written in house once city government officials become more familiar with the process and can tailor the plan to local experiences.

### Style and Length

Because the energy and emissions plan is meant to be public and easily comprehended by a wide variety of stakeholders, it should be a graphically compelling document that illustrates data and policies using figures and diagrams that make energy data interesting and easy to read. (See the technical assistance opportunity in box 13.1.) It should contain comparisons with other cities and should refer to aspects of everyday life that help nontechnical readers understand the targets and how the goals can be reached. Its description of the urgent need to change the course of the city's energy consumption should be inspirational and should showcase the thought and efforts put into setting the goals, targets, and strategies. These elements of the document should be tailored to the city so that readers believe that they are achievable. Most energy plans are between 50 and 100 pages and are produced in both digital format and hard copy.

---

**Box 13.1  Technical Assistance Opportunity: Graphic Design and Report Writing**

Some city governments have graphics departments that can publish a compelling document, but even large cities such as Chicago and London hire external consultants to put the SUEEP information into a format and style that is attractive and readable. Conveying technical data in graphically simple ways requires graphic design software and unique experience. This may also be a resourcing requirement because permanent staff may already be busy with existing obligations and the SUEEP process is a biannual activity.

---

**Box 13.2  Resources: Sample Energy and Emissions Plans for Reference**

Examples of how other cities have developed their plans are in the references below:
Singapore: *Climate Change and Singapore: Challenges. Opportunities. Partnerships.*
New York, New York, USA: *PlaNYC 2030: A Greener, Greater New York*
Birmingham, UK: *Birmingham 2026, Our Vision for the Future*
Toronto, Canada: *Climate Change, Clean Air and Sustainable Energy Action Plan: Moving from Framework to Action*
Huntington Beach, California, USA: *City of Huntington Beach Energy Action Plan*
Dublin, Ireland: *Dublin City Sustainable Energy Action Plan 2010–2020*
Tbilisi, Georgia: *Sustainable Energy Action Plan—City of Tbilisi for 2011–2020*
Cape Town, South Africa: *Moving Mountains: Cape Town's Action Plan for Energy and Climate Change*
Christchurch, New Zealand: *Sustainable Energy Strategy for Christchurch 2008–18*

---

## Contents of an Energy and Emissions Plan

All SUEEP processes are different, and as it develops, the contents of the energy and emissions plan will evolve to fit the process. See the resources in box 13.2 and the case study in box 13.3. Some energy and emissions plans will be produced earlier, during the data collection and analysis steps, and some cities will have done significant background research and planning before writing the document. The example in box 13.4 provides a general template of the contents of a typical energy and emissions plan. Although a city's first energy and emissions plan may not be a fully comprehensive document, it should at the very least incorporate the results of the energy balance and greenhouse gas (GHG) inventory to profile the city's energy use, outline its targets and projects to change energy use patterns, and flesh out an action plan to demonstrate the steps the city will take to begin to implement the energy and emissions plan.

## Chapter Summaries
### Inventory and Benchmarking

The first chapter of the energy and emissions plan (after an executive summary) should summarize the energy balance and GHG emissions inventory and highlight the major energy users by sector. Data should be put into context using

**Box 13.3  Case Study: Dublin, Ireland—Sustainable Energy Action Plan**

Dublin published its "Sustainable Energy Action Plan 2010–2020" in a 45-page document. The document includes the city's energy vision, a summary of the GHG inventory, and the energy planning actions the city will take to reduce its GHG emissions. The document is full of graphs, tables, charts, and examples of ongoing energy projects, and it reads easily and provides information to a wide variety of readers.

*Source:* Dublin City Council and Codema 2010.

**Box 13.4  Example: Table of Contents of an Energy and Emissions Plan**

All SUEEP processes are different, but a starting place for the contents of an energy and emissions plan follows:

1. Inventory and Benchmarking
2. Energy and Emissions Growth Projections
3. Energy and Emissions Goals
4. Priority Projects
5. Financial Resources
6. Action Plan

historical trends and previous goals set by the mayor relating to the use of energy. Progress indicators for specific sectors should also be compared with those for similar cities and peer cities. Data should be illustrated in an interesting and informative manner using graphs and charts.

This chapter should also describe likely opportunities for improvement.

(Collecting data for and calculating the energy balance and the GHG inventory are described in Step 4 of this Guidebook [see chapter 11 of this volume].)

*Energy and Emissions Growth Projections*

This chapter describes the wide variety of future energy consumption scenarios that a city could face. Using the diagnostics of the current inventory year, it makes a number of assumptions about how energy consumption will change in the coming 10–20 years. Such assumptions should take into account trends that accompany economic growth in the East Asia and Pacific region, such as the growth of energy consumption as a result of shifts from two-wheel vehicles to cars; from industrial manufacturing to tourism and knowledge work; from multigenerational housing to higher quality single family housing; and from traditional naturally ventilated buildings that use relatively little energy to higher quality, international-style office space with air conditioning, computers, and overhead lighting.

Because growth and trends for the future are uncertain, a number of alternative scenarios for high and low growth should be examined. Assumptions should also include trends such as densifying urban centers versus city expansion via low-density development on the outskirts of town.

(The process for developing growth projections and scenarios is described in Step 8 of this Guidebook [see chapter 12 of this volume].)

### Energy and Emissions Goals

The energy and emissions plan should state the city's goals and priorities for an alternative energy future to clarify the vision and to set targets for energy and emissions reductions. These goals should coincide with the mayor's agenda, helping to prioritize projects and make decisions about which investments will create the most significant change in pursuit of the desired goals. Goals and targets are specific and should apply to citywide performance as well as to each sector.

(Setting goals is described in Step 8 of this Guidebook [see chapter 12 of this volume]).

### Priority Projects

The final three chapters set out the specific steps for achieving the city's goals. This chapter summarizes how all possible projects were evaluated and which were given priority. The mix of high-priority projects should include some quick and easy projects for quick wins, some long-term big picture projects, and projects that cut across all sectors. This chapter will briefly describe each project. Projects can range from energy efficiency to organizational or institutional development projects. Some projects may also be the establishment of policies that set the stage for the viability of future energy projects.

(Prioritizing and selecting projects can be found in Step 9 of this Guidebook [see chapter 12 of this volume].)

### Financial Resources

This chapter summarizes the costs and benefits of proposed high-priority projects. It also shows how the city plans to pay for each project, but it is not meant to be a detailed cost exercise with extensive figures or analyses. Because costs and benefits of various projects accrue to multiple stakeholders, including those outside the city, the energy and emissions plan should address this issue in a compelling way. This chapter should also make a strong financial case for the plan's benefits to the city's macroeconomic situation and how it will boost rather than hinder economic growth.

(Financing mechanisms for energy projects can be found in Step 14 of this Guidebook [see chapter 14 of this volume].)

### Action Plan

The final chapter outlines the short- and long-term actions required to implement the plan using a time-based sequence of activities. The action plan will show the institutional and policy changes necessary as a result of the SUEEP process. Seeing the 1-year action plan as well as a 10-year outlook helps to put the process into perspective for readers who want to know what is happening and when it is happening.

An action plan must be time-based and a schedule must be set for meeting major milestones. The energy and emissions plan is only one step along the path, but it should include an action plan summary.

The action plan has two major components:

- The institutional or organizational activities required for implementation
- Actual project implementation phasing, annual monitoring and reporting activities, and cycles of financial planning and recalibration of implementation plans

The example in box 13.5 shows an action plan with major milestones set out over a 10-year horizon along with regular GHG inventories, budget allocations, and status reports.

**Box 13.5  Example: Sample 10-Year Energy and Emissions Action Plan**

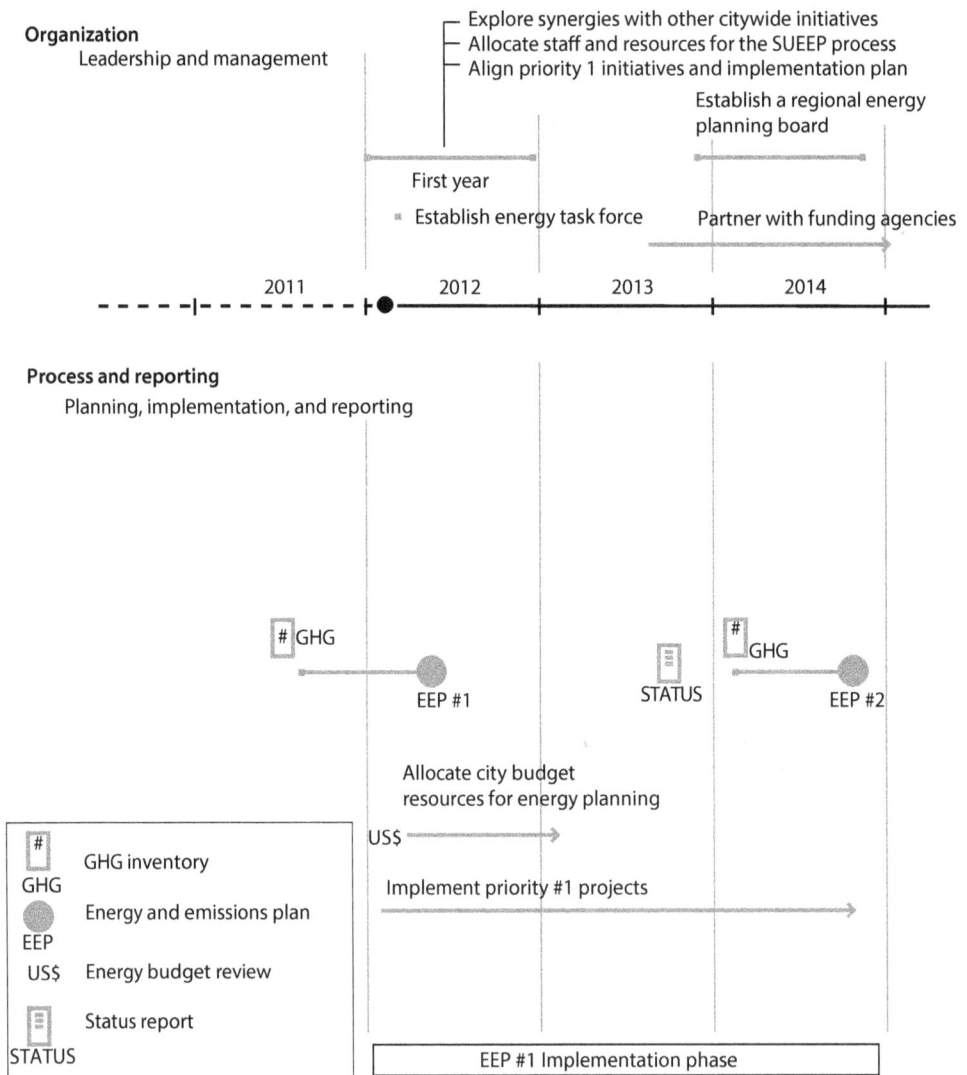

*box continues next page*

## Step 11: Finalize and Distribute the Plan

Once the energy and emissions plan has been drafted, it must go through a stake-holder review and input process. This review has many purposes, including getting technical corrections and clarifications. Most important, the review serves to get buy-in from stakeholders. It may be difficult to get support for the plan if a stakeholder who is expected to implement high-priority projects has not seen the plan or agreed to implement the projects. In contrast, stakeholders who have been in regular dialogue with the energy task force and have been involved in

**Box 13.5  Example: Sample 10-Year Energy and Emissions Action Plan** *(continued)*

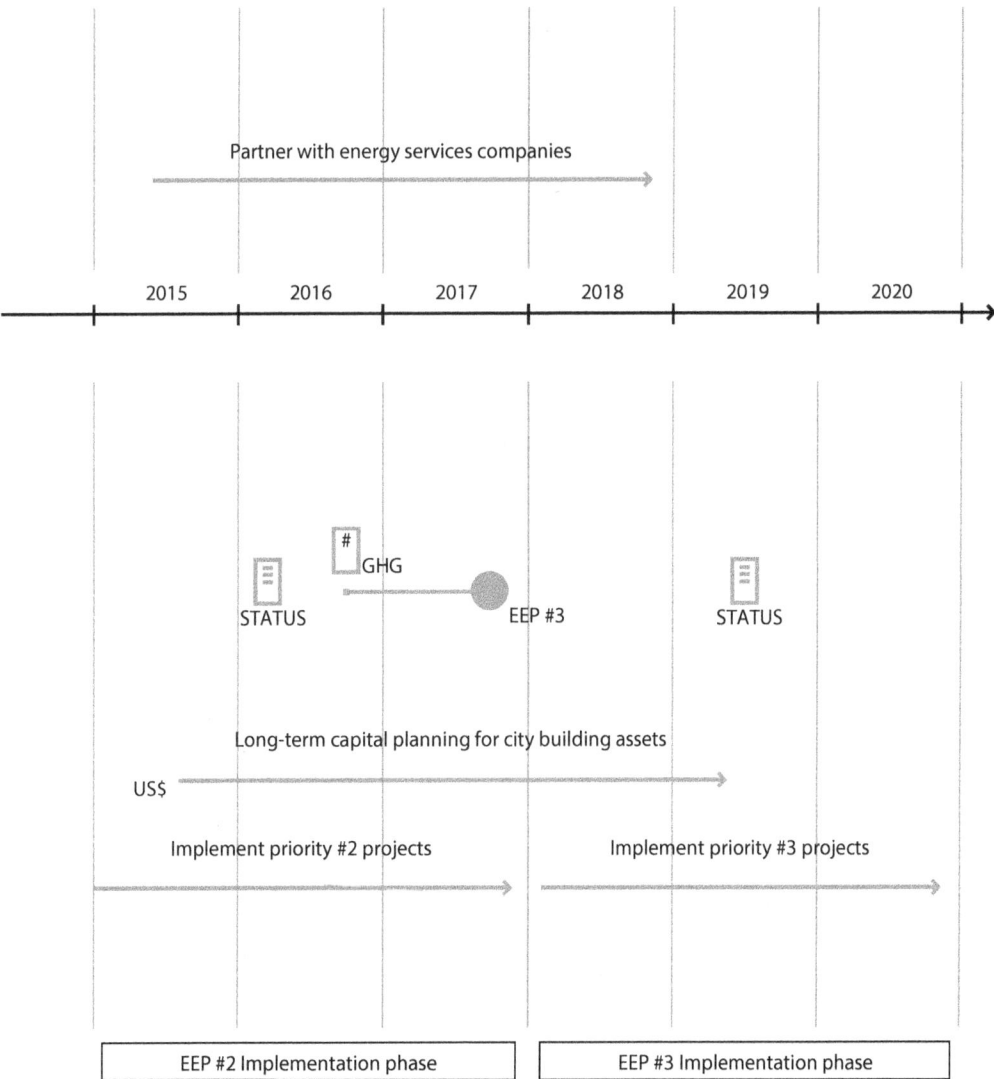

*Note:* GHG = greenhouse gas; SUEEP = Sustaining Urban Energy and Emissions Planning.

project review and assessment would likely support initiatives that are related to or need to be implemented by them.

### Receive Input on the Draft Plan

Once the draft energy and emissions plan is completed and distributed to all stakeholders for review, it is important to meet with each stakeholder to walk through project prioritization and selection. This is particularly important if the stakeholder is expected to implement one or more of the high-priority projects. If a formal meeting is not possible or if a stakeholder will not have direct responsibility for implementing a high-priority project, written input will suffice.

### Revise and Synthesize the Plan

All comments should be collected, recorded, and evaluated to ensure that all inputs are considered. Comments should be prioritized based on the relevance, the technical or experiential basis of the comment, the priority of the stakeholder making the comment, and the compatibility of the comment with the city's broader goals. Once the changes to the plan have been determined, it can be revised to incorporate the most important comments.

### Finalize the Plan

The final step in developing the energy and emissions plan is to gain approval from the mayor and the energy task force. Finalizing the document should be straightforward once the diagnostics have been completed, goals have been set, projects have been prioritized, and stakeholders have been consulted. However, if the mayor has not been involved in the day-to-day SUEEP process, time should be taken to explain the process and its outcomes to the final decision makers.

Most energy and emissions plans begin with a personal letter from the mayor introducing and supporting the plan. Once the plan has received final approval, it should be printed for distribution and published digitally for uploading to the city's website to be accessible to a wider audience.

### Get the Word Out—Locally

Once the energy and emissions plan has been published and posted on the city's website, the public, private businesses, local nongovernmental organizations, and all other stakeholders who will be interested in participating and benefiting from the SUEEP projects should be informed. News articles in local papers, press releases, events to introduce aspects of the plan, and other publicity-generating activities should be pursued. (See the technical assistance opportunity in box 13.6.) One of the main reasons that mayors develop energy plans is that they believe such plans will support the public's expectations for improved quality of life and environmental stewardship of the city. Therefore, if the plan is written and implemented, but the public remains unaware of it, much of its benefit will be lost.

---

**Box 13.6 Technical Assistance Opportunity: Public Relations Press Release**

Once the energy and emissions plan is complete, professional support may be needed for getting the word out. Some cities have strong public relations and community outreach capabilities, but these are often confined to the city itself. Professional public relations firms can help to expand the reach of publicity for the energy and emissions plan regionally, nationally, and internationally. This is particularly important if a city is looking for international donor funding and needs to show how serious the SUEEP process is to the mayor. Publicity may even attract funding that was not in the picture before the press release.

---

### Get the Word Out—Internationally

Because some of the stakeholders in the SUEEP process are international businesses, donor agencies, and peer cities, the energy and emissions plan needs to be publicized beyond the city and even the country. The commitments made and the thoughtful planning demonstrate the city's capability to implement a progressive agenda and could make the city more economically competitive. If the energy and emissions plan is credible, it can attract financial, technical, and political support for implementing the projects. A well-written energy and emissions plan will also attract interest from international organizations, which are on the lookout for good projects and cities' capacity and ability to implement the projects.

### References

Birmingham, U.K. 2010. *Birmingham 2026, Our Vision for the Future*. Birmingham, U.K. http://www.bebirmingham.org.uk/uploads/scs%20draft(1).pdf?phpMyAdmin=b599 8cc58dff68a4b03a480ef59038da.

Cape Town, South Africa. 2011. *Moving Mountains: Cape Town's Action Plan for Energy and Climate Change*. Cape Town: Environmental Resource Management Department.

Christchurch City Council. undated. *Sustainable Energy Strategy for Christchurch 2008–18*. Energy Team and Strategy and Planning Group, Christchurch, New Zealand. http://resources.ccc.govt.nz/files/EnergyStrategy-docs.pdf.

Dublin City Council and Codema. 2010. *Dublin City Sustainable Energy Action Plan 2010–2020*. Dublin. http://www.dublincity.ie/WaterWasteEnvironment/Sustainability/Documents/SEAP-FINAL%20version%20for%20website.pdf.

Huntington Beach, California. 2011. *City of Huntington Beach Energy Action Plan*. Huntington Beach, CA. http://www.huntingtonbeachca.gov/residents/green_city/hb-eap-adopted.pdf.

New York, New York. 2007. *PlaNYC 2030: A Greener, Greater New York*. New York: The Mayor's Office. http://nytelecom.vo.llnwd.net/o15/agencies/planyc2030/pdf/full_report_2007.pdf.

Singapore. 2012. *Climate Change and Singapore: Challenges. Opportunities. Partnerships*. National Climate Change Secretariat, Singapore.

Tbilisi, Georgia. 2011. *Sustainable Energy Action Plan—City of Tbilisi for 2011–2020.* Tbilisi. http://helpdesk.eumayors.eu/docs/seap/1537_1520_1303144302.pdf.

Toronto Environment Office. 2007. *Climate Change, Clean Air, and Sustainable Energy Action Plan: Moving from Framework to Action.* Toronto. http://www.toronto.ca/changeisintheair/pdf/clean_air_action_plan.pdf.

# CHAPTER 14

# Stage V: Implementation

*The implementation stage is where good planning pays off. The strong foundation for city action established in the previous steps enables the city to take on the challenges associated with Sustainable Urban Energy and Emissions Planning (SUEEP). The information presented in this stage will help address the basics of overcoming policy and financing barriers. Because this Guidebook maintains a high-level overview of the issues associated with implementation, numerous external resources are referenced in the text. The references provide valuable information and would be useful sources of information during project implementation.*

## Step 12: Develop Content for High-Priority Projects

This section outlines issues to consider when developing the most common types of energy efficiency and energy planning projects. The high-priority projects were selected in Step 9 of the SUEEP process to be implemented first. The details of every energy project are different, so a considerable amount of experience and time are required to turn a project idea into a project that can be fully implemented. See the technical assistance opportunity in box 14.1.

The following descriptions provide greater detail for four common types of projects:

- Incentive projects
- Major single projects
- Organizational development
- Policy projects

(See the resources in box 14.2 for more information on developing a policy-based energy and emissions plan.)

### Incentive Projects

Many energy efficiency projects encourage uptake of better, more expensive equipment by paying the purchaser the difference between the low cost–high energy product and the high cost–low energy one. Examples include

---

**Box 14.1  Technical Assistance Opportunity: Project Development**

The variety of energy projects in this Guidebook show the range of expertise required to develop any particular project. Most of the projects described in the Project Assessment and Prioritization Toolkit have been implemented somewhere in the world, but probably not in your city. It is helpful to bring in technical expertise in either a peer review role or an advisory role to help identify pitfalls and develop the details of specific aspects of a given energy project. For example, a bus rapid transit manufacturer or transport consulting firm could be brought in to support the design and feasibility steps.

---

**Box 14.2  Resources: Policy and Energy and Emissions Plans**

*Energy Efficiency Governance—Handbook*
    IEA 2010.
*Eco² Cities: Ecological Cities as Economic Cities*
    Suzuki and others 2010.
"Energy Efficiency Indicators: Best Practice and Potential Use in Developing Country Policy Making"
    Phylipsen 2010.
*Cities and Climate Change*
    OECD 2010.
*Pathways to a Low-Carbon Economy: Version 2 of the Global Greenhouse Gas Abatement Cost Curve.*
    McKinsey and Company 2009.
*Shanghai Manual: A Guide for Sustainable Urban Development in the 21st Century.*
    United Nations, Bureau International des Expositions, and Shanghai 2010 World Exposition Executive Committee, 2011.

---

high-efficiency air conditioners, boilers, lighting, and pump motors, or fuel-efficient cars.

*Technology Assessment*
The first step is to rigorously assess the technologies to find the ones that are better suited to the circumstances of the city. The advantages of, for instance, particular lights, pump motors, or air conditioners over typical equipment should be determined.

*Target Customers*
Businesses, residents, or manufacturers that would be interested in an incentive project should be identified. Interviews to understand what would motivate them to engage in the project, and factors that would dissuade them, should be conducted. The number of customers that could plausibly participate in the project should be estimated and targets for the uptake of the project set

(for example: 1,000 incentives [rebates, discount coupons] for very-high-efficiency scooters will be distributed every year for five years).

### Set Incentive Levels
Incentive levels can be set once the available energy efficient technologies are understood, including how much they cost and how much energy can be saved. Typically, incentive levels are set at 50–100 percent of the difference in cost between baseline equipment and high-efficiency equipment. The energy and emissions task force will have to set the level of incentives based on the group of consumers being targeted. If the time consumers spend in the application and validation process is greater than the energy cost savings they expect to accrue, they will have no incentive to become more energy efficient. For example, if the incentive for a dimmable T5 light bulb does not completely cover the cost differential with a typical T12 light bulb, consumers will need to believe that the energy savings will pay for the additional cost within a short time.

### Allocate Funding
Once incentives, target rollout volumes, and project administrative staffing levels have been estimated, a total project budget can be developed. Armed with a solid plan that identifies the budget, the projects, and the potential energy or greenhouse gas (GHG) savings, you can approach the city council, donor agencies, and specialized financing bodies (for example, energy efficiency funds) for financing support. Step 14 provides details on financing options.

### Develop Application and Selection Process
A good incentive project could be oversubscribed, so a process should be developed to select recipients based on need, speed of implementation, and other characteristics that make them attractive.

### Validate Installation
Most incentive projects require that the purchase, installation, and correct use of the equipment be validated. This follow-up also ensures that the technology is appropriate and delivers the energy savings predicted in the initial technical assessments.

## Major Single Projects
Some projects are potent enough to change the way a city uses energy. Examples include a large-scale district-level combined heat and power facility, a bus rapid transit line, or a citywide water network leak detection and reduction program. These projects do not need to engage a large number of businesses or residents but do require the involvement of many intergovernmental agencies and funding sources, as well as substantial planning, approvals, and political support.

### Conceptual Design
The ideas behind a project with the potential to achieve one or more of the city's energy goals should be refined to provide a high-level understanding of its

primary concepts. A firm grasp of major characteristics—overall cost of construction and operation; annual revenue potential; stakeholders involved (including property owners, local businesses, residents, nongovernmental organization [NGOs], and city agencies); and timeline for full design, construction, and implementation—is critical.

### Project Feasibility
With a conceptual design, a project feasibility study can be undertaken. Analyses should include costs and benefits, technical components, environmental and social considerations, political roadblocks, and financing issues.

### Project Approvals
If a project is deemed feasible, a more detailed design should be formulated and approved by the regional or national electrical, regulatory, and environmental bodies. For example, a large-scale renewable power generation facility (wind, solar, geothermal, biomass, or the like) should be approved by the regional or national electrical regulatory body and should gain environmental and legislative approvals.

### Project Financing
Throughout the development of a major project, the project leader should be aware of different financing structures, and engage lenders, partners, and donor agencies. Many of the major project examples shown in this Guidebook were implemented using innovative financing methods such as public-private partnerships (PPPs), design-build-transfer, design-build-operate-transfer, and other methods to bring in private sector technical and financial expertise and risk sharing.

### Project Bidding
Once the project is approved and the procurement strategy has been designed, the project should be put through a competitive bid process. The request for bids should be publicized as widely as possible to bring in a large pool of potential bidders. The bidding process may require multiple stages; the initial stages might request only statements of capabilities and team structure, with subsequent stages requesting more detail about finances and implementation plans.

### Project Implementation Plan
Before a final bidder is selected, a rough implementation plan should be drawn up by the city and circulated to all stakeholders to ensure they have been consulted on implementation hurdles.

### Organizational Development
Successful energy planning goes beyond identifying and developing incentive projects or major impact projects. Good energy planning increases the capacity within the city government to implement future projects. Increasing

energy-related capacity includes building knowledge of successful projects from throughout the region, building a network of contacts for technical support and advice, and changing the mindset of city employees to work collaboratively toward the city's wider energy goals.

### New or Improved Organization

Whether formation of a new agency or group to undertake responsibility for the SUEEP process and status reports is required or an existing organization could simply be improved to be more effective should be determined. The city government should ensure that the new or existing agency, group, or organization is properly funded and provided with the authority to make decisions and implement projects; otherwise, the initiative will be ineffective.

### Staffing Requirements

The minimum number of staff required to make the organizational improvement should be determined. Keeping staffing cost low is critical to minimizing project administration overhead.

### Training Requirements

The most cost-effective available training (including conferences for key staff, professional consultancy, training programs for energy services, and donor-funded capacity-building activities) should be identified. Many donors look for opportunities to provide technical assistance through capacity building rather than direct consultancy, so cities should take advantage of these opportunities.

## Policy Projects

Finally, policy projects can lay the groundwork to ensure that tactical energy projects are successful. For example, fuel subsidies for private vehicles could be decreased or eliminated in tandem with access and affordability improvements in public transportation.

### Policy Analysis

The cost of the new policy or regulation to stakeholders should be assessed; this analysis should be used to ensure that its benefits outweigh the costs. Finding the right balance between regulation or policy changes and economic development and improvement in citizens' quality of life, safety, and security is critical.

### Stakeholder Consultation

Changes to the existing policy environment will affect residents and businesses, so it is important to gain buy-in on the changes before they are implemented.

### Legislation, Regulation, and Enforcement

Sometimes incentives are insufficient to bring the energy efficiency agenda to fruition. In these cases, city governments may choose to put in place legislation or regulations; for example, building codes may need to be formalized to make it compulsory for buildings to be designed in an energy efficient manner.

A regulatory policy must be both enforceable and enforced to create change, so ensure resources are allocated and a reasonable process is set up to monitor compliance.

## Step 13: Improve Policy Environment

Cities need to recognize the close relationship between the policy environment and the success of the SUEEP process. Each city is shaped by its unique political environment, which means that the intricacies of adapting policies to the SUEEP process will be similarly specific. This step aims to help a city understand how its current policy environment will affect the SUEEP process, the policy options that are potentially beneficial to the SUEEP process, and how policy recommendations can be established. This step also describes the process a city can use to analyze how current policies can be improved to streamline energy and emissions–related policy, remove potential bottlenecks or conflicts with the SUEEP process, and preempt issues arising from mismatched policy. This is achieved through a three-stage process:

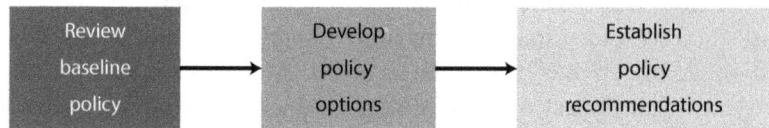

### Review Baseline Policy
A baseline review will clarify the strengths, weaknesses, and gaps in existing policies and policy instruments related to the SUEEP process. The review provides an understanding of current policies enacted at national, regional, and local levels; the way in which policies can complement or conflict with each other; the city's role as it relates to energy; and the capacity of the city to act.

### Current Policy and Project Review
The first step in the review is to take stock of existing policies, initiatives, projects, and programs at the national, regional, and city levels (see the example in box 14.3). Initiatives and projects developed by utilities, NGOs, and other organizations should be included in the review.

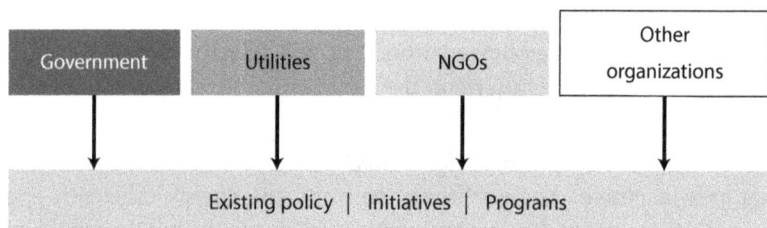

---

**Box 14.3  Example: Level of Government and Policy Type**

National level: Renewables policy
Regional level: Mass transit
City level: Building codes

---

**Box 14.4  Case Study: Curitiba, Brazil—Transportation Planning**

Curitiba is the capital of the state of Parana in the south of Brazil and is home to 1.83 million people. The city occupies a strategic location along Brazil's economic corridor, which includes Brasilia, Porto Alegre, Rio de Janeiro, and São Paolo, and other major South American cities such as Buenos Aires and Montevideo.

Curitiba has encouraged a lively and thriving downtown area through the city's efficient transportation system and a holistic approach to city planning. Curitiba's city planning integrates strategic transportation corridors, land-use zoning, and its comprehensive master plan to encourage high-density growth in close proximity to public transportation.

This integrated approach has resulted in numerous benefits. Bus ridership has reached 45 percent, the city's air quality has improved, traffic congestion has decreased, and green space has been added to the city. The center of the city has been pedestrianized, resulting in a vibrant downtown area and a reduction in crime.

The Institute for Research and Urban Planning of Curitiba (IPPUC) is responsible for monitoring, implementing, and updating Curitiba's master plan. The delegation of this authority to the IPPUC is one of Curitiba's key successes. The organization is a largely independent institute and therefore less susceptible than a government department to political pressures and changes.

*Source:* Suzuki and others 2010.

---

The results of the review can be presented as a list, table, memo, presentation, or any other suitable format. This review will serve as the basis for policy analysis, so the information should be clear and easy to analyze.

### Policy Support and Conflict

Once the policy and project review is completed, it is important to identify how the policies interact. A supportive policy relationship allows different policies to augment each other's desired effect. For example, for a densely populated city, Singapore has remarkably low traffic congestion and good air quality. These attributes resulted from the implementation of a number of policies that work together, including a high tax on gasoline, congestion charging, and stringent automobile standards. See the case study in box 14.4 for another example of mutually reinforcing policies.

However, policies also have the potential to conflict with each other if they have not been formulated strategically. For example, a city might want to improve the fuel efficiency of its vehicle fleet and has identified fuel mileage as a key concern. However, existing procurement guidelines might preclude the use of fuel efficiency as a product selection criterion. In this case, the city may need to consider updating its vehicle procurement guidelines to enable achievement of its goal.

In addition, when considering policy support and conflict, pointing out the many additional benefits that come along with energy and emissions policies is important. For example, energy and emissions policies can result in the following:

- Public health improvements
- Cost savings and increased efficiency
- Reliability of energy supply and infrastructure improvements
- Improved quality of life

Using the baseline policy review as a way to cross-check policies against each other will help the city understand the relationships between different policies. Some instances of policy symbiosis and conflict may arise with policies that are not related to energy and emissions, and these relationships should be noted for reference because they may prove to be important later on. See the example in box 14.5 for a suggested framework for assessing policy support and conflicts.

### Gap Analysis
A gap analysis builds on the policy and project review to understand the areas in which further policy action is required for the effective implementation of projects identified by the SUEEP process. A gap analysis consists of mapping current policies and projects against a set of categories or areas that need to be addressed to plan for effective energy and emissions management.

The tip in box 14.6 lists areas that are typically not fully covered by energy and emissions policies. This list is useful as a starting point, but an individual analysis is essential for each city given the wide variability in activity.

### Develop Policy Options
The outcome of the baseline policy review sets the stage for the development of policy options with the potential to achieve the city's goals. The aim is to identify a wide range of alternatives and then narrow them down to those that are most suited to the city's situation.

### Consider Multiple Policy Approaches
Establishing multiple policy approaches to achieve a desired outcome is a good strategy for policy reinforcement. For example, if the city has identified reduced water use in buildings as a policy objective, then a two-prong policy approach that includes fixture flow rate requirements in building codes and public education initiatives on water savings is an effective way to achieve the city's goals. Although this is a simplistic illustration, all the potential avenues available to the city to establish, reinforce, and support change should be considered.

---

**Box 14.5  Example: Policy Support and Conflict Analysis**

| Existing city level (CL) energy-related policies | Supported by existing CL policies? Which? | Supported by existing regional and national (RNL) policies? Which? | Conflict with existing RNL policies? Which? | Conflict with existing CL policies? Which? | Existing RNL energy-related policies |
|---|---|---|---|---|---|
| Rule making and enforcement | | | | | Rule making and enforcement |
| 1. … | ☐———— | ☐———— | ☐———— | ☐———— | 1. … |
| 2. … | ☐———— | ☐———— | ☐———— | ☐———— | 2. … |
| 3. … | ☐———— | ☐———— | ☐———— | ☐———— | 3. … |
| 4. … | ☐———— | ☐———— | ☐———— | ☐———— | 4. … |
| 5. … | ☐———— | ☐———— | ☐———— | ☐———— | 5. … |
| Direct capital expenditure | | | | | Direct capital expenditure |
| 1. … | ☐———— | ☐———— | ☐———— | ☐———— | 1. … |
| 2. … | ☐———— | ☐———— | ☐———— | ☐———— | 2. … |
| 3. … | ☐———— | ☐———— | ☐———— | ☐———— | 3. … |
| 4. … | ☐———— | ☐———— | ☐———— | ☐———— | 4. … |
| 5. … | | ☐———— | ☐———— | ☐———— | 5. … |
| Financial incentives | | | | | Financial incentives |
| 1. … | ☐———— | ☐———— | ☐———— | ☐———— | 1. … |
| 2. … | ☐———— | ☐———— | ☐———— | ☐———— | 2. … |
| 3. … | ☐———— | ☐———— | ☐———— | ☐———— | 3. … |
| 4. … | ☐———— | ☐———— | ☐———— | ☐———— | 4. … |
| 5. … | ☐———— | ☐———— | ☐———— | ☐———— | 5. … |
| Awareness and knowledge sharing | | | | | Awareness and knowledge sharing |
| 1. … | ☐———— | ☐———— | ☐———— | ☐———— | 1. … |
| 2. … | ☐———— | ☐———— | ☐———— | ☐———— | 2. … |
| 3. … | ☐———— | ☐———— | ☐———— | ☐———— | 3. … |
| 4. … | ☐———— | ☐———— | ☐———— | ☐———— | 4. … |
| 5. … | ☐———— | ☐———— | ☐———— | ☐———— | 5. … |

In developing multiple policy approaches, potential measures should be aligned with the desired outcome. This will allow policy planners to match the policy's goal with the means available to achieve it.

### Stakeholder Engagement

As discussed in Step 3: Identify Stakeholders and Links, gaining policy insights from stakeholders and securing stakeholder buy-in are essential to developing energy and emissions policies. Stakeholders contribute to the city's understanding of the necessary policies and enable a multidisciplinary approach to policy development. They also play a variety of roles related to energy efficiency and emissions policy development, including the following:

- Developing and writing policies (for example, department of energy)
- Enabling project delivery (for example, energy services companies [ESCOs])

---

**Box 14.6 Tip: Common Policy Gaps**

- City governance structures, for example, institutions not organized to address energy and emissions
- Procurement policy, for example, procurement guidelines preclude selection of goods based on energy efficiency
- Barriers to partnerships with the private sector
- Data availability
- Public health, for example, policy does not recognize the benefits of energy and emissions planning on human health
- Economic development, for example, policy does not link the SUEEP process and energy and emissions plan to growth strategies

---

- Coordinating strategic planning (for example, city chamber of commerce)
- Implementing projects (for example, building operators)
- Receiving services from the city and participating in public consultation (for example, citizens)
- Enabling knowledge sharing (for example, local academic institutions)

In particular, engagement with the national government is key, given that national policies, especially those that cut across sectors, affect policies implemented at the city level. For example, electricity tariffs, which are usually determined by the national government, could potentially impede the city's efforts to enhance energy efficiency should the price of electricity be subsidized or set particularly low.

(Further discussions of stakeholders and how they can be engaged can be found in Step 3.)

### Establish Policy Recommendations

Establishing policy recommendations for the energy and emissions plan requires that the list of policy options identified in the previous steps be reduced. This reduction will be based on an analysis of the city's capacity to act, the partnerships that may enable policy implementation, and the empirical database underpinning development of the energy and emissions plan.

#### Capacity to Act

The first consideration in establishing policy recommendations is the city's capacity to act. Capacity to act refers to the city's scope of influence with respect to energy and emissions policy. For example, cities generally have the power to regulate, enforce regulations, invest in infrastructure upgrades, provide subsidies, and educate the public. See the example in box 14.7 for more information on how cities can classify their powers to accomplish the goals of the energy and emissions plan.

---

**Box 14.7  Example: Capacity to Act**

A city's capacity to act and its level of influence determine how it can affect the energy and emissions plan. This matrix can be used to classify initiatives and clarify how they can best be implemented.

| Type of Mayoral Lever | Local authority | Provincial or state government | National government | Private sector | Households and individuals | NGOs and others |
|---|---|---|---|---|---|---|
| Rule making | | | | | | |
| Regulatory oversight | | | | | | |
| Direct expenditures and procurement | | | | | | |
| Financial incentives | | | | | | |
| Information gathering, dissemination, convening, facilitation, advocacy | | | | | | |

*Source:* Hammer 2009.
*Note:* NGO = nongovernmental organization.

---

**Box 14.8  Example: Policies According to City Role**

*Energy Consumer*
Air conditioning turned off during certain hours
City building retrofit project

*Regulator*
Building codes
City planning requirements

*Energy Producer and Supplier*
Tariff structure
Fuel procurement policy

*Motivator*
Energy efficiency publicity campaigns
Energy efficiency pilot projects

*Source:* Management of Domains Related to Energy in Local Authorities (MODEL), 2010 (http://energy-cities.eu/MODEL).

---

There are four major roles that a city can take with respect to energy and emissions (see the example in box 14.8):

- Energy consumer
- Regulator
- Energy producer and supplier
- Motivator

**Box 14.9  Case Study: Mexico—Green Building Codes**

Mexico has had a mandatory building energy standard for commercial buildings since 2001 that was developed by CONUEE, the national energy conservation agency, with support from the Lawrence Berkeley National Laboratory. The code has not yet been incorporated into the country's construction regulations, but it is recognized that this is necessary to encourage its effective implementation.

For additional detailed information, see Feng, Meyer, and Hogan (2010).

For each role, the city's capacity to act is limited in a specific way. For example, the city should have significant control over its own energy consumption and can introduce policies to retrofit city building stock, develop procurement policies that prioritize energy efficiency, and educate civil servants on energy efficient behavior.

However, as a regulator, the city's control is more limited. For example, if the city aims to reduce per capita energy consumption in homes, the available policy levers are generally information dissemination through educational campaigns or regulation through building codes. (See the case study in box 14.9.) These are not likely to be as effective as policies that are outside the city's authority, such as a progressive electricity tariff structure.

Each policy in the baseline review will relate to one of the city's roles. Identifying the extent of the city's influence and its policy levers in the proposed projects will allow it to determine if it is using its full capacity to act within each role, or whether some policy levers should be favored over others. (See the example in box 14.10 for a description of available policy levers.) This review will also enable a city to consider indirect methods of enacting change if the analysis shows that some outcomes it desires are not within its scope of influence.

See the case study in box 14.11 for an illustration of many of the concepts in this section.

### External Partnerships

Some policies cannot be implemented without the help and support of external partnerships. ESCOs and PPPs are good examples of external partnerships that have enabled energy efficiency policies to be successful in cities across the world.

In addition, development banks, international organizations, and NGOs can potentially help plan for a rollout of the energy and emissions plan. The potential for external partnerships is discussed in more detail in Step 3: Identify Stakeholders and Links.

### Empirical Base

The foundation of the SUEEP process is an empirical base of periodic energy and emissions data collection and analysis, enabling systematic measurement and monitoring of policy successes and failures. This information allows a city to

---

**Box 14.10   Example: Available Energy Efficiency and GHG Policy Levers**

Below are examples of policy levers available to city leadership:

*Rule making and enforcement*

Building performance standards and green building codes

Industrial efficiency standards

Green procurement policy

*Direct capital expenditure*

Improved vehicle testing

Audit and upgrade for different building types

Efficient technologies program

Efficiency in government operations

Demonstration projects

*Financial incentives*

Subsidies, tax deductions, or loans with favorable rates for energy efficient products, for example, roof insulation

Renewable technology rebates

*Awareness and knowledge sharing*

Online information portal

Consumer guide to energy efficient products

Energy efficiency, GHG mitigation awards

Energy efficiency partnerships

Training for energy efficiency professionals

*Funding*

Energy efficiency and GHG funds

---

adjust policies where required and develop an efficient approach to energy and emissions management by eliminating unsuccessful or redundant policies. The empirical base of the SUEEP process can be leveraged by a city to analyze its policy structure, especially after one iteration of the SUEEP process has been completed.

## SUEEP Policy Process

This section covers the key factors that enable the development and implementation of the energy and emissions plan to be successful.

### Transparency

Transparency is a critical part of policy development because a process that is communicated well to all stakeholders builds support and ensures widespread understanding of the city's intent. Transparency is also strongly linked to perceived regulatory risk from the perspective of potential investors. By improving transparency, the city is reducing this perceived risk and improving its position to attract financing.

Energizing Green Cities in Southeast Asia • http://dx.doi.org/10.1596/978-0-8213-9837-1

**Box 14.11  Case Study: Seattle, Washington, USA—Green Building Program**

Through a collection of successful regulatory standards, measures, and incentives for the building industry, Seattle now has one of the highest concentrations of sustainable buildings in the United States and a powerful sustainable building industry worth US$671 million.

Having initially established a Green Building Team in 1999, Seattle regrouped its green building experts to form a single business unit called City Green Building in 2005. Its main program is funded through interdepartmental resources and staffed by green building experts in residential, commercial, institutional, and city capital projects. Using its strong relationships with the city's water and energy utilities and their incentive programs, it connects developers, design teams, and building permit applicants with green building resources and helps eliminate code barriers to building green.

A fundamental element of the city's green building program is the promotion and measurement of the environmental impact of buildings and third-party verification. Seattle's successful programs include the following:

*Sustainable Building Action Plan.* The action plan identified key strategies for promoting green buildings in the marketplace. The two most important strategies identified were to lead by example and to develop a standard for green building.

*Sustainable Building Policy.* This policy requires new municipal buildings of more than 5,000 square feet to meet a minimum Leadership in Energy and Environmental Design (LEED) Silver standard. Through 2011, an investment in state-of-the-art sustainable buildings of more than US$500 million has resulted in 10 LEED Certified projects owned by the city (5 Gold, 3 Silver, 2 Certified), with a further 28 projects planned or in development.

*City LEED Incentive Program 2001–05.* The city of Seattle provided support to green buildings through its City LEED Incentive Program, with incentives of more than US$2 million for energy conservation, more than US$2 million for natural drainage and water conservation, and more than US$300,000 for design and consulting fees for LEED projects. The program was launched in 2001 as a joint program of Seattle City Light and Seattle Public Utilities. It provided upfront soft-cost assistance to projects committing to LEED. Funds can be used for additional design and consulting fees and for participation in the LEED program. Funding levels were US$15,000 for LEED Certified and US$20,000 for LEED Silver or above.

*Density Bonus.* The density bonus offers downtown commercial, residential, and mixed-use developments greater height or floor area (or both) if a green building standard of LEED Silver or higher is met. Projects must also contribute to affordable housing and other public amenities. Three projects have so far registered, and five projects are currently considering registration as of 2011.

*Source:* C40 Large Cities Climate Summit 2007.

## Codification of the Energy and Emissions Plan

Formalizing the legal status of the energy and emissions plan embeds it into the long-term citywide strategy. The projects are no longer at the mercy of political cycles and the responsibility to follow through on the plan must be taken seriously. For example, New York's PlaNYC has been codified, cementing the city's commitment to take on the actions identified in the plan. Although codification of PlaNYC has been effective in New York, cities in the East Asia and Pacific region will have to consider if a similar approach would be effective in entrenching the SUEEP process in city planning.

## Step 14: Identify Financing Mechanisms

This section provides a high-level overview of the basics of energy efficiency and emissions project financing. Energy efficiency and emissions reduction projects tend to suffer from a financing viability gap when compared with conventional projects. The information presented here will help you address this challenge. However, because energy efficiency and emissions projects cover a diverse range of sectors, stakeholders, and technologies, developing financing strategies for these projects is complex and cannot be fully addressed in this Guidebook. To augment the information presented here, a selection of supplementary information is presented in the resources in box 14.12.

This section is structured according to a general process a city can use to assess a project's financial viability as shown in the diagram below:

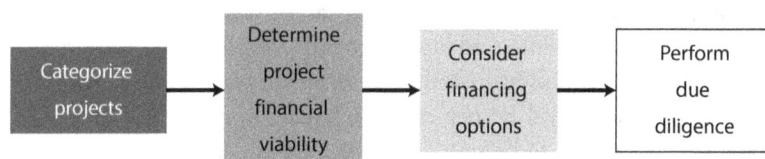

---

**Box 14.12  Resources: Further Reading**

*Financing Energy Efficiency: Lessons from Brazil, China, India, and Beyond.* Robert P. Taylor, Chandrasekar Gavindarajalu, Jeremy Levin, Anke S. Meyer, and William A. Ward. World Bank. 2008.

"Financing Energy Efficiency: Forging the Link between Financing and Project Implementation." Silvia Rezessy and Paolo Bertoldi. Joint Research Centre of the European Commission. 2010.

*Energy Efficiency and the Finance Sector: A Survey on Lending Activities and Policy Issues.* A report commissioned by UNEP Finance Initiative's Climate Change Working Group. 2009.

*Public Procurement of Energy Efficiency Services: Lessons from International Experience.* Jas Singh, Dilip R. Limaye, Brian Henderson, and Xiaoyu Shi. World Bank. 2009.

"Energy Efficiency Indicators: Best Practice and Potential Use in Developing Country Policy Making." G. J. M. Phylipsen, commissioned by the World Bank. 2010.

*Shanghai Manual: A Guide for Sustainable Urban Development in the 21st Century.* United Nations, Bureau International des Expositions, and Shanghai 2010 World Exposition Executive Committee. 2011.

*Green Infrastructure Finance: Framework Report.* AusAID and the World Bank. 2012.

---

### Categorize Projects

Once a collection of projects has been identified and prioritized by a city, the projects must be categorized according to specific criteria that will streamline the approach to financing. Projects in different categories may be eligible for different forms of investments and incentives. The matrix shown here is a generic approach to categorizing projects by size (small or large) and nature (centralized or decentralized); however, if a city's prioritized projects tend to fall into the same category, more detailed levels of categorization may be required, for example, breaking down projects by infrastructure capital investment versus operational measures.

|  | Large scale | Small scale |
|---|---|---|
| Decentralized | Green building codes<br>Improved public transportation | Efficient lighting<br>Household solar hot water |
| Centralized | Renewables development<br>Water treatment system location | Biomass energy<br>Landfill gas capture |

Using a categorization matrix is helpful for mapping projects to a city's investment environment. This will enable viable projects to be matched to available financing and ensure that a financial analysis can be undertaken within each category's market segment. Unless additional energy-specific incentives are provided to investors, projects will be evaluated head-to-head against non-energy projects—but the energy projects will be perceived as having a higher risk profile because financial institutions tend to be unfamiliar with energy efficiency projects.

### Financing Barriers

From a banking perspective, financial attractiveness boils down to risk and return on investment (ROI). Financing decisions are based on comparisons of investment options and an analysis of the trade-offs between the risks and returns expected from those projects. However, the public sector must also incorporate socioeconomic goals into the decision-making process when considering the financial viability of a project. It is important that this additional layer of complexity be acknowledged when pursuing project financing.

Because sustainable energy and emissions management is still novel, streamlined mechanisms for financing associated projects are not yet available. Planners should consider the impacts of the most common financial barriers to energy efficiency initiatives (Rezessy and Bertoldi 2010). These barriers include the following:

- High development and transaction costs for small projects
- Lack of awareness of energy efficiency projects and technologies on the part of investors
- Lack of energy efficiency financing experience
- High perceived end-user credit risks
- Long marketing cycles
- Low collateral asset value

- Energy savings not considered a revenue source
- High up-front costs
- Short payback period requirements
- Budgetary rules that make it difficult to finance projects from energy savings
- Energy efficiency financing coming from the investment budget whereas savings are credited to the operational budget
- Lack of consideration of life-cycle costs
- Ambiguous ownership and operation of major energy assets

Not all of these barriers will apply to every project category, and conversely, some categories may experience additional challenges. However, addressing these potential impediments during the planning process will reduce the risk that they later prevent the implementation of an energy efficiency or emissions mitigation project.

Using the categorization matrix, a like-for-like comparison of financial risks and attractiveness is possible within each project category. Comparing projects across categories does not necessarily provide insight into the best potential project options because the criteria for financial viability will differ.

### Determine Project Financial Viability

Broadly speaking, projects are financially viable if the ROI reaches an agreed-on threshold, or "hurdle rate," and the identified risks are tolerable and might be mitigated during implementation. Therefore, before financing options are identified, a thorough risk assessment and financial analysis must be undertaken.

### Risk Assessment

The level of risk associated with a project influences the hurdle rate required to make the project viable. A city can encourage private sector investment by mitigating certain risk elements of a prioritized project, for example, by providing loan guarantees for initial seed funding of investments or by fast-tracking regulatory permits. In developing-world economies, the perceived risks are elevated, especially for newer technologies in the energy efficiency sector, and can lead to higher hurdle rates. Therefore, a city would be well served by developing favorable investment policies early in the project assessment phase that will mitigate key risks and thereby help to channel private sector capital into prioritized projects.

Risk assessment is a critical component of project financing, but generic methodologies may not apply to energy efficiency projects. The risk profile of the different categories of projects will also vary considerably, so a custom approach will be required for each category.

Generic risks commonly associated with energy efficiency projects that should be considered during the SUEEP process include (Covenant of Mayors 2010; and author's experience) the following:

- *Project-related risks:* cost and time overruns, poor contract management, contractual disputes, delays in tendering and selection procedures, poor communication between project parties

- *Government-related risks:* inadequate approved project budgets, delays in obtaining permits, changes in government regulations and laws, lack of project controls, administrative interference
- *Technical risks:* inadequate design or technical specifications, technical failures, poorer-than-expected performance, higher-than-expected operating costs
- *Contractor-related risks:* poorer-than-expected performance, higher-than-expected operating costs
- *Market-related risks:* increases in wages, shortages of technical personnel, materials inflation, shortages of materials or equipment, variations in the price of energy carriers

Unique risks, by category, might include the following:

- *Technology risk:* Is the technology proven? Technologies that appear in the early stages of the adoption curve will incur higher risk premiums and inflated hurdle rates (see the tip in box 14.13).
- *Timing risk:* Is the project planning and construction time too long or too short to attract the type of financing required? The availability of financing must meet the timeline requirements of project cash flows.
- *Country risk:* How does the perceived risk of deploying capital in the host country impact investment decisions? The host country risk profile (credit ratings, gross domestic product [GDP], consumer price index, corruption perceptions index, and the like) will have an impact on the hurdle rate.
- *Regulatory risk:* If incentives are offered, are they sustainable? What is the likelihood that regulation will be consistent into the future?

**Box 14.13  Tip: Technology Diffusion Curve**

**Figure B14.13.1  Technology Diffusion Curve**

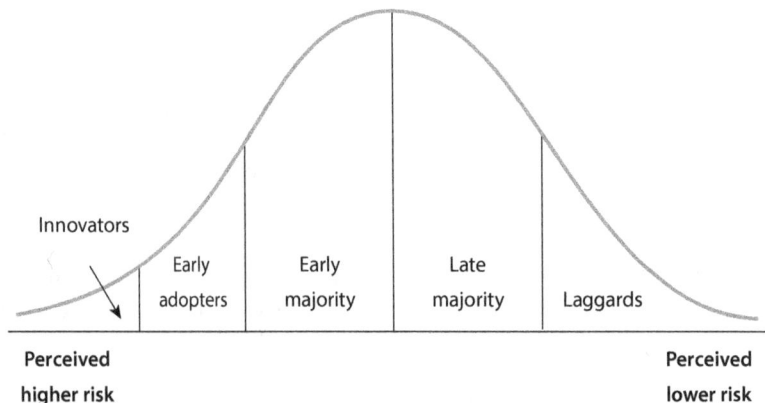

*Source:* Based on Rogers 1962.

Risk assessments will incorporate a multitude of project-specific elements that are difficult to generalize. Elements of the assessment identified above will give a reasonable picture of the level of risk attributable to each project type and therefore an understanding of the likely hurdle rate for each project. Once this is known, project-specific ROI analyses should be undertaken to enable benchmarking of project ROI rates versus investors' required hurdle rates. This will determine the proposed project's financial viability.

### *Return on Investment*
ROI analysis is typically undertaken to determine whether investment capital deployed will deliver a return that meets an agreed-on hurdle rate. The ROI should be analyzed for every proposed project. The analysis will map project cash flows (costs vs. receipts) over the project's lifetime to determine its overall profitability.

Hurdle rates acceptable in the private sector, to a bank or financial institution, for example, are often different from those acceptable to a city. A private financial institution would not typically accept an ROI lower than a government bond rate, whereas a city authority might consider investments with very low or even negative returns if the socioeconomic benefits are significant.

If a project exceeds market-wide benchmark ROI, project financing options will be more plentiful. If it does not, additional incentives will be required to attract investments (see the example in box 14.14). In the majority of cases, the ROI for energy efficiency projects will not meet market-wide hurdle rates

---

**Box 14.14  Example: Energy Efficiency Project Considerations**

**Figure B14.14.1  The Use of Incentives**

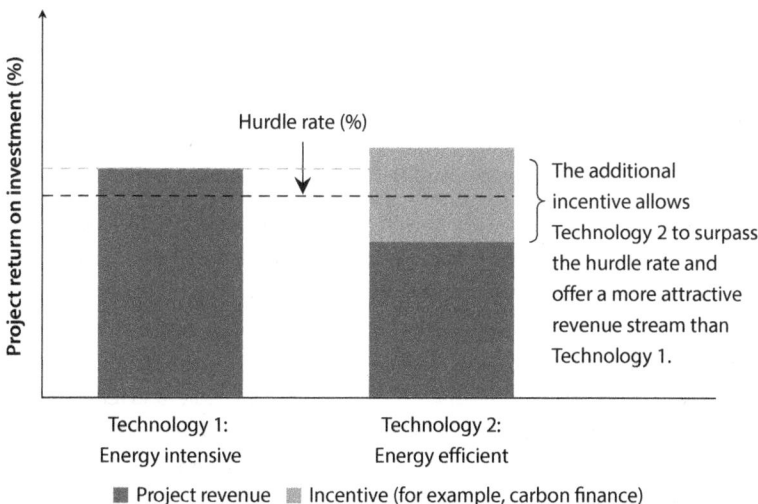

Hurdle rate (%)

The additional incentive allows Technology 2 to surpass the hurdle rate and offer a more attractive revenue stream than Technology 1.

Project return on investment (%)

Technology 1: Energy intensive          Technology 2: Energy efficient

■ Project revenue   ▨ Incentive (for example, carbon finance)

---

because of the inherently high levels of perceived risk. However, additional incentives might be available to these projects that do not exist in the wider market. These incentives may include structured and nonstructured climate or energy efficiency schemes promoted by regional, national, or supranational governments, for example, carbon credits or subsidies. Further information on bridging the hurdle rate gap can be obtained from the Australian Agency for International Development (AusAID) and World Bank (2012).

Once the ROI analyses are complete, like-for-like comparisons of projects' financial attractiveness within categories may be undertaken, and the financing options can be considered.

### Consider Financing Options

City decision makers will use risk and ROI analyses to assess proposed projects and determine how projects can best be financed. In some cases, the analysis will show that certain projects are not viable. For those that are, cities must determine the most attractive sources of financing available to them, plus the incentives that may be available (see the case study in box 14.15). The categorization of projects is important for subsequent identification of sources of financing. Although not comprehensive, and varying from country to country, examples of types of financing available for certain project categories are shown in the following diagram.

|               | Large scale                                     | Small scale                          |
|---------------|-------------------------------------------------|--------------------------------------|
| Decentralized | Government intervention                          | Development funding                  |
|               | Multistage or mezzanine financing               | Carbon finance (Clean                |
|               | Initial public offering                          | Development Mechanism)               |
| Centralized   | Structured finance (internal investment,        | Private equity                       |
|               |     bond sale, or the like)                      | Carbon finance                       |
|               | Traditional project finance (debt or equity)    |                                      |

### Self-Funding

In many circumstances, self-funding by a city may appear simpler and less time consuming than seeking outside funding. But if sources of price-competitive

---

**Box 14.15  Case Study: Maldives—Public-Private Partnership (PPP)**

In May 2011, the government of Maldives, with the help of the World Bank as lead transaction adviser, implemented a solid waste management PPP. The transaction was structured as a 20-year build-operate-transfer project and mobilized US$50 million in private investment that will improve waste collection, transportation, and disposal; reduce marine and air pollution; and generate power through a 2.7 megawatt waste-to-energy plant. The project will benefit 120,000 people, process up to 70 percent of the country's solid waste, and reduce annual GHG emissions by 16,000 tons. This project supports the Maldivian government's goal of becoming carbon neutral by 2020, while helping it to comply with good global practices for the treatment and disposal of solid waste.

---

---

**Box 14.16  Case Study: Tokyo, Japan—Waterworks Maintenance Planning**

Fees and charges are important to revenue-generating enterprises, such as water companies, in considering an appropriate level of reserve funding. (Reserve funds are used to meet unexpected costs that may arise, for example, emergencies, and for future needs.) The Tokyo Waterworks, which serves 12.5 million people in metropolitan Tokyo, finances operating expenses and capital expenditures by relying on water tariff revenues.

Various reserve funds have been set aside to cover fluctuations in these costs. In 2011, the utility was facing the daunting task of replacing old water pipes. The project was to begin in nine years. The total investment is estimated to be ¥1 trillion (US$10 billion), which represents 40 percent of the utility's total assets of ¥2.5 trillion (US$25 billion) in current yen.

To meet this challenge, Tokyo Waterworks started identifying ways to level out the ¥1 trillion (US$10 billion) planned investment over a reasonable period by planning for maintenance and rehabilitation well ahead of the project and establishing a detailed construction plan. Meanwhile, the utility has already started accelerating debt repayments so that outstanding debt can be maintained at the current level of ¥0.5 trillion (US$5 billion) even after project financing has been undertaken. The accelerated repayments are being covered by water tariff revenues even though the Tokyo metropolitan government lowered the water tariff on January 1, 2005. The utility plans to finance the ¥1 trillion (US$10 billion) replacement project by implementing a reasonable tariff adjustment.

*Source:* Suzuki and others 2010.

---

financing are available on the open market supported by risk-mitigating incentives, it is often more efficient to take advantage of these funds. Then, city budgetary funds can be used for high-priority projects that were demonstrated to be financially untenable.

If a city decides to implement energy efficiency and emissions reduction projects using its own funds, financing most often will be drawn from revenues derived from fees and taxes, business activities, privatization of city property, and state budget subsidies. See the case study in box 14.16.

Self-funded projects give a city the greatest level of control over implementation, but also require the highest level of involvement and use of internal resources. See the case study in box 14.17.

### Generating Alternative Funds

City funds and resources are often not sufficient for energy efficiency projects requiring high capital expenditure or risk. In these cases, external funding will be required and other means of financing must be sought, such as credits, PPPs, leasing and concessions deals, third-party financing, donations, and so forth. A key consideration for generating alternative or external funds is the time horizon for the availability of financing. Timing may be established in the early analysis of funding options. If funding timelines do not match project timelines, these

---

**Box 14.17  Case Study: Various Cities—Lamp Rollout**

Many governments simply use their own funds to implement domestic compact fluorescent lamp (CFL) projects. These projects can be a "quick win" under energy efficiency and climate change policies. Cuba and South Africa, to name just two, have adopted this approach.

However, international carbon finance can also be leveraged without the need self-funding. CFL projects fall under the Clean Development Mechanism (CDM) methodology "AMS-II.C. ver. 9—Demand-side energy efficiency activities for specific technologies." So CFL projects are eligible for registration with the United Nations Framework Convention on Climate Change and can deliver Certified Emissions Reduction credits (CERs) equivalent to the emissions reductions achieved.

OSRAM lighting company and RWE Power registered three small-scale CFL CDM projects in 2009 using this methodology. The projects are all located in India and will result in the issuance of approximately 27,000, 40,000, and 30,000 CERs per year.

The development of the CDM Programmes of Activity (PoA) mechanism has further encouraged CFL projects. The first PoA registered with the CDM Executive Board, CUIDEMOS Mexico (Campaña de Uso Inteligente de Energía México, or Smart Use of Energy Mexico), was undertaken by a U.K. and Australian partnership and will lead initially to the issuance of 24,283 CERs per year. The PoA includes within the umbrella-design document the capacity to deliver 520,365 CERs per year.

---

sources should be discarded at an early stage unless guarantees can be obtained from project implementers that project-related timelines will be adhered to.

If an ROI analysis shows that returns from a project will exceed a market-linked hurdle, private sector financing may be sought at prevailing market rates. Private sector financing is often preferable to energy efficiency–specific financing that has links to incentives, which may prove more time consuming and costly to attain.

If an ROI analysis demonstrates that the proposed project is not competitive with projects in nonenvironmental sectors, tailored financing may be sought from investors experienced in environmental project risk and who have obligations to direct capital into certain market segments. Socially responsible investing (SRI) funds are a good example. SRI funds deploy capital according to criteria that include environmental and social performance indicators. The managers of these funds, although still hoping to achieve a good ROI, are more cognizant of project risk in the environmental sector (see SRI Fund Portal Asia: http://www.asria.org). The SRI sector includes climate-specific and ecology funds (for example, Jupiter Ecology Fund, HSBC Global Investment Fund—Climate Change), and fund managers have a more detailed understanding of energy efficiency project risk, which can save valuable time. See the resources in box 14.18 for additional options.

Although it is not possible to list all options in this Guidebook, a few examples of alternative arrangements are identified in the example in box 14.19.

---

**Box 14.18  Resources: Finding Financing Resources**

Numerous climate funds are available for carbon mitigation projects but no single portal contains links to them all. However, some good resources can be found on the Internet:

    http://www.climatefinanceoptions.org

    http://www.climateinvestmentfunds.org

---

## Criteria for Assessing Funding Arrangements

A wide variety of potential funding mechanisms are available for energy efficiency projects. A city must understand the criteria important to its assessment of financing options, which may include the following:

- Cost of capital
- Transaction costs
- Funding timing and project milestones
- Required project delivery vehicle
- Credit ratings of potential investors and lenders
- Securities required
- Equity demands
- Ethical considerations

## Project Delivery Vehicle

Thought should be given to the appropriate vehicle for delivery of each project. This entity will be linked to the method of financing; for example, if funds are sought for decentralized, small-scale projects, a project-specific NGO might make the most efficient use of the funds because it would be able to take advantage of particular tax-efficient benefits. A cost-benefit analysis should be made of each delivery vehicle to ascertain the most efficient mechanism for the project category. Vehicles might include the following:

|               | Large scale                                 | Small scale              |
| ------------- | ------------------------------------------- | ------------------------ |
| Decentralized | ESCO                                        | NGO                      |
|               | Limited liability company                   | Development organization |
| Centralized   | Special purpose vehicle                     | ESCO                     |
|               | (linked to utility or city)                 | Private company          |

## Incentives

To counter any time or cost impacts of obtaining financing specifically for environmental projects, available incentives—which will vary by project category—should be fully assessed. These incentives can mitigate project risk and increase the likelihood of attracting additional external financing. Incentives often take the form of mezzanine financing that ensures the project's development while

## Box 14.19 Example: Alternative Funding Schemes

Financing strategies commonly used to fund energy efficiency and emissions projects include the following:

- Debt financing: Funds are acquired by borrowing.
- Equity financing: Funds are acquired by issuing shares of common or preferred stock in anticipation of income from dividends and capital gains.
- Subordinated debt financing (mezzanine financing): Financing using capital that sits midway between senior debt and equity and has features of both kinds of financing.
- Project financing: Project financing relies on a project's cash flow expectations and spreads risk between different actors involved in the project.
- Supplementary mechanisms: Governments close financing gaps, catalyze private investment, and accelerate energy efficiency market uptake via financial and nonfinancial interventions.
- Revolving funds: These can include loans or grants and aim to be self-sustaining after the first capitalization.
- Third-party financing schemes: An external party funds the energy efficiency scheme and takes on the risk.
- Leasing: The city makes payments of principal and interest to the lending financial institution. This type of scheme includes capital leases and operating leases.
- ESCOs: ESCOs finance energy savings projects and recover the investment through the contract period energy savings.
- Public internal performance commitments: A department in the public administration acts similarly to an ESCO for another department.
- Public-private partnerships: The public authority uses a concession scheme to enable the private sector to contract public projects.
- Bonds: Energy efficiency and emissions projects are funded through city-issued bonds.
- Commodity or commercial credits: A delay of payment is accepted in exchange for raw materials or goods.
- Leasing of equipment: Equipment is leased to generate additional funds.
- Climate Investment Funds: A US$6.4 billion facility draws on the expertise of several multilateral development banks to help developing countries pilot low-emission and climate-resilient development (http://www.climateinvestmentfunds.org/cif/).
- Clean Development Mechanism: See the example in box 14.20 and the resources in box 14.21.
- Global Environment Facility: This facility provides grants to developing countries and those with economies in transition for projects related to biodiversity, climate change, international waters, land degradation, the ozone layer, and organic pollutants.
- Policy-oriented private equity fund-of-funds: These are funds such as the Global Energy Efficiency and Renewable Energy Fund, which provides global risk capital through private equity investments for energy efficiency and renewable energy projects in developing countries.

*box continues next page*

**Box 14.19  Example: Alternative Funding Schemes** *(continued)*

- Partial risk guarantees (PRGs): PRGs are particularly helpful in supporting energy efficiency projects that do not fall under traditional lending categories. This instrument is offered by the World Bank and covers private lenders or investors against the risk of a government (or government-owned) entity failing to perform its obligations with respect to a private project. In the case of default resulting from the nonperformance of contractual obligations undertaken by governments or their agencies in private sector projects, PRGs ensure payment and thereby significantly reduce the risk assumed by investors. PRGs are available for projects with private participation dependent on certain government contractual undertakings such as build-operate-transfer and concession projects, PPP projects, and privatizations. PRGs are available for both greenfield and existing projects.

*Sources:* Adapted from Covenant of Mayors 2010 (http://www.eumayors.eu/actions/sustainable-energy-action-plans_en.html); Rezessy and Bertoldi 2010; MODEL 2010 (http://www.energy-cities.eu.MODEL); AusAID and World Bank 2012.

---

longer-term, more substantial project financing is sought. Incentives can originate from regional, national, and supranational sources and be on a structured or non-structured basis. Structured incentives might include carbon credits, such as those from the Clean Development Mechanism (CDM) (see the example in box 14.20 and the resources in box 14.21), whereas nonstructured incentives might include ad hoc loan guarantees from development organizations.

Incentives may be available at many levels, ranging from local tax incentives to national government subsidies and international development seed funding. Technical assistance for accessing financing is also available, such as the CTI Private Finance Advisory Network (http://www.cti-pfan.net), supported by the UN Framework Convention on Climate Change (UNFCCC). Such support can be quantified financially in ROI assessments. The level of incentive will also vary by project category and potentially by market forces that dictate pricing.

Incentives may also be shaped to some extent by the city itself to widen the availability of private financing. City policy development should consider what levels of incentives would attract private sector financing into prioritized projects to achieve the goals set out in the sustainable urban energy and emissions plan. Limited city financial resources can thus be leveraged to draw larger external private sector investment. The diagram illustrates the incentives a city might offer.

| *Large scale* | *Small scale* |
|---|---|
| Grant | Soft loan |
| Subsidy | Loan guarantee |
| | Environmental credits |

## Perform Due Diligence

Once financing options have been considered and a finance provider engaged, due diligence will likely be required by investors. Transparency in documentation and process is essential to give a potential investor confidence in claims that were made at the project financial viability stage.

**Box 14.20  Example: Clean Development Mechanism (CDM) Project Development**

The CDM is a specific alternative financing approach developed for projects that address climate change in developing nations. The CDM is an arrangement under the UNFCCC allowing industrial countries with GHG reduction commitments under the Kyoto Protocol to invest in projects that reduce emissions in developing countries as an alternative to more expensive emission reductions in their own countries (ICLEI 2009). (CDM is addressed separately from other alternative funding sources because it is applicable to many of the priority projects identified in energy and emissions plans.)

The CDM allows emissions reduction projects in developing countries to earn Certified Emission Reduction credits (CERs), each equivalent to one ton of carbon dioxide. These CERs can be traded and sold, and can be used by industrial countries to meet a part of their emissions reduction targets under the Kyoto Protocol (UNFCCC at http://cdm.unfccc.int/faq/index.html). The CDM provides an additional incentive for financing environmental projects by allowing CERs to be traded at a market-determined price, as well as indirect advantages such as publicity value and lower risk perception, hence, lower cost of capital.

Procuring CDM funding takes a number of stages, each involving different stakeholders. The CDM Executive Board (CDM EB), under the UNFCCC, plays a vital role at the registration and CER

**Figure B14.20.1  Project Development under CDM**

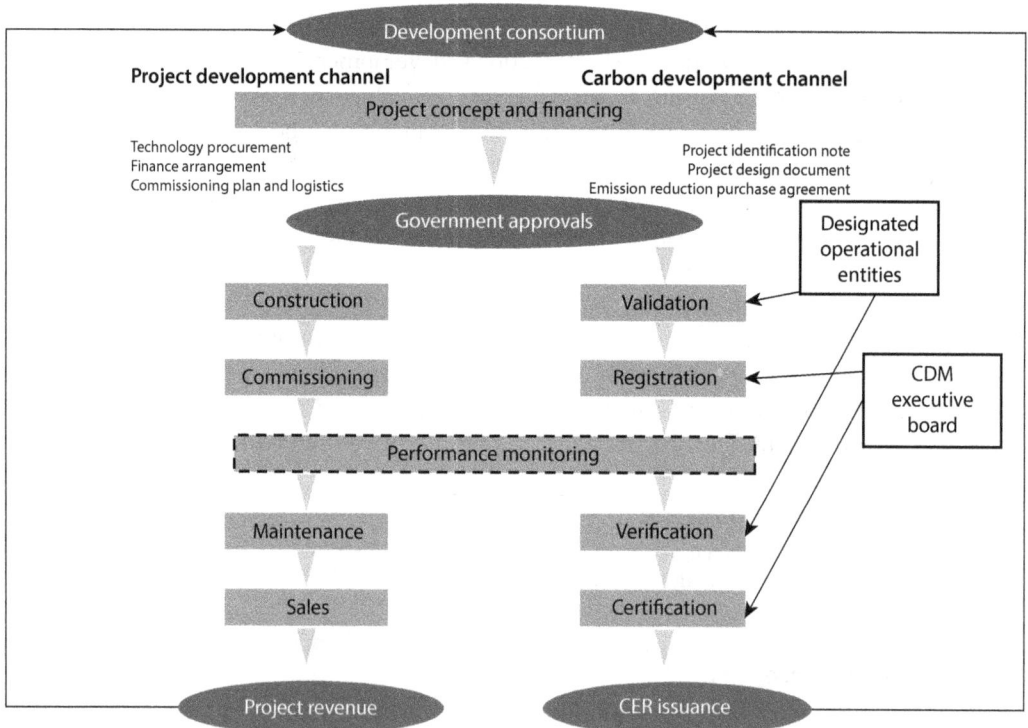

*Note:* CDM = Clean Development Mechanism; CER = Certified Emission Reduction.

*box continues next page*

**Box 14.20  Example: Clean Development Mechanism (CDM) Project Development** *(continued)*

issuance stages. Projects must apply a preapproved project methodology (see figure B14.20.1), obtain an approval from the host party government, and undergo a series of independent audits by UN-approved designated operational entities (DOEs) before registration by the CDM EB. This process is exceptionally transparent—all project design documents, including ROI analyses, must be published on the UNFCCC's public website for review and comment by global stakeholders.

The ongoing monitoring of emissions reductions is emphasized in CDM projects, and monitoring plans are core to any project design. The application of monitoring plans is central to the issuance of CERs. Verification of emissions reductions by DOEs are completed at the end of every monitoring period, and it is the DOE's verification report that forms the basis of CER issuance requests made to the CDM EB.

Additionally, CDM Programs of Activity (PoA) allow project concepts to be registered with preapproved approaches to additionality assessments,[a] monitoring plans, emissions reductions calculations, and so forth. During the period in which the PoA is accruing credits, individual program activities can be included as separate small-scale projects. PoAs are intended to reduce transaction costs and development time for smaller-scale projects. Different program activities can be coordinated and managed by separate entities on the ground, but included in the same PoA. This is an efficient approach to implementing decentralized, small-scale energy efficiency projects while taking advantage of available carbon finance.

Typically, carbon finance available for CDM projects will hinge on the project developer's ability to secure an Emissions Reduction Purchase Agreement with a CER buyer. These agreements often consist of an agreed-on price for the future delivery of a stated volume of CERs during the project's crediting period. If the agreement contains either a fixed price or a floor price for CERs, cash flows can be extrapolated for the project's ROI analysis because emissions reduction estimates are available from the project design documents (taking into account monitoring and issuance risks). Many CER buyers are willing to make risk-adjusted, up-front payments for future CER deliveries, depending on the forward CER price curve.

At present, the majority of CER demand originates in the European Union's Emissions Trading System. The third phase of this scheme will begin in 2013 and conclude in 2020. The number of CERs eligible to enter the scheme is limited, but during the third phase any new CDM project that delivers CERs into the scheme must originate in a developing country.

When considering CDM, cities must be aware of the exposures inherent in the process:

- *Conventional project exposures:* cost overruns, market risks, counterparty credit risk, underperformance, currency risk, and force majeure
- *Host country exposures:* confiscation, expropriation, and nationalization; civil war; contract repudiation or frustration; host country sovereign risk; administrative barriers; lack of institutional capacity in host country
- *CDM process exposures:* CDM EB nonapproval; timing and delays; CER supply-demand dynamics; monitoring and verification risk; institutional barriers; CER legal ownership

a. Additionality assessments demonstrate that a carbon reduction project actually reduces carbon emissions and that it would not have already been performed without the project's intervention.

---

**Box 14.21  Resources: CDM Funding Sources**

The United Nations Adaptation Fund: The Adaptation Fund is financed through CDM project activities and other sources of funding. The share of proceeds amounts to 2 percent of Certified Emission Reductions issued for a CDM project activity.

Pure carbon funds: World Bank Prototype Carbon Fund; Certified Emission Reduction Unit Procurement Tender; GTZ fund (German government development fund)

Carbon equity funds: FE Clean Energy Group, Inc.'s Asian Clean Energy Services Fund; Asia Pacific Carbon Fund of the Asian Development Bank

---

For example, investors are usually unfamiliar with energy efficiency projects. If the methodology used to establish potential energy savings is not transparent and well documented, investors will not be able to understand the risks of the project and might well respond by withdrawing funds or requiring a higher hurdle rate to mitigate the perceived risk.

Similar documentation will be necessary in applying for incentives if the incentive is linked to energy savings or GHG emissions reduction. Without clear documentation and application of reliable methodologies, incentives are unlikely to be granted.

Parameter values used in ROI analyses must be supportable with clear documentary evidence, preferably audited by reliable third parties. This will help to educate investors who are unclear on the specifics of energy efficiency projects and thereby help to mitigate perceived investment risks.

Due diligence is the final step in identifying financing mechanisms and will be used by potential investors to scrutinize management control over the project and the concomitant risk profile.

## Step 15: Roll Out Projects

This step describes the process of rolling out projects once they have been identified, developed, and funded. This is the last step in the implementation phase and may require months or years to accomplish, particularly for a major project. Although implementation processes for the various types of projects may differ (for example, incentive projects, major projects, organizational development programs, and policies and regulations), several standard factors should be considered as projects are being rolled out. These are explored in Step 15. See the example in box 14.22 for a high-level rollout plan.

### Identifying Needed Skills

It is important to acknowledge that new skills and time are critical to roll out and deliver the energy and emissions projects. City government departments should not be expected to undertake new projects on top of their existing tasks without training and additional manpower. The skills needed to implement each project

---

**Box 14.22  Example: Typical Incentive Project Rollout Plan**

**Publicize Project Availability**
Set a level of publicity appropriate to the attractiveness of the incentive and demand for the project. Conduct radio ads for residential projects with large funding, for example. Or simply reach out directly to large industrial customers that might best take advantage of a small industrial energy project.

**Select Applicants**
Develop an application submission and selection process that complies with city procurement rules. Be careful not to make the application process too onerous—you do not want the cost to apply to outweigh the benefit of the incentive.

**Confirm Specifications**
Review final designs or purchase orders to confirm that the energy efficiency equipment conforms to the list of equipment for which incentives are provided.

**Perform Audits**
If the incentive project is based on a series of audits, for example, hospitals or electrical substation transformer upgrades, perform the audits quickly and efficiently. You do not want to lose the best customers and you do want to capture the benefits.

**Validate Installation and Distribute Incentive**
Most incentive projects require a validation process in which the managing organization verifies that the new equipment was installed correctly so that the energy savings will actually occur.

---

should be identified. For example, the personnel required to inventory street lamps, research lamp types, and update a database to implement a street lighting audit and retrofit project should be identified. Afterward, additional manpower requirements—beyond the standard street lighting maintenance team—to replace old lamps with new high-efficiency lamps should be estimated. Personnel requirements should also consider the qualifications and experience needed for each position.

## Developing Skills and Manpower

The planning and analysis for the SUEEP process up to this point may have been accomplished by a few people in the energy task force, or by technical consultants. However, city government staff will have primary responsibility for implementation and validation of projects—they are experts in their own systems and may need only minimal training to become project champions. More important, they know how to get the bureaucracy to run.

Staff members may need one-on-one training from experts in the region who have implemented similar projects. They may also attend relevant conferences and embark on study trips to learn about similar projects in the region.

Energizing Green Cities in Southeast Asia  •  http://dx.doi.org/10.1596/978-0-8213-9837-1

In-house training also provides an opportunity for city government staff from different disciplines and agencies to network, serving as a platform for future collaboration.

### Bringing in Skills and Manpower

Hiring new staff with special energy programming or sector experience may be the least expensive way to build capacity. Part of the new staff members' assignments could be to train those around them in new ways of thinking, procurement methods, or technologies.

To keep a lid on budgets, contract staff may be the lowest-cost option for when a large number of employees are needed for a short time to execute a portion of a project, such as an audit or equipment inventory.

### Identifying Needed Resources

Skilled and motivated people are the most important factors for the success of a rollout plan. However, if they lack equipment, funding, or time to implement projects, even the best staff will not be effective. It is therefore important to acknowledge that both resources and time are necessary to undertake new projects, and to make provisions to enable adequate staffing to implement the sustainable urban energy and emissions plan.

### Project Management

An execution plan for each project and an oversight process that will ensure each step is delivered on time and on budget should be developed. In particular, managers should be made responsible for the success of the project to ensure that agencies prioritize its implementation, rather than leave it on the back burner while administering important day-to-day city services.

### Communication Plan

Communication with stakeholders occurs throughout the project development and implementation stages. Affected stakeholders should be informed of their roles before projects are rolled out. To increase buy-in for the policy or to increase the uptake of an incentive project, public outreach efforts can be extended even while the project is midway through the implementation phase. Just as important, projects that are successfully completed should be publicized.

### Monitoring and Reporting

Project monitoring should be part of project design and should commence, together with the collection of data and assessment of performance, when the project is rolled out. Stakeholders should be aware of the metrics used to measure success.

### Stay Positive

Defining the "successful implementation" of the project will be something only you can do, as the leader of the city's energy and emissions plan. Stay optimistic about the ways these projects can affect your city and that optimism will

---

**Box 14.23  Tip: Critical Success Factors for Projects**

Be sure to consider the following aspects when rolling out each energy project:

- Adequate skills and experience of staff
- Adequate funding
- Clear success metrics
- Clear line of responsibility for delivery
- Time-based performance targets
- Political support and commitment
- Communication plan for sharing successes and experiences

---

**Box 14.24  Resources: Further Reading on Implementing Projects**

*Financing Energy Efficiency: Lessons from Brazil, China, India, and Beyond.* Robert P. Taylor, Chandrasekar Gavindarajalu, Jeremy Levin, Anke S. Meyer, and William A. Ward. World Bank. 2008.

---

resonate with the stakeholders and community. Remember, although this is an "energy" plan, the wide reach and positive impacts of successful projects will be felt throughout your city, well beyond the realm of energy and emissions.

The tip in box 14.23 summarizes the "must haves" of successful projects.

See the resource in box 14.24 for additional information on implementing projects.

## References

AusAID (Australian Agency for International Development) and World Bank. 2012. *Green Infrastructure Finance: Framework Report.* Washington, DC: World Bank.

C40 Large Cities Climate Summit. 2007. "Case Study, Seattle Sets the Standards for Green Buildings." London. http://www.c40cities.org/c40cities/seattle/city_case_studies/seattle-sets-the-standards-for-green-buildings.

Covenant of Mayors. 2010. "How to Develop a Sustainable Energy Action Plan (SEAP)—Guidebook." Publications Office of the European Union, Luxembourg.

Feng, Liu, Anke S. Meyer, and John F. Hogan. 2010. "Mainstreaming Building Energy Efficiency Codes in Developing Countries: Global Experiences and Lessons Learned from Early Adopters." Working Paper 204, World Bank, Washington, DC.

Hammer, Stephen A. 2009. "Capacity to Act: The Critical Determinant of Local Energy Planning and Program Implementation." Columbia University Center for Energy, Marine Transportation and Public Policy, New York, NY.

ICLEI (Local Governments for Sustainability). 2009. "International Local Government GHG Emissions Analysis Protocol (IEAP)." Bonn, Germany. http://carbonn.org/fileadmin/user_upload/carbonn/Standards/IEAP_October2010_color.pdf.

IEA (International Energy Association). 2010. *Energy Efficiency Governance—Handbook.* Paris: OECD/IEA.

McKinsey and Company. 2009. *Pathways to a Low-Carbon Economy, Version 2 of the Global Greenhouse Gas Abatement Cost Curve.* Washington, DC: McKinsey and Company. https://solutions.mckinsey.com/ClimateDesk/default.aspx.

OECD (Organisation for Economic Co-operation and Development). 2010. *Cities and Climate Change.* Paris: OECD Publishing.

Phylipsen, Gerardina Josephina Maria. 2010. "Energy Efficiency Indicators: Best Practice and Potential Use in Developing Country Policy Making." Phylipsen Climate Change Consulting, commissioned by the World Bank, Washington, DC.

Rezessy, Silvia, and Paolo Bertoldi. 2010. "Financing Energy Efficiency: Forging the Link between Financing and Project Implementation." Joint Research Centre, European Commission, Ispra, Italy. http://ec.europa.eu/energy/efficiency/doc/financing_energy_efficiency.pdf.

Rogers, Everett M. 1962. *Diffusion of Innovations.* New York: The Free Press.

Singh, Jas, Dilip R. Limaye, Brian Henderson, and Xiaoyu Shi. 2009. *Public Procurement of Energy Efficiency Services: Lessons from International Experience.* Washington, DC: World Bank.

Suzuki, Hiroaki, Arish Dastur, Sebastian Moffatt, Nanae Yabuki, and Hinako Maruyama. 2010. *Eco² Cities: Ecological Cities as Economic Cities.* Washington, DC: World Bank.

Taylor, Robert P., Chandrasekar Gavindarajalu, Jeremy Levin, Anke S. Meyer, and William A. Ward. 2008. *Financing Energy Efficiency: Lessons from Brazil, China, India, and Beyond.* Washington, DC: World Bank.

United Nations, Bureau International des Expositions, and Shanghai 2010 World Exposition Executive Committee. 2011. *Shanghai Manual: A Guide for Sustainable Urban Development in the 21st Century.* Shanghai, China. http://sustainabledevelopment.un.org/content/documents/shanghaimanual.pdf.

UNEP (United Nations Environment Program). 2009. *Energy Efficiency and the Finance Sector: A Survey on Lending Activities and Policy Issues.* Nairobi: UNEP Finance Initiative's Climate Change Working Group.

CHAPTER 15

# Stage VI: Monitoring and Reporting

*Regular collection, compilation, and understanding of the progress of the plan is critical to its success. The tasks outlined in this stage focus on engagement with the leaders of the projects, who are working together to collect various levels of data and to understand what the data are revealing about the projects and the status of the city beyond just "energy." Using the data and the feedback from leaders of the projects and industry stakeholders, a city evaluates the progress of its program and decides on the stories to communicate in a status report to the city, stakeholders, and all interested parties.*

## Step 16: Collect Information on Projects

The lead agency for the energy and emissions plan should continuously monitor the city's overall energy program and evaluate each project annually. Coordination with implementing agencies and stakeholders is no small task, so successfully gathering performance data will require planning and organization. Although progress will be monitored in a variety of ways across projects and cities, the information will consist of hard data, anecdotes from end users, and stakeholders' observations of outcomes as a result of the implementation of specific projects.

This section describes a series of tasks and organizational actions that will lead to successful collection of data from the various projects. The examples can be referred to for clarity. See the example in box 15.1 to start.

### Collect Project Data

Specific hard data will have to be collected in each project operations phase. In some cases, data will be collected in the natural course of running the project (requiring little effort) and the annual data for that project can be collected in one meeting or report. In other cases, a variety of resources and additional data collection processes will be required to ensure that project data are sufficient to measure progress.

### Comparability of Data

Stage II described the need to establish data trends and to ensure clarity about the source of data. This step builds on Stage II and requires the collection of data

---

**Box 15.1  Example: Three Sample Projects**

Three examples are used throughout this stage to demonstrate how particular projects can be monitored and reported.

**Project 1. Commercial Building Energy Code**
> The rollout, adoption, and enforcement of a commercial building energy code to mandate energy efficient construction practices for new buildings and retrofits

**Project 2. Bus Rapid Transit (BRT) System**
> The phased installation of a BRT line from a popular suburban residential neighborhood into the downtown, financial district of the city

**Project 3. Street Light Efficiency Project**
> The phased installation of more efficient and well-designed (spaced and implemented) street lighting throughout the city

---

**Box 15.2  Example: Data Collection for the Three Sample Projects**

**Project 1. Commercial Building Energy Code**
> The electrical utility provides total electricity sold to commercial customers. The Department of Buildings provides the number of new buildings that have been issued Building Code Compliance Forms and the total square footage of floor space.

**Project 2. Bus Trapid Transit (BRT) System**
> The Department of Transportation provides annual ridership counts, total annual passenger-kilometers, annual fuel consumption and expenditure, and kilometers of new BRT operating in the year.

**Project 3. Street Light Efficiency Project**
> The Street Lighting department provides an annual inventory of all street lights, lamp types, total electricity consumption and expenditure, percentage of city streets lit, and total kilometers of streets lit.

---

over similar time scales, boundaries, and users so that trends can be established to enable a city to assess its progress in implementing energy and emissions projects. See the example in box 15.2.

### Creating Key Performance Indicators (KPIs)

Stage II describes how KPIs are established to track energy performance over time. Project-specific KPIs should also be established to demonstrate the performance of individual projects.

KPIs may be energy specific, but they should also be related to the underlying drivers of energy or other city qualities, such as health or economic development.

---

**Box 15.3  Example: KPIs for the Three Sample Projects**

**Project 1. Commercial Building Energy Code**

- Total megajoules per city occupant
- Consistent census data for city population

**Project 2. Bus Rapid Transit (BRT) System**

- BRT riders per day, wait times, energy consumption per BRT passenger-kilometer

**Project 3. Street Light Efficiency Project**

- Megajoules saved (compared with business as usual)
- Total electric site lighting (calculated estimates in kilowatt-hours of electricity per year)

---

The challenge for the energy task force is to acknowledge these links between various metrics and understand how they measure progress in the context of the unique circumstances of the city. See the example in box 15.3.

Sometimes KPIs will require that data be combined in a way that ensures the appropriate metrics are used to track success, for instance, the denominator of a KPI, which allows for consistency over time, such as area (gross square meters) or occupants (riders). To be consistent with other KPIs that use similar denominators, the energy task force should ensure that the denominator (for example, occupant or area) is consistent for various metrics.

### External Data

Although a city's primary efforts will be to collect useful and accurate data and information from its constituencies and stakeholders, data from external parties will also be needed. Priority projects could affect not only the city, but regional or national policies and actions. Thus, the city will also have to work with utility providers, national organizations, and external stakeholders to ensure that regional data (for example, electricity fuel mix, transportation trends, and larger-scale economic growth and trends beyond the city's boundary) are collected as well.

### Assess KPIs against Targets

Assessing KPIs regularly is critical to understanding the results of either a specific project or the energy and emissions plan overall. See the case study in box 15.4. KPIs should be assessed to determine whether they meet the targets for a particular year. If the KPI suggests that the targets will not be met, an effort should be made to find out why. Alternatively, if KPIs are easily met, then perhaps more aggressive targets should be set.

### Remember the Context

Data alone do not indicate success or failure. Data and KPIs have to be put into context. For example, external factors such as population increases or boundary

**Box 15.4  Case Study: London, the United Kingdom—Congestion Charging**

In 2003, London introduced a daily congestion fee for vehicles traveling in the city's central district during weekdays. This fee was meant to ease traffic congestion, improve travel time and reliability, and make central London more attractive to businesses and visitors. According to the city's analysis, the program largely met its objectives. After four years of operation, traffic entering the charge zone was reduced by 21 percent; congestion, measured as a travel rate (minutes per kilometer), was 8 percent lower; and annual fuel consumption fell by approximately 3 percent. These changes translated into annual reductions of 110,000–120,000 tons of carbon dioxide, 112 tons of nitrogen oxides, 8 tons of particulate matter, and some 250 fewer accidents. The identified benefits exceeded the costs by more than 5 percent.

In addition, the scheme brought a steady net revenue stream for transportation improvements, of which 80 percent was reinvested in improving public bus operations and infrastructure. The city proved to be innovative and resourceful by ensuring key elements of the congestion-charging project were in place, including technical design, public consultation, project management, an information campaign, and impact monitoring. These factors led to the successful implementation of the project.

*Source:* http://www.esmap.org/esmap/node/1279.

changes may have prevented targets from being met, affecting the KPI of a specific project. In such a case, insufficient efforts from the implementing agency were not the cause of the failure to make progress. Hence, establishing connections between targets and the city's context is key to assessing progress. See the example in box 15.5.

### Engage Stakeholders

Although data are necessary for an understanding of performance levels and trends, qualitative information and feedback from stakeholders are crucial to assessing the success of a project. A city needs to consider which stakeholders to engage and to what extent, based on the stakeholders' influence on projects and on the effect projects have had on them. Feedback from a good sample of users, organizations, and industry leaders will help a city to understand and implement measures to improve projects. See the example in box 15.6.

### Summarize and Learn

A city selects projects not only because the projects can reduce energy use or carbon emissions, but also because the projects have the potential to contribute to realization of the city's vision. It is important to emphasize the links between energy projects and the city's wider goals and objectives. With data and feedback from stakeholders in hand, key lessons can be learned and projects could be revamped to contribute to the overall success of the energy and emissions plan.

---

**Box 15.5  Example: What Data May Reveal**

Data alone may show trends, but putting it into context for stakeholders and end users will help to provide a clearer understanding of the drivers behind the data.

**Project 1. Commercial Building Energy Code**

Increased energy demand without an increase in the number of commercial customers may be symptomatic of an increase in energy intensity for the city. Growing economic development leads to lower-quality commercial space being upgraded to higher-quality (and higher-energy-using) commercial office space.

A moderate increase in energy demand accompanied by an increase in the number of customers suggests that lower-energy-using commercial clients are being attracted.

**Project 2. Bus Rapid Transit (BRT) System**

Rider surveys may reveal whether an increase in public transportation has resulted in reduced energy use per person.

Air quality reports in areas in which the BRT was implemented show that emissions are lower, based on records for the period before the BRT, suggesting that car usage has been reduced.

**Project 3. Street Light Efficiency Project**

As more street lights are retrofitted with the new lamps, energy usage decreased. As an added benefit, areas with new lighting show safety increases.

---

## Find the Lessons Learned

The energy leadership should now have a good idea of the project's challenges (planning, implementation, and enforcement) as well as of the factors that contributed to its success. Such information, including key messages from the information collected, should enable a city to draw lessons from each project.

Although experiences differ across projects, lessons learned from experiences in specific projects are at times relevant to other projects and different audiences. Thus, sharing these lessons across relevant organizations will enable the city to benefit from an all-encompassing view of the lessons learned across projects. Compiling the lessons learned and communicating them clearly to current and future leaders will drive future successes.

## Highlight Key Success Stories

A well-executed plan to implement priority projects is likely to bear fruit. Although accomplishments during the first year of the energy and emissions plan may be limited, sharing success stories will encourage parties implementing projects to intensify their efforts in attaining targets and goals. And success stories need not be based exclusively on the reduction of a city's energy use and carbon emissions—contributions to a healthier and more economically prosperous city can also be highlighted. See the case studies in boxes 15.7 and 15.8.

## Box 15.6  Example: Stakeholder Engagement for Sample Projects

Examples of stakeholders to interview when understanding the impacts of these projects:

### Project 1. Commercial Building Energy Code
- Real estate owners who have adopted the code or advocacy groups who have agreed to support the code
- Department of Buildings, which manages the project
- Utility providers
- Local associations (American Society of Heating, Refrigerating, and Air-Conditioning Engineers [ASHRAE], local professional engineering organizations, and others)

### Project 2. Bus Rapid Transit (BRT) System
- Department of Transportation, which manages the project
- Users (commuters)
- Local businesses on the streets through which the BRT lanes run

### Project 3. Street Light Efficiency Project
- Parks department
- Police department
- Businesses on streets
- General public

## Box 15.7  Case Study: Curitiba, Brazil—Bus Rapid Transit (BRT) Network

The popularity of Curitiba's BRT has effected a modal shift from automobile travel to bus travel. Based on 1991 traveler survey results, it was estimated that the introduction of the BRT had caused a reduction of about 27 million auto trips per year, saving about 27 million liters of fuel annually. Other policies have also contributed to the success of the transit system. Land within two blocks of the transit arteries is zoned for high density, because high density generates more transit ridership per square foot.

Compared with eight other Brazilian cities of similar size, Curitiba uses about 30 percent less fuel per capita, resulting in one of the lowest rates of ambient air pollution in the country. As of 2010, about 1,100 buses were making 12,500 trips every day, serving more than 1.3 million passengers—50 times the number of 20 years ago. Some 80 percent of travelers use the express or direct bus services. Best of all, Curitibanos spend only about 10 percent of their income on travel—much lower than the national average.

All data were provided by 10 private bus companies in partnership with the local department of transportation. This illustrates a multidisciplinary, comprehensive approach to an energy- and carbon-saving solution.

*Source:* http://urbanhabitat.org/node/344.

---

**Box 15.8  Case Study: Seattle, Washington, USA—Street Lighting Efficiency Program**

The publicly owned utility Seattle City Light successfully installed more than 6,000 street lights. The project is part of a plan to replace 41,000 residential street lights in Seattle by the end of 2014, a program that is already saving the city US$300,000 per year (http://www.ledsmagazine .com/news/8/7/11). Once completed, the city council estimates a US$2.4 million reduction in operating costs will be achieved.

Edward Smalley, manager of street light engineering at Seattle City Light, said the decision to install light-emitting-diode-based street lighting was the result of the technology's demonstrated illumination performance, controllability, and operational efficiency (48 percent energy savings), all needed to satisfy the city's lighting needs. Council members were also swayed by the tremendous savings in maintenance costs. "Every two years, we would pay workers overtime to quickly replace the high-pressure sodium lamps before the winter came," Smalley said. "Now that cost has been essentially eliminated."

*Source:* http://www.ledsmagazine.com/features/8/9/4.

---

### Be Transparent and Honest, but Find the Wins

Stakeholders respect transparency and honesty. Packaging the data will require significant effort, and assumptions on boundaries, time scales, and conversion factors should be clearly stated either in the body of the report or in an appendix. Industry technical leaders and even international groups referring to the energy and emissions plan and status reports will notice inconsistencies or alterations to the data that falsely show successes, resulting in the loss of credibility of the city's efforts in the Sustainable Urban Energy and Emissions Planning (SUEEP) process.

### (Re)Defining Success

Although data and information on projects in their first year of implementation may be limited, efforts should be made to collect as much as possible from all projects to provide snapshots of progress and to evaluate the status of the overall energy and emissions plan implementation efforts.

KPIs should be reviewed in conjunction with the wider city projections of growth and economic development, and vision and goal statements should be revisited. This may not be the time to actually amend them—this task is a better fit for an upcoming version of the larger energy and emissions plan. The data and information collection process will be a time to understand the congruity of "reality" (the current and trending quantities and qualities of the city) with the exercise of projecting energy use based on the impact of the SUEEP team's actions.

With a better understanding of the city's capability to achieve its target KPIs, previous definitions of success should be revisited to determine whether the data, trends, and information collected are sufficient and whether the original targets had been too ambitious or not ambitious enough. It may be too early to redefine success until more data and information are obtained—it is recommended that the vision or goal statements *not* be altered in haste.

Now with the data and feedback in hand, the story of the energy and emissions plan's success can be told.

## Step 17: Publish Status Report

Data and information have been collected and successes have been outlined. This step now describes how to write and release an SUEEP status report.

### Identify Reporting Entity

Compiling information into a concise, well-formatted, and organized document requires significant resources. To ensure responsibility for delivery of the document, a group or an individual should be assigned to prepare the report.

The most appropriate entity to lead preparation and production of this report will depend on a city's internal resources, skill sets, and structures. This status report is jointly owned by the energy task force and the person or agency that has been assigned to prepare it, with the latter acting as "lead" to oversee all contributions and resourcing.

Sometimes a city hires third-party organizations to compile the data, information, and key success stories and produce a report that attracts stakeholders' (including financial institutions') attention. If a third party is used, it will have to report to a representative on the energy task force or high-level mayoral staff member to ensure that the report is accurate and conveys messages that the city government wants to spread. Production of the report commences once there is clarity on the party responsible for delivery of the document.

### *Scheduling the Report*

Many cities publish an annual status report for the energy and emissions plan that reviews the outcomes resulting from implementation of selected projects. If a city does not have the resources to publish an annual report, it could vary the format and timing for the release of data to still provide useful indicators of progress.

### Select Data and Draft the Report
#### *Use Illustrative Data*

The task force should use appropriate metrics to communicate the progress of the energy and emissions plan. An example may be the communication of data at a project level (the results of a street lighting project, for instance). Alternatively, the task force may want to provide overarching information on the energy and emissions plan (such as citywide reductions in carbon emissions).

This Guidebook suggests the type of data to report, but an appropriate set of city-specific indicators should be used to communicate progress of the energy and emissions plan and its component projects and their impact on the city.

## Acknowledge Lessons Learned

The report should showcase success stories but also highlight the hard lessons learned and hurdles the city had to overcome. The report should then propose actions to remediate shortcomings in the plan.

## Pull It All Together

The tip in box 15.9 gives an indicative list of contents generally expected in a successful status report. Although formatting and style make the report more attractive, ensuring that the messages are clear and understood by the general public is key. Technical terms should be explained and jargon should be minimized. The more important messages should be clear enough for a grade school student to comprehend.

---

### Box 15.9  Tip: Typical Contents of an SUEEP Status Report

Every SUEEP status report is different, but the following is a typical table of contents covering the primary components of the report.

**Overview**
Background and executive summary of the status of the plan. Introduce goals, objectives, and important messages.

**Energy Balance and Greenhouse Gas (GHG) Inventory**
Include the numbers, and tie them to the overarching goal statement. Details of the inventory should be included in an appendix.

**Key Performance Indicators (KPIs)**
Link the GHG inventory and data collected for each project to the sector and project KPIs (as you see fit).

**Major Highlights**
Document the fun facts, links to successes in other sectors, and qualities of the city that are changing for the better as a result of the energy and emissions plan.

**Project Updates**
Elaborate on the success of the priority projects and perhaps discuss future actions.

**Lessons Learned**
Share challenges and obstacles that were overcome, as well as key messages and "stories from the trenches."

**Appendixes**
- Inventories of processes and data
- Acknowledgment of reporting team and stakeholders who contributed to the report

---

## Release, Follow-Up, and Future Actions

### Finalize and Release the Status Report

The draft report should be reviewed and approved before it is published. After all reviewers' comments have been dealt with, the report can be made available in a variety of forms (see the tip in box 15.10).

### Follow Up with Stakeholders

Stakeholders are crucial to the SUEEP process, and the release of the status report can be used as an opportunity to strengthen relationships with them. Special invitations to events, or photo opportunities with high-level staff, will reinforce the city's appreciation of their participation in the SUEEP process.

### Give Credit Where Credit Is Due

Appreciation should be shown to city staff, including the members of the energy task force, who have spent countless hours on the SUEEP process and implementation of the plan. Although city governments may not be able to match private sector salaries, acknowledging the efforts of public employees helps to motivate the individuals who have chosen to drive change.

### Plan for Future Actions

The release of the first status report is a solid step, but it is only the beginning of the city's long and continuing journey toward attaining its vision of a sustainable

---

**Box 15.10  Tip: Media Outlets for Distribution of the Status Report**

There are several ways to communicate the results of the report and to release it:

**Post report on website**
Prepare a PDF version of the report and post it on the city website. The website can highlight important messages and stories.

**Hard copy reports**
Published copies should be sent to major stakeholders.

**Public announcements**
Messages can be shared through press conferences, public announcements, and on-site public engagement with stakeholders.

**Press releases**
Important stories, using testimonials about the impact of these projects, should be released to newspapers and local radio and TV stations.

**Social media**
Popular social media platforms such as Facebook and Twitter can be used to spread important messages.

---

future. City governments should maintain the momentum and develop programs and projects to ensure that future challenges can be tackled. See the case study in box 15.11.

See the resources in box 15.12 for more information on monitoring and reporting.

---

### Box 15.11  Case Study: Portland, Oregon, USA—Reporting One Year after the Climate Action Plan

Portland released a report showing the city's progress toward reducing local carbon emissions and the status of efforts made in the first year of implementing the Climate Action Plan. The report outlines improvements in several sectors, and in the specific focal area of buildings and energy, the city's Climate Action Plan contains four objectives for 2030:

1. Reduce total energy use of all buildings.
2. Achieve zero net greenhouse gas emissions in all new buildings.
3. Produce some energy from on-site renewable and clean district energy systems.
4. Ensure that buildings can adapt to a changing climate.

The "Highlights" section of the plan describes successes and ties program status to the objectives.

The city also created an easy way to track the status of all the projects (or "Actions") by developing a rating system, using colored dots to signify the following:

- RED: Action has not yet been initiated and/or little progress has been made.
- YELLOW: Action is under way, but may face obstacles.
- GREEN: Action is on track for completion by 2012.
- BLUE: Action is completed.

*Source:* Portland, Oregon, Bureau of Planning and Sustainability 2009.

---

### Box 15.12  Resources: Monitoring and Reporting References

The following resources can help you to determine effective performance metrics and reporting processes. Examples of other cities' annual status reports are given below.
*Singapore:* Singapore Green Plan 2012
Measurement, Reporting and Verification (MRV)
(http://app.mewr.gov.sg/data/ImgCont/1342/sgp2012.pdf)
*Berkeley, California:* Climate Action Plan
Metrics and Website Communication
(http://www.cityofberkeley.info/climate/)
*New York, New York:* PlaNYC
Greenhouse Gas Inventory and Status Reports
(http://www.nyc.gov/html/planyc2030/html/publications/publications.shtml)

*box continues next page*

---

**Box 15.12  Resources: Monitoring and Reporting References** *(continued)*

*Fort Collins, Colorado:* Climate Action Plan 2009 Status Report
(http://www.fcgov.com/airquality/pdf/2009capstatus-sept2010.pdf)

**Papers**

"Measurement, Reporting and Verification (MRV) of GHG Mitigation," OECD. undated. (OECD, Paris) (http://www.oecd.org/document/50/0,3746,en_2649_34361_42546674_1_1_1_1,00.html) "Mitigation Actions in China: Measurement, Reporting and Verification," Fei Teng, Yu Wang, Alun Gu, Ruina Xu, Hilary McMahon, and Deborah Seligsohn. 2009. (Institute of Energy, Environment and Economy, Tsinghua University, Beijing, China; and World Resources Institute, Washington, DC).

**A Protocol**

The GHG Protocol is the most widely used accounting tool for government and business leaders for understanding, quantifying, and managing greenhouse gas emissions. (http://www.ghgprotocol.org)

---

## The Beginning

This Guidebook outlines a pathway, a framework, and relevant tools that a city could use to effect changes to promote a sustainable future.

The steps summarized here are based on experiences gained through pilot studies and on industry knowledge. This Guidebook serves as a platform for cities to commence their own SUEEP processes, but the Guidebook alone is insufficient—each city must tailor its program to its own needs.

As this Guidebook reaches your desk, projects within your city are probably already ongoing—some version of an inventory may have been undertaken and leadership frameworks might have been outlined. Only you, as a leader in your city, will be able to outline a process, schedule, and an overall energy and emissions plan that is compatible with your city's needs and aspirations. It is a big task, and help is available.

## References

Berkeley, California. 2009. *Climate Action Plan.* Berkeley, CA. http://www.cityofberkeley .info/climate/.

Fei Teng, Yu Wang, Alun Gu, Ruina Xu, Hilary McMahon, and Deborah Seligsohn. 2009. "Mitigation Actions in China: Measurement, Reporting and Verification." Institute of Energy, Environment and Economy, Tsinghua University, Beijing; World Resources Institute, Washington, DC.

Fort Collins, Colorado. 2010. *Climate Action Plan 2009 Status Report.* Fort Collins, CO.

New York, New York. 2007. *PlaNYC 2030: A Greener, Greater New York.* New York: The Mayor's Office. http://nytelecom.vo.llnwd.net/o15/agencies/planyc2030/pdf/full_ report_2007.pdf.

OECD (Organisation for Economic Co-operation and Development). undated. "Measurement, Reporting and Verification (MRV) of GHG Mitigation." Paris. http://www.oecd.org/document/50/0,3746,en_2649_34361_42546674_1_1_1_1,00.html.

Portland, Oregon, Bureau of Planning and Sustainability. 2009. *City of Portland and Multnomah County Climate Action Plan 2009*. Portland, OR. http://www.portlandonline.com/bps/index.cfm?c=49989.

Singapore, Ministry of the Environment and Resources. 2002. "Singapore Green Plan 2012." Singapore. http://app.mewr.gov.sg/web/Contents/Contents.aspx?ContId=1342.

# Approach and Methodology

The Sustainable Urban Energy and Emissions Planning (SUEEP) project team applied new techniques for the data collection and analysis that were used to identify and inform priority energy system interventions in each city. This report presents the findings of the project, which was structured in three phases:

- Four weeks of preliminary data gathering by a local consultant
- A two-week mission to each city involving international consultants from Happold Consulting
- A six-week follow-up period for final data gathering and report writing

A discussion of the methodological frameworks applied during each mission and the resulting outputs follows.

## Energy Balance Study

The energy balance study maps primary and secondary energy supply and use in each city. Energy balance studies highlight the relative importance of different fuel supplies through their contribution to the city's economy; demonstrate the efficiency of different energy conversion technologies; establish benchmarks that allow a city to assess the effectiveness of different policies; and potentially highlight the sectors to which energy efficiency, security, and conservation efforts can be targeted by clearly showing how energy is used in a city. Energy balance studies are further used as a check on the completeness of the available data, as a high-level check on data accuracy, and as a starting point for construction of various indicators of energy consumption and energy efficiency.

In 2011, the World Bank developed an energy balance assessment technique that breaks down information by the type of fuel source, the sector consuming the energy, and the way energy is consumed within that sector. Notably, the breakdown is limited to the sector level and all data and calculations are contained within a single, practical worksheet. Data for the energy balance studies were compiled from published data, local electric utility companies, and officials from different government departments in each city.

In most cases, data were collected before and during the two-week mission. The energy balance study contributes to the compilation of background information for the Tool for Rapid Assessment of City Energy (TRACE), although the TRACE model also requires specialized data that go beyond that collected in a traditional energy balance study.

## Greenhouse Gas (GHG) Emissions Inventory

The GHG emissions inventory establishes the principal sources of GHG emissions and may identify policy or program initiatives not highlighted by a purely energy-based analysis. The GHG emissions inventory may result in, for example, a heightened focus on renewable power or steps to reduce fugitive landfill gas emissions. This analysis provides the basis for emissions information that may help local government agencies or utilities obtain carbon finance funding for different renewable or energy efficiency projects.

The GHG inventories were compiled using the Urban Greenhouse Gas Emissions Inventory Data Collection Tool, developed by the World Bank in 2011. The tool covers emissions from transportation, solid waste management, water and wastewater treatment, and stationary combustion (power generation, building energy usage, industry, and the like). (Urban forestry considerations may be included in the tool in the future.) The tool aims to provide a simple, user-friendly, and easily replicable approach, and places significant emphasis on providing descriptions of data sources, years, implied boundaries, and quality of the data.

## Analysis Using the TRACE

The TRACE is the project component that most directly links local energy use data with the identification of policy alternatives city authorities may use to reduce energy use. Developed by Happold Consulting for the World Bank's Energy Sector Management Assistance Program (ESMAP) team in 2010, the TRACE was specifically designed for application at the city government level and was piloted in Quezon City, the Philippines, and in Gaziantep, Turkey.

The TRACE is a decision-support diagnostic that compares one city's energy performance with that of other cities through a custom benchmarking system. Benchmarking results are used to identify priority sectors to which a city should target its energy efficiency efforts. Policy recommendations built into the TRACE are linked to each sector, triggering field interviews and further data gathering by the TRACE project consultant to assess which recommendations are most appropriate for the local city context. In some cases, the city may have already implemented or begun work on these policy options, whereas in other cases local political, policy, or market circumstances make a recommendation unviable. Ultimately, the TRACE model is designed to produce tailored, actionable recommendations that will guide or support local policy efforts and ideally be of interest to development banks and other project funders.

The energy balance and GHG emissions components of the study were carried out using frameworks created by the World Bank to quickly assess these two issues, with emphasis on city government services provision and citywide energy-consuming or GHG-creating activities. The application of more detailed, international standards would have required further data gathering and analysis and was not considered necessary to provide the macro-level insights afforded by the methodological frameworks used.

## Data Validity

Data used in the study were provided by local sources, either city government office records or other agencies such as utility companies. The accuracy, reliability, and completeness of the data were examined and queried where necessary, but were not verified by the consultant team because of the rapid nature of the study. Happold Consulting does not, therefore, assure or guarantee these aspects of the study or acknowledge any liability arising from their future reliance or use by others.

## Mission Activities

The major activities undertaken during the premission and mission phases of the project included the following:

- Training of local consultants on the logic and techniques used in the TRACE
- World Bank–initiated contact between the consultancy team and pilot city governments and other key local utilities (water, electricity) to introduce the project, establish high-level project engagement and support for the release of relevant data, and obtain agreement to follow the proposed mission schedule
- Initial data collection by local consultants through a literature review and telephone and in-person contact with relevant agency and utility staff
- Data review and data input into the TRACE by the consulting firm
- Meetings between the mission team and pilot city government and utility officials
- Final presentation on TRACE findings to pilot city government officials
- TRACE training for relevant pilot city government officials
- Follow-up data collection and preparation of a final report

# Additional Reading

ADB (Asian Development Bank). 2011. *Toward Sustainable Municipal Organic Waste Management in South Asia: A Guidebook for Policy Makers and Practitioners.* Manila: ADB.

ALMEC Corporation/JICA (Japan International Cooperation Agency). 2010. *The Study on Integrated Development Strategy for Danang City and Its Neighboring Area in the Socialist Republic of Vietnam (DaCRISS).* Tokyo: JICA.

Barry, Judith. 2007. "Watergy: Energy and Water Efficiency in Municipal Water Supply and Wastewater Treatment—Cost-Effective Savings of Water and Energy." Alliance to Save Energy, Washington, DC.

City of Cebu Planning and Development Office. 2008. "Cebu City, Philippines: Profile." Planning and Development Office, Cebu, the Philippines.

Climate Alliance. 2006. "The Climate Compass Compendium of Measures for Local Climate Change Policy." Climate Alliance, Frankfurt, Germany. http://www.climate-compass.net/fileadmin/cc/dokumente/Compendium/CC_compendium_of_measures_en.pdf.

Da Nang Department of Finance. 2010. *Statistical Yearbook.* Da Nang, Vietnam: Department of Finance.

Enova SF. 2008. *Municipal Energy and Climate Planning: A Guide to the Process.* Trondheim, Norway: Norwegian Ministry of Energy and Petroleum.

ESMAP (Energy Sector Management Assistance Program). 2010. *Rapid Assessment Framework: An Innovative Decision Support Tool for Evaluating Energy Efficiency Opportunities in Cities, Final Report.* Report 57685. Washington, DC: World Bank.

European Energy Award. 2011. http://www.european-energy-award.org.

Global Environment Facility/United Nations Development Programme. 2004. *Municipal Energy Planning: Guide for Municipal Decision Makers and Experts.* Sofia, Bulgaria: EnEffect. http://docs.china-europa-forum.net/doc_748.pdf.

IEA (International Energy Agency). 2011. *Key World Energy Statistics 2011.* Paris: OECD Publishing.

Natural Capitalism Solutions. 2007. *Climate Protection Manual for Cities.* Eldorado Springs, CO: Natural Capitalism Solutions. http://www.climatemanual.org.

Norling, Malin. 2009. "Energy Planning Guidance: An Introduction." Deliverable 4.2 of the IEE Pepesec project. http://www.pepesec.eu/archives/306.

UNEP (United Nations Environment Programme). 2006. *Geo Yearbook 2006: An Overview of Our Changing Environment*. New York: UNEP.

United States Department of Energy. 2011. "Regional Energy Efficiency Workshop: Handout on Energy Efficiency Action Plan Approaches and Resource." U.S. Department of Energy National Renewable Energy Lab for the International Partnership for Energy Efficient Cooperation. http://www.iea.org/work/2011/IPEEC_WEACT/Day1_SessionI/Benioff.pdf.

World Bank. 2003a. *Indonesia Environment Monitor 2003: Special Focus—Reducing Pollution*. Jakarta: World Bank Indonesia Office.

———. 2003b. *Vietnam Environment Monitor 2003: Special Focus—Water*. Washington, DC: World Bank.

———. 2007. *Philippines Environment Monitor 2006: Environmental Health*. Washington, DC: World Bank.

———. 2009a. "Reducing Technical and Non-Technical Losses in the Power Sector." Background Paper for Energy Sector Strategy, World Bank, Washington, DC.

———. 2009b. *World Development Indicators*. World Bank, Washington, DC.

———. 2010. "Mainstreaming Building Efficiency Codes in Developing Countries: Global Experiences and Lessons from Early Adopters." Working Paper 204, World Bank, Washington, DC.

———. 2011a. "Building Sustainable and Climate-Smart Cities in Europe and Central Asia." Europe and Central Asia Region, Sustainable Development Department, and World Bank Institute, Washington, DC.

———. 2011b. *One Goal, Two Paths: Achieving Universal Access to Modern Energy in East Asia and Pacific*. Washington, DC: World Bank.

green press INITIATIVE

www.ingramcontent.com/pod-product-compliance
Lightning Source LLC
Chambersburg PA
CBHW080811280326
41926CB00091B/4177